Electrical Product Compliance and Safety Engineering

Volume 2

For a listing of recent titles in the
*Artech House Technology Management and
Professional Development Series,* turn to the back of this book.

Electrical Product Compliance and Safety Engineering

Volume 2

Steli Loznen
Constantin Bolintineanu

ARTECH
HOUSE
BOSTON | LONDON
artechhouse.com

Library of Congress Cataloging-in-Publication Data
A catalog record for this book is available from the U.S. Library of Congress.

British Library Cataloguing in Publication Data
A catalog record for this book is available from the British Library.

ISBN-13: 978-1-63081-838-8

Cover design by Charlene Stevens

685 Canton Street
Norwood, MA 02062

10 9 8 7 6 5 4 3 2 1

For Ruthy, Abigail, Albert and the memory of my parents
—S. L.

For all of our readers, for Danuta, for Steli,
and all my beloved ones,
and the memory of my parents
—C. B.

Contents

Foreword

This second volume of *Electrical Product Compliance and Safety Engineering* is an exciting update to the first volume with content not readily known to most compliance engineers. The topics presented in this book are not taught in academia. When asking university students what their lifelong career choice will be and the academic degree sought after, product safety engineering is generally at the bottom of their list. There are, however, universities with strong EMC programs recognized worldwide. Safety and EMC are tightly coupled under the category of compliance engineering.

Graduating students generally do not intentionally interview for a lucrative high-paying position with a company as their safety or EMC engineer. They want to be a designer or programmer, working with hardware/software, manufacturing, test, quality, or other aspects of engineering that are fun and exciting, not compliance. Most compliance engineers are unexpectantly assigned to work on getting a product certified or approved to regulatory standards, safety, and/or EMC. They are invited into their manager's office and told "Congratulations, you are now the compliance engineer...get us international safety and EMC approvals."

There was never a professional society worldwide for product safety engineers until 2004 when the IEEE created the Product Safety Engineering Society. Both authors of this book were early members of this society, with one eventually being elected to a vice president position. The goal of creating this society, on an international basis, was to disseminate intellectual property (IP) on topics never before published in the field of product safety. IP includes textbooks and technical papers, the results of research into many categories associated with safety, networking, and learning from others who may have knowledge on a specific topic and their desire to share this information in the public domain.

There has been in recent years a shift in the compliance approach from meeting proscriptive standards to risk hazard analysis. In other words, in the past we "designed to meet" regulatory requirement and pass specific tests for the elusive safety logo. Today we must now "design to comply" to essential requirements to ensure products cannot and do not cause harm to users and domestic animals when used within a defined environment of use, including abnormal operation.

This new compliance process is based on risk hazard analysis. Included in the arena of risk hazard analysis is "frequency of occurrence" and "hazardous level or likelihood," clearly defined within a matrix. To ensure due diligence is performed in accordance to the matrix, testing must occur to levels appropriate for the desired environment of use such as residential, commercial, military, medical, transportation, appliances, light and heavy industrial manufacturing, and the like. In addition, an EMC event such as a high-power radiated field may cause functional disruption to hardware that may result in a product entering a mode of operation through software that could cause injury or death. Engineers must "design to comply" to unknown environmental conditions and unauthorized alteration that may trigger an event.

The first volume of this book focused on standards, design philosophy, methods for failure analysis, construction requirements, component selection and implementation, markings and labeling, human factors, testing, and manufacturing for compliance and education. There is no other book that discusses these topics in such involved detail, making this book a classic.

This second volume presents an enhanced approach toward product safety compliance that includes safety aspects of various product approvals, energy management, environment concerns, material science, radiation, hazardous location use, and global market access. This book does not replace the outstanding content of the first volume—it enhances it with advances in safety compliance.

For those who are now responsible for product safety compliance with limited knowledge on "design to comply," one will find essential aspects of safety compliance engineering not found in any other publication. Both authors are well known internationally as experts, sharing their knowledge in an easy-to-read book that will make the role of safety compliance engineering easier.

Mark I. Montrose
Founder and First President—IEEE Product Safety Engineering Society
Professional EMC Consultant and Educator
September 2021

Preface

Electrical product compliance and safety engineering may help with the development of a more profitable product, contribute to better satisfied customers, and reduce the risk of liability by building confidence in meeting the requirements of standards and regulatory bodies. We believe that this second volume of *Electrical Product Compliance and Safety Engineering* provides a continuity of the first volume published in 2017, and will help readers develop more effective professional mindsets, approaches, and insights.

This book represents a multiyear effort by the authors; we hope that you, our distinguished readers, will benefit and learn from the content in this book. We want this book to be a guide for professionals. As the technology and standards are evolving and changing nonstop, the relevance of the content in this book may change over time: new updates, new standards developed, and discoveries of new safety concerns in different applications. Therefore, users should always use both volumes as a guideline and always verify the relevance of a standard and that it is the most current, and determine whether additional standards exist or new standards for a specific product have been developed.

The content of this second volume was chosen to provide additional background on why you need to know about compliance and safety engineering for electrical products and how to use the information provided.

Product compliance and safety science is a broad and multidisciplinary field (electrical, electronics, mechanical, chemical, materials, general engineering, environmental, etc.) governed by a well-established philosophy. This second volume analyzes concepts, principles, and methods and their influences for proper understanding of product compliance and safety, highlighting the ways in which these are applied.

This book is intended for compliance and safety professionals who are responsible for designing, implementing, managing, testing, manufacturing, marketing, installing, maintaining, servicing, and regulating electrical products. The primary audience for the book includes design, quality assurance and control, testing, regulatory, manufacturing, service, sales, and marketing practitioners. No major background in electrical and electronics engineering is required, but a familiarity with specific topics will get you started right away. This book is also intended for instructors and students in electrical and electronics departments of engineering universities, to whom we suggest adding the training syllabus on issues that have not been covered until now. Due to a lack of formal education in the field of product compliance and safety, many employees will benefit from the information provided here.

The book is structured as follows:

Chapter 1 examines compliance and safety aspects pending on specific application need, by referring to the following categories of products: information technology Equipment and audio-video (multimedia), medical electrical, household, laboratory, measurement, control, luminaires (including LED lamps) and lamp control, industrial machinery, semiconductors manufacturing (SEMI), electrical tools, alarm, and security.

Chapter 2 provides additional information on energy management addressing the smart grid, stability of power sources, energy efficiency, stored energy systems, DC/DC conversion, inverters DC/AC, UPS, and the use of photovoltaic and solar energy, fuel cells, and wind turbines.

Chapter 3 covers environmental aspects, analyzing how temperature, atmospheric pressure, relative humidity, vibration, shock, and packaging/transportation affects product safety. Information about testing (environmental, HALT, and HASS) is also provided. Special attention is paid to compliance of electrical products with dedicated environmental regulations (RoHS, REACH, WEEE, etc.).

Chapter 4 details the materials considerations: metal corrosion, adhesives, insulation materials, information about safety (safety data sheet (SDS)) of hazardous materials.

Chapter 5 details the safety of electronic product radiation sources, detailing non-ionizing and ionizing sources as: electrical and magnetic field (EMS), magnetic resonance, Wi-Fi, mobile phones, microwaves, lasers, LEDs, photo-biological (IPL), infrared, ultraviolet, X-Ray, and gamma radiations. Specific hazards and means of protection are also specified for sonic waves (acoustic and ultrasound exposure).

Chapter 6 focuses on the safety for hazardous locations: flammable environment, hazardous zones, flammable and combustible substances, sources of ignition, equipment and protection systems, components and construction requirements, installation of electrical product in hazardous locations, standards and codes, documentation and marking, and similarities and differences between ATEX and IECEx.

Chapter 7 discusses the practical aspects related to global market access: specific documentation and local labeling requirements, languages used for safety instruction and user manual, internal nonaccredited testing versus an ISO/IEC 17025 accreditation, post-production surveillance, follow-up services, market surveillance, and recalls.

Throughout this book we have included some personal opinions resulting from more than 30 years of activities in the new science of compliance and product safety engineering. These opinions refer especially to the importance of an in-house accredited test laboratory, a proposal for continuous improvement of follow-up services, and a fundamental need for developing a compliance and product safety culture encompassing all stakeholders involved in the process of compliance and product safety.

Although every effort has been made to ensure that the contents of this book were correct and up to date at the time of writing, errors sometimes arise from the mass of technical details. We have relied upon documents and materials and cannot guarantee the accuracy of this source material. We have exercised the usual and customary care in our presentation of the contents of this book; however, the responsibility and usage remains fully with the user.

It is our hope that this book will increase awareness about product safety hazards and, consequently, will help to avoid dangerous situations and meet safety requirements, for the benefit of manufacturers, users, and service personnel.

If every reader would follow the advice in this book, our world would become that much safer.

We are grateful to Dr. Merlin Fox, Soraya Nair, and the staff at Artech House for providing us with helpful insights and reviews. Because the reviewers remain anonymous, we can only thank them collectively for their professionalism and accuracy.

Acknowledgments

After more than four decades of working with many people, I am amazed at the spirit of goodwill and cooperation.

A special consideration I have to a number of generous individuals, from which I learned a great deal. Each of whom knows the importance of developing and implementing a new science such as compliance and product safety engineering. I want to thank them for their help, including (alphabetically) Michel Brossoit, Professor Vlad Cehan, Professor Alf Dolan, Frank O'Brien, Peter Merguerian, Charles Sidebottom, Michael Sippl, Alex Vilenski, Ph.D, and Acad. Professor Horia N. Teodorescu.

For these special people and all my readers, clients, and students who have shared their hopes, dreams, and problems with me over the years, thank you for your help. The best parts are yours—all the mistakes are mine. I hope all the readers will profit from wisdom and generosity acquired.

Steli P. Loznen
Tel Aviv, Israel

So many years that amount to a table of contents' worth of a career. A career in which I was surrounded by many amazing people. I want to thank Bob Davison, Siva Pillai, Jose Tillett, Richard Nute, Pete Perkins, John Woodgate, Robert Miller, Thomas Dickten, Mark Montrose, and my colleague and mentor Kevin Harris, and all the people who are working nonstop to expand Digital Security Controls. It was an honor to be with you all throughout this journey. We hold everything that we did inside a box of dedication and hard work. This book would not exist without you.

Constantin Bolintineanu
Toronto, Canada

CHAPTER 1

Compliance and Safety Aspects Depending on the Specific Application

This chapter is intended as a practical guide with references to the designation of the application, regulatory requirements, characteristic concepts, differences between categories, and main safety aspects for certain electric and electronic products categories.

The following sections are organized by the target's product categories.

1.1 Information Technology and Audio-Video/Multimedia

Audio/video, information, and communication technology equipment is defined as electrical and electronic equipment dedicated to areas of audio, video, information, and communication, as well as businesses and offices, with a rated voltage not exceeding 600V. This equipment was previously covered by separate standards (IEC 60950-1 for information and communication technology equipment and IEC 60065 for audio/video), and since December 2020 this equipment is covered by only one main standard for safety, the IEC 62368-1 Audio/Video, Information and Communication Technology Equipment—Part 1: Safety Requirements, which was more or less adopted worldwide with some specific national deviations.

Existing equipment that is already in the market as per previous standards may be kept as is on any market. After December 20, 2020, IEC 60065 and IEC 60950-1 certified components are no longer permitted in equipment that should be covered by IEC 62368-1 without additional investigation.

In the European Union (EU), the CE Mark regulatory requirements for audio/video, information, and communication technology equipment are specified in the LVD, EMC Directive (2014/30/EU), Energy Efficiency (2017/1369/EC) (see Chapter 2, Section 2.2), and RoHS, REACH, and WEEE Directives (see Chapter 3, Section 3.3). Depending on the specificity of the product, this can also apply to other directives such as the Radio Equipment Directive (RED), or the General Product Safety Directive (GPSD 2001/95/EC) for equipment with a supply voltage below 50V for AC or below 75V for DC. For electromagnetic compatibility (EMC), the EN 55032 (emissions), EN 55035 (immunity), EN 61000-3-2 (harmonics), and EN 61000-3-3 (voltage fluctuations and flicker) standards apply.

In North America, the binational standard CAN CSA C22.2 60950 and the UL 60950 are still valid, and even the UL 62368-1 was adopted. Suppliers of the equipment may still use these standards for a while. The UL Mark is used by a few Category Control Numbers (CCNs) for the product categories covered by the scope of UL 62368-1, UL60950, and UL 60065 standards: AZOT (for UL 62368-1), NWGQ (for UL 60950), AZSQ (for UL 60065), QQGQ (Power Supplies for ITE), and so on. For EMC aspects the Federal Communications Commission (FCC) Part 15 and Part 18 requirements apply.

In China, the IEC 62368-1 standard is still not enforced and a formal transition schedule remains under development.

The following material will delineate several aspects regarding this large category of electrical and electronic equipment to highlight the change that brought about the new hazard-based safety engineering (HBSE) approach.

Traditional safety standards define the safety requirements, informing the designers on what they should do about construction, selection of components, markings, and user instructions.

A step-by-step process delineating HBSE may be included in a flowchart as follows:

1. Identification of the energy source(s) used within the equipment.
2. Perform the measurements of those energy levels they can generate.
3. Consider the decision if the energy from the sources is or not at a hazardous level.
4. Classify the energy sources accordingly.
5. Identify if and how the energy can be transferred to a body part or to a property.
6. Determine the appropriate necessary safeguard in a way that:
   - Persons are protected against pain and injury from the classified energy sources
   - Property damage due to the energy (originated within the equipment) transfer will be avoided
7. Measure and verify by testing the effectiveness of the considered safeguards [1].

No concepts—in terms of judging the equipment from the point of view of the product safety evaluation—were maintained from the old standards. However, some of the definitions were kept along with "classic" terminology used in most of the product safety standards, but many new concepts were implemented in each hazard area that falls under the scope of the IEC 62368-1, observing the hazards and pointing to the use of an appropriate safeguard to protect property and the human body from injuries and/or damages [2].

The IEC 62368-1 standard specifies three classes of energy sources defined by magnitudes and durations of source parameters relative to the human body and the effect on combustible materials (Table 1.1).

All the equipment that falls within the scope of the IEC 62368-1 standard is judged based on this approach. Energy sources are addressed by this standard, together with the pain or injury that results from a transfer of the energy to the body

Table 1.1 Effects of Energy Source Classes per IEC 62368-1 Standard

Energy Source	*Effect on the Body*	*Effect on Combustible Materials*
Class 1	Not painful but detectable	Ignition not likely
Class 2	Painful, but does not generate an injury	Ignition possible, but limited growth and spread of fire
Class 3	Causing an injury	Ignition likely, rapid growth, and spread of fire

From: [3].

and the likelihood of generating property damage as a result of fire escaping from the equipment.

The philosophy of the standard that covers all these categories of equipment is based on the classification of the energy sources and prescribes safeguards against those energy sources as necessary, and provides guidance on the application of, and requirements for, those safeguards.

The prescribed safeguards are intended to reduce the likelihood of pain, injury, and, in the case of fire, property damage.

Injuries are regarded for each category of persons who may have contact with the equipment during its life, starting with users up to skilled persons.

The standard IEC 62368-1 specifies the levels corresponding to each class of energy source (1, 2, or 3) associated with the following hazards [3]:

- Electrical energy hazards: ES1, ES2, ES3;
- Power sources: PS1, PS2, PS3;
- Potential ignition sources: PIS1, PIS2, PIS3;
- Chemical hazards: CS1, CS2, CS3;
- Mechanical hazards: MS1, MS2, MS3;
- Radiation hazards: RS1, RS2, RS3;
- Thermal burn hazards: TS1, TS2, TS3.

There is no doubt that the progress of the technology within the last 25 years, along with the use of information technology (IT) elements within each type of equipment, makes clear that combining the following equipment under the umbrella of one single standard is necessary in terms of protection of end users:

- Consumer electronic equipment (e.g., professional and home audio and video, musical instruments, digital cameras, amplifiers, and personal music players);
- Electrical and electronic office machines, (e.g., calculators, copying machines, document shredding machines, micrographic office equipment, paper trimmers such as punchers, cutting machines, and separators, and typewriters);
- Data network equipment (e.g., data circuit terminating equipment, data terminal equipment, routers, and expanders);
- Banking equipment;
- Data and text processing machines;

- Telecommunication network infrastructure and telecommunication terminal equipment;
- Postage equipment (mail processing machines, postage machines);
- Electrical and electronic retail equipment (e.g., cash registers, point-of-sale terminals including associated electronic scales);
- Multimedia equipment (e.g., webcams, digital cameras, musical instrument digital interface-MIDI ports, microphones, sound cards, and video capture cards);
- Public terminal equipment;
- Power supplies for all above products that are currently part of the same family of electronic and electrical equipment covered by the IEC 62368-1 standard [3].

During the last 20 to 25 years, many organizations struggled to provide unified acceptance criteria for these products, products which we can find now almost everywhere. In our opinion, this process started in 1994 when Ecma[1] was born, and over time, this organization promoted globalization of information technology equipment (ITE) and some consumer electronics equipment.

As of today, Ecma International facilitates the timely creation of a wide range of global information and communications technology (ICT) and consumer electronics (CE) standards. We have many reasons to believe that people who used to be part of the ECMA Standard for Safety contributed in a significant manner to the creation of the new standard [4].

This standard put together product safety design requirements for this category of equipment by using a new concept, the hazard-based approach.

The IEC 62368-1 standard (which covers everything that several previous standards did) is not an equipment-testing and criteria-oriented standard; it is a hazard-based and performance-oriented standard.

It focuses on the energy sources available inside the equipment and how these sources can be kept under control in the process of energy transfer to the human body or the surrounding environment. This makes the standard easier to apply to new technologies by using a general frame for a wider range of electrical and electronic equipment.

HBSE visualizes the hazards and safeguards in terms of three-block models that describe the links between energy source, the energy-transfer mechanism to the human body, and/or to property (Figure 1.1).

The authors of the standard considered that three categories of persons may come in contact with the equipment that is the subject of the standard as follows [3]:

Ordinary persons are defined as not only the end users of the equipment but all persons who may have unrestricted access to the equipment or even who "may be in the vicinity of the equipment," in normal operating conditions and/or in abnormal operating conditions. Ordinary persons "should not be exposed to parts

1. Ecma International initially was the European Computer Manufacturers Association (ECMA) but changed its name to reflect the organization's global presence and activities; their name is no longer considered an acronym and thus they no longer use full capitalization.

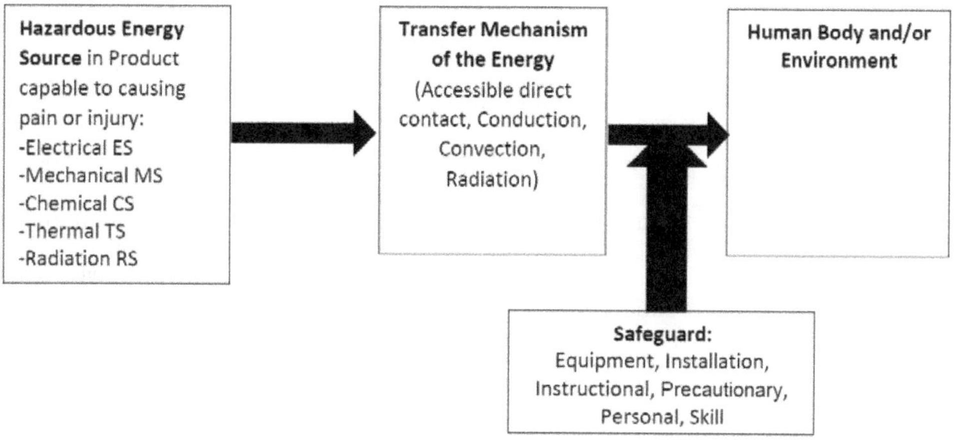

Figure 1.1 The three-blocks model.

comprising energy sources capable of causing pain or injury." Under a single-fault condition, ordinary persons should not be exposed to parts comprising energy sources capable of causing injury. The term "ordinary person" is applied to all persons other than instructed persons and skilled persons.

Instructed persons are the persons who have been instructed and trained by a skilled person to identify energy sources that may cause pain or injury and to take precautions to avoid unintentional contact with or exposure to those energy sources (the training and experience or supervision of an instructed person by a skilled person is considered a precautionary safeguard). Under normal operating conditions, abnormal operating conditions, or single fault conditions, instructed persons should not be exposed to parts comprising energy sources capable of causing injury.

Skilled persons (e.g., service persons) are the persons who have the necessary training or experience in multiple areas, such as equipment technology, and know the various energies and energy magnitudes used within the equipment. It is expected that the skilled persons will use their training and experience to recognize energy sources capable of causing pain or injury and to take action for protection from injury from those energies. The education, training, knowledge, and experience of the skilled person that is employed to protect the skilled person against class 2 and class 3 energy sources are considered a skill safeguard. Despite their knowledge and experience, skilled persons should also be protected against unintentional contact or be exposed to energy sources capable of causing injury. Table 1.2 shows a few examples of safeguards characteristics.

This category of electronic and electrical equipment shares common energy hazards and the standard constructively addresses each of them. Considering how the IEC 62368-1 standard was designed, we want to make it clear that a major contribution to the success of it was due to the specialists who have dedicated themselves to the field of product safety, including (alphabetically): Morten Andersen, Thomas M. Burke, Bob Griffin, Richard Nute, Pete Perkins, Grant Schmidbauer, and Robert A Taylor.

We want to emphasize that unlike the previous standards (IEC 60950, IEC 60065), the IEC 62368-1 places almost all of the tests into annexes instead of in

Table 1.2 Examples of Safeguard Characteristics

Safeguard	*Basic*	*Supplementary*	*Reinforced*
Equipment	Normal temperatures below ignition temperatures	Fire enclosure	Not applicable
Installation	Wire size	Overcurrent protective device	Socket outlet
Personal	Glove	Insulating floor material	Electrically insulated glove for handling live conductors
Instructional	Instruction to disconnect telecommunication cable before opening the cover	After opening a door, instruction to protect against hot parts	Instruction to not touch hot parts in an office photocopier or a continuous-roll paper cutter on a commercial printer

the body of the standard. There are also several changes regarding the testing and acceptance criteria:

- Changes to the test voltages (as a value) regarding the electric strength;
- Changes to the values of the touch temperature;
- Changes regarding the compliance criteria for pluggable type A equipment, and pluggable type B equipment regarding the capacitor discharge test;
- Changes regarding testing tools for devices that contain lithium batteries;
- Changes regarding mechanical energy hazards, wall mounting test, stability test, etc. [5].

This standard assumes that a person will not intentionally create conditions or situations that could cause pain or injury. At the same time, even if does not directly address any manufacturing-related aspects, it provides requirements for safeguards for subassemblies, for enclosures, and in this manner indirectly provides a safer manufacturing environment.

This category of equipment uses the principle of safeguards (equivalent to a type of insulation and means of protection as referred to in other safety standards). Many products use energy levels capable of causing pain or injury. Product design does not try to eliminate the use of energy, and thus designers are faced with having to use a scheme that reduces the likelihood of such energy being transferred to a body part. The means that can reduce the likelihood of the total energy transfer to a body part is a safeguard. The safeguards may be categorized into three main categories:

1. Equipment safeguards (acting as basic, supplementary, double, or reinforced safeguards);
2. Installation safeguards (acting as a supplementary safeguards);
3. Instructional safeguards (requesting a personal safeguard, such as personal protective equipment, or avoidance behavior) (acting as basic or supplementary safeguards).

See Table 1.2 for examples of safeguard characteristics.

In practice, safeguard selection will take account of the nature of the energy source, the intended user, the functional requirements of the equipment, and similar considerations [3].

Insulation is part of the first category of safeguards, equipment safeguards that could be basic safeguards (e.g., basic insulation), supplementary safeguard (e.g., supplementary insulation), and reinforced safeguard (e.g., reinforced insulation).

Means of protection from other standards are accepted safeguards for each category mentioned above, depending on the means.

Safeguards that need to be interposed between a specified class of the energy source and ordinary, instructed, or skilled person are indicated in Table 1.3.

It is clear that differences between the categories of equipment will require specific testing and may also require specific marking. At the same time, many of the instructional safeguards are similar because the equipment shares similar responsibilities in terms of the hazards they may present during their lifespans. A specific instructional safeguard, such as the warnings, cautions, safety signs, and symbols, shall be placed on the equipment label, on the packaging, or in the instruction manual. Table 1.4 summarizes a few of these.

Alternatively, an instructional safeguard may be shown on the equipment display during use. Interactive menus are available that offer live assistance to the user.

The goal to ensure that failures during the manufacturing process or unacceptable tolerances in the manufacturing process and/or the materials or components used to build some equipment is accomplished by performing routine tests in production. General information about routine tests was included in the Volume I of this book in Section 15.5 [6].

The standard IEC 62911 Audio, Video and Information Technology Equipment–Routine Electrical Safety Testing in Production [7] states that it covers the routine safety tests performed on the equipment powered by an AC mains supply or DC mains supply and that claim to be compliant with IEC 62368-1.

In our opinion, each manufacturer should design the routine tests in a way that the tests will not affect the product in any form (aesthetically and/or the functionality). We know that using a high current source may involve powerful clamps or the involved wiring may be stressed by limit. A good approach would be to analyze the best structure that is relevant to the routine tests and to perform these tests as the standard allows: "during or at the end of manufacture."

Table 1.3 Relation between Classes of Energy Sources, Safeguards, and Persons

Class of Energy Source	*Ordinary Person*	*Instructed Person*	*Skilled Person*
Class 1	Does not need safeguard	Does not need safeguard	Does not need safeguard
Class 2	One basic safeguard	Precautionary safeguard	Skill safeguard
Class 3	An equipment basic safeguard and an equipment supplementary safeguard (double safeguard) or a piece of equipment reinforced safeguard	An equipment basic safeguard and an equipment supplementary safeguard (double safeguard) or a piece of equipment reinforced safeguard	Skill safeguard

Table 1.4 Examples of Instructional Safeguard Marking

Symbol	*Meaning*	*Symbol*	*Meaning*
	Avoid listening to high volume levels for long periods		Sharp edges
	Risk of electrical shock		Caution: Moving parts
	Caution: Very bright light		Connection to protective earthing (PE)
	Class II equipment with functional earthing	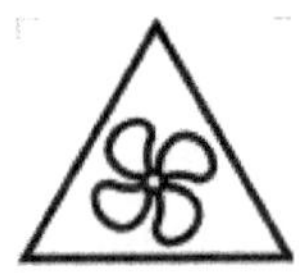	Caution: Moving fan blades
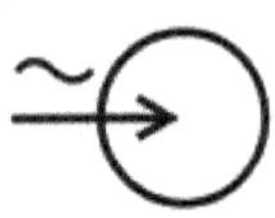	Rated power input AC		Rated power output DC

Table 1.5 summarizes the most important standards applicable to audio/video, information, and communication technology equipment.

1.2 Household and Consumer

As a generic designation, "household" refers to the appliances that are encountered by all persons in and around the home. The household appliance is included in the consumer equipment, which covers a wider area. According to the 15 US Code 2052, the term "consumer product" means any article, or part thereof, produced or distributed (i) for sale to a consumer for use in or around a permanent or temporary household or residence, or otherwise said, or (ii) for the personal use, consumption or enjoyment of a consumer in or around a permanent or temporary household or residence, or otherwise. But such a term does not include any article that is not customarily produced or distributed for sale to, or use of, a consumer; tobacco products; motor vehicles; aircraft; boats; food; drugs; medical devices; or cosmetics [8].

Table 1.5 Standard Applicable to Audio/Video, Information, and Communication Technology Equipment

Standard	*Title of Standard*
IEC/EN 62368-1	Audio/Video, Information and Communication Technology Equipment–Part 1: Safety Requirements
IEC/TR 62368-2	Audio/Video, Information and Communication Technology Equipment–Part 2: Explanatory Information Related to IEC 62368-1
EN 55032 (CISPR 32)	Electromagnetic Compatibility of Multimedia Equipment–Emission Requirements
EN 55035 (CISPR 35)	Electromagnetic Compatibility of Multimedia Equipment–Immunity Requirements
IEC/EN 61000-3-2	Electromagnetic Compatibility (EMC)–Part 3-2: Limits–Limits for Harmonic Current Emissions (Equipment Input Current ≤16 A per Phase)
IEC/EN 61000-3-3	Electromagnetic Compatibility (EMC)–Part 3-3: Limits–Limitation of Voltage Changes, Voltage Fluctuations and Flicker in Public Low-Voltage Supply Systems, for Equipment with Rated Current ≤ 16 A per Phase and Not Subject to Conditional Connection
IEC Guide 112	Guide on the Safety of Multimedia Equipment
IEC 62911	Audio, Video and Information Technology Equipment–Routine Electrical Safety Testing in Production
IEEE 802.3bt	IEEE Standard for Ethernet Amendment 2: Physical Layer and Management Parameters for Power over Ethernet over 4 Pairs
IEEE 1680.1	Environmental and Social Responsibility Assessment of Computers and Displays
UL 2755	Outline of Investigation Covers Modular Data Centers (MDC)
Title 47 CFR Part 15	Radio Frequency Devices
Title 47 CFR Part 18	Industrial, Scientific, and Medical Equipment
IEC 60950-21	Information Technology Equipment–Safety–Part 21: Remote Power Feeding
IEC 60950-23	Information Technology Equipment–Safety–Part 23: Large Data Storage Equipment
IEC 60950-22	Information Technology Equipment–Safety–Part 22: Equipment to Be Installed Outdoors
IEC 60728-11	Cable Networks for Television Signals, Sound Signals and Interactive Services–Part 11: Safety
IEC/TS 62393	Portable and Hand-Held Multimedia Equipment–Mobile Computers–Battery Run-Time Measurement

The U.S. Consumer Product Safety Commission (CPSC) regulates the sale and manufacture of more than 15,000 different consumer products. CPSC has also maintained a public database of public complaints of safety problems and recalls connected with any of the consumer goods regulated by the CSPC (www.saferproducts.gov). This database provides a growing and potentially rich database for understanding trends in consumer product safety. Authority for European consumer product safety is provided under the framework of the General Product Safety Directive. A rapid alert system, RAPEX, allows for the rapid exchange of information on dangerous consumer products between the member countries and the European Commission, except for food, pharmaceutical, and medical devices, which are covered under other mechanisms [9]. In many countries additional requirements are specified by the national health authorities, the national authorities responsible for the protection of labor, the national water supply authorities, and other similar authorities.

In the EU, the CE Mark regulatory requirements for household products are specified in Low Voltage Directive (LVD 2014/35/EU), EMC Directive (2014/30/EU), Energy Efficiency (2017/1369/EC) (see Chapter 2, Section 2.2), RoHS, REACH, and WEEE Directives (see Chapter 3, Section 3.3). Depending on the specificity of the product, other directives can also apply, such as machinery and radio. The Low Voltage Directive (LVD) covers electrical equipment within certain voltage limits (50–1,000V AC supplied or 75–1,500V DC supplied). Household products can also be on the General Product Safety Directive (GPSD) 2001/95/EC that refers to products with a voltage below 50V for AC or below 75V for DC and applies in the absence of specific EU regulations on the safety of certain product categories and complements the provisions of sector legislation, which do not cover certain matters in relation to producers' obligations.

The most used safety series of standards that cover household equipment is IEC/EN/UL/ AS/NZS 60335 Household and Similar Electrical Appliances–Safety [10].

From point of view of protection against electric shock, household products are classified in the IEC 60335-1, as specified in Table 1.6.

Table 1.6 Classification of Household Equipment Depending on the Protection Against Electric Shock

Class of Equipment	*Description*
Class 0 Appliance	Appliance in which protection against electric shock relies upon Basic Insulation only, there being no means for the connection of conductive accessible parts, if any, to the protective conductor in the fixed wiring of the installation, reliance in the event of a failure of the basic insulation being placed upon the environment.
Class 0I Appliance	An appliance having at least Basic Insulation throughout and incorporating an earthing terminal but having a supply cord without earthing conductor and a plug without earthing contact.
Class I Appliance	Appliance in which protection against electric shock does not rely on Basic Insulation only but which includes an additional safety precaution, in that conductive accessible parts are connected to the protective earthing conductor in the fixed wiring of the installation in such a way that conductive accessible parts cannot become live in the event of a failure of the Basic Insulation. This provision includes a protective conductor in the supply cord.
Class II Appliance	Appliance in which protection against electric shock does not rely on Basic Insulation only but in which additional safety precautions are provided, such as Double Insulation or Reinforced Insulation, there being no provision for protective earthing or reliance upon installation conditions.
Class III Appliance	Appliance in which protection against electric shock relies on supply at safety extra-low voltage and in which voltages higher than those of safety extra-low voltage are not generated.
Class II Construction	Part of an appliance for which protection against electric shock relies upon Double Insulation or Reinforced Insulation.
Class III Construction	Part of an appliance for which protection against electric shock relies upon safety extra-low voltage and in which voltages higher than those of safety extra-low voltage are not generated.

The household product also has a specific classification, in addition to the above classification included within Table 1.6, regarding the usual types of appliances (portable, hand-held, stationary, and fixed) as follows:

- *Built-in appliance:* Fixed appliance intended to be installed in a cabinet, in a prepared recess in a wall, or in a similar situation.
- *Heating appliance:* Appliance incorporating heating elements but without any motor.
- *Motor-operated appliance:* Appliance incorporating motors but without any heating element. Note that magnetically driven appliances are considered to be motor-operated appliances.
- *Combined appliance:* Appliance incorporating heating elements and motors.

In general, the safety requirements and tests applied to household products are similar to those applicable for other electrical and electronic products. However, some specific requirements apply to the following household products [10]:

- Appliances for heating liquids and appliances causing undue vibration shall not be provided with pins for insertion into socket outlets.
- Appliances shall be constructed so that their electrical insulation cannot be affected by water which could condense on cold surfaces or by liquid that could leak from containers, hoses, couplings, and similar parts of the appliance. The electrical insulation of Class II Appliances and Class II Constructions shall not be affected if a hose ruptures or a seal leaks.
- Appliances containing liquid or gases in normal use or provided with steam-producing devices shall incorporate adequate safeguards against the risk of excessive pressure.
- Non-self-resetting thermal motor protectors shall have a trip-free action unless they are voltage maintained.
- The fixing properties of snap-in devices used in parts that are likely to be removed during installation or servicing shall be reliable.
- Handles, knobs, grips, levers, and similar parts shall be fixed reliably so that they will not work loose in normal use if loosening could result in a hazard. If these parts are used to indicate the position of switches or similar components, it shall not be possible to fix them incorrectly if this could result in a hazard.
- Spacers intended to prevent the appliance from overheating walls shall be fixed so that it is not possible to remove them from the outside of the appliance by hand or using a screwdriver or a spanner.
- Driving belts shall not be relied upon to provide the required level of insulation unless they are constructed to prevent inappropriate replacement.
- Direct contact between live parts and thermal insulation shall be effectively prevented unless such material is noncorrosive, nonhygroscopic, and noncombustible.

- Wood, cotton, silk, ordinary paper, and similar fibrous or hygroscopic material shall not be used as insulation, unless impregnated.
- Asbestos and oils containing polychlorinated biphenyl (PCB) shall not be used in appliances.
- Bare heating elements shall be supported so that the heating conductor is unlikely to come into contact with accessible metal parts if they rupture.
- Appliances, other than those of Class III, shall be constructed so that sagging heating conductors cannot come into contact with accessible metal parts.
- Appliances having parts of Class III construction shall be constructed so that the insulation between parts operating at safety extra-low voltage and other live parts complies with the requirements for double insulation or reinforced insulation.
- Parts connected by protective impedance shall be separated by double insulation or reinforced insulation.
- For Class II appliances connected in normal use to the gas mains or the water mains, metal parts conductively connected to the gas pipes or in contact with the water shall be separated from live parts by double insulation or reinforced insulation.
- Parts of natural or synthetic rubber used as supplementary insulation shall be resistant to aging or be located and dimensioned so that creepage distances are not reduced below the values specified even if cracks occur.
- Ceramic material which is not tightly sintered, similar materials or beads alone shall not be used as supplementary insulation or reinforced insulation.
- Conductive liquids that are or may become accessible in normal use shall not be in direct contact with live parts. Electrodes shall not be used for heating liquids.
- Shafts of operating knobs, handles, levers, and similar parts shall not be live unless the shaft is not accessible when the part is removed.
- Capacitors shall not be connected between the contacts of a thermal cut-out.
- Lamp-holders shall be used only for the connection of lamps.
- Motor-operated appliances and combined appliances, which are intended to be moved while in operation, or which have accessible moving parts, shall be fitted with a switch to control the motor. The actuating member of this switch shall be easily visible and accessible.
- Appliances shall not have an enclosure that is shaped and decorated so that the appliance is likely to be treated as a toy by children. Examples are enclosures representing animals or persons or resembling scale models.
- Software used in protective electronic circuits shall be software class B or software class C, as specified in the IEC 60335-1 standard.
- Appliances intended to be connected to the water mains shall withstand the water pressure expected in normal use and shall be constructed to prevent back-siphonage of nonpotable water into the water mains.
- Appliances, other than stationary appliances for multiple supplies, shall not be provided with more than one means of connection to the supply mains.

Stationary appliances for multiple supplies may be provided with more than one means of connection provided the relevant circuits are adequately insulated from each other.

- Plugs shall not be fitted with more than one flexible cord.

The annexes of the IEC 60335-1 provide some normative tests for motors: aging (Annex C), thermal motor protectors (Annex D), and motors having basic insulation that is inadequate for the rated voltage of the appliance (Annex I).

The IEC 60335 series include a dedicated standard for testing the chargers for secondary batteries: the IEC 60335-2-29. This standard deals with the safety of electric battery chargers for household and similar use having an output not exceeding 120V ripple-free direct current, their rated voltage being not more than 250V. Battery chargers intended for charging batteries in a household end-use application outside the scope of the IEC 60335 series of standards are within the scope of this standard. Battery chargers not intended for normal household use, but which nevertheless may be a source of danger to the public, such as battery chargers intended for use in garages, shops, light industry, and on farms, are within the scope of this standard.

The supply cord in household equipment can have one of the following types of attachment:

- Type X attachment: Method of attachment of the supply cord such that it can easily be replaced. The supply cord may be specially prepared and only available from the manufacturer or its service agent. A specially prepared cord may also include a part of the appliance.
- Type Y attachment: Method of attachment of the supply cord such that any replacement is intended to be made by the manufacturer, its service agent, or a similar qualified person.
- Type Z attachment: Method of attachment of the supply cord such that it cannot be replaced without breaking or destroying the appliance.

EMC requirements for household products are included in the EN 55014-1 (Emission) and the EN 55014-2 (Immunity). Noise levels for household and related types of electrical equipment can be measured by following the tests that are included in the IEC 60704 series of standards.

Household and related types of electrical equipment shall be marked by the information provided by the relevant product safety standard.

Some specific symbols used in the household products are summarized in Table 1.7.

The safety requirements for specific types of household and similar types of electrical equipment are primarily covered, as specified above, by the IEC/EN/UL/AS/NZS 60335, household and similar electrical appliances (series) [10]. See Table 1.8 for a list of the main standards applied to household products. Production line tests (routine tests) as recommended in Annex A of IEC 60335-1 standard should be performed by the manufacturer on every produced item to ensure the right construction of the household equipment (for details on routine tests refer to Section 1.1).

Table 1.7 Specific Symbols Applicable to Household Equipment

Symbol	*Meaning*	*Symbol*	*Meaning*
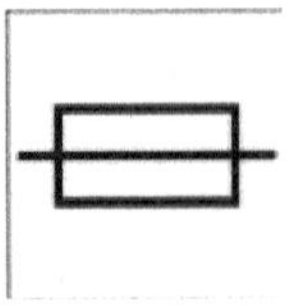	Fuse (rated current may be indicated with the symbol)		Lamp (rated voltage may be indicated with the symbol)
	Hot surfaces	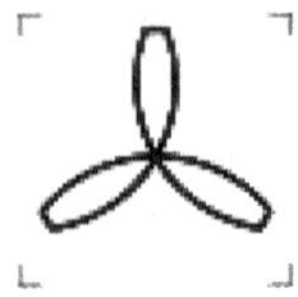	Ventilator; general
	Remote control		Manual control
	Automatic control (closed loop)		Open (a container)
	Frozen food storage compartment		Steam

1.3 Measurement, Control, and Laboratory Use

Measurement and control systems, including the instruments, transducers, and sensors used along within them, are important in a wide variety of applications, such as industrial, medical, and even in domestic activities. The safety requirements for specific types of measurement, control, and similar types of electrical systems are primarily covered by the IEC/EN/UL/AS/NZS 61010, Safety Requirements for Electrical Equipment for Measurement, Control, and Laboratory Use(series) standards [11].

In general, measurement equipment is defined as a device that tests, measures, indicates, or records one or more physical quantities. The science of measurement called metrology is probably the oldest science in the world, and includes the following categories: scientific, industrial, and legal. We will focus here on industrial meterology, which has to ensure adequate and traceable measurements in such areas as industry and medicine, as well in production and testing processes.

The IEC 61010 series of standards also applies to auxiliary equipment used on the measurement systems, such as signal generators, measurement standards,

Table 1.8 Standards for Household Equipment

Standard	*Title of Standard*
EN 50564	Electrical and Electronic Household and Office Equipment: Measurement of Low Power Consumption.
EN 55014-1	Electromagnetic Compatibility: Requirements for Household Appliances, Electric Tools, and Similar Apparatus. Emissions.
EN 55014-2	Electromagnetic Compatibility: Requirements for Household Appliances, Electric Tools, and Similar Apparatus. Immunity.
IEC 60335-1	Household and similar electrical appliances–Safety:
	Part 1: General requirements
	Part 2-2: Specific requirements for vacuum cleaners and water-suction cleaning appliances
	Part 2-3: Specific requirements for electric irons
	Part 2-4: Specific requirements for spin extractors
	Part 2-5: Specific requirements for dishwashers
	Part 2-6: Specific requirements for stationary cooking ranges, hobs, ovens
	Part 2-7: Specific requirements for washing machines
	Part 2-8: Specific requirements for shavers, hair clippers
	Part 2-9: Specific requirements for grills, toasters, and similar portable cooking appliances
	Part 2-10: Specific requirements for floor treatment machines and wet scrubbing machines
	Part 2-11: Specific requirements for tumble dryers
	Part 2-12: Specific requirements for warming plates
	Part 2-13: Specific requirements for deep-fat fryers, frying pans
	Part 2-14: Specific requirements for kitchen machines
	Part 2-15: Specific requirements for appliances for heating liquids
	Part 2-16: Specific requirements for food waste disposers
	Part 2-17: Specific requirements for blankets, pads
	Part 2-21: Specific requirements for storage water heaters
	Part 2-23: Specific requirements for appliances for skin or hair care
	Part 2-24: Specific requirements for refrigerating appliances, ice cream appliances, and ice-makers
	Part 2-25: Specific requirements for microwave ovens, including combination microwave ovens
	Part 2-26: Specific requirements for clocks
	Part 2-27: Specific requirements for appliances for skin exposure to ultraviolet and infrared radiation
	Part 2-28: Specific requirements for sewing machines
	Part 2-29: Specific requirements for battery chargers
	Part 2-30: Specific requirements for room heaters
	Part 2-31: Specific requirements for range hoods and other cooking fume extractors
	Part 2-32: Specific requirements for massage appliances
	Part 2-34: Specific requirements for motor-compressors
	Part 2-35: Specific requirements for instantaneous water heaters

Table 1.8 (continued)

Standard	*Title of Standard*
IEC 60335-1	Part 2-36: Specific requirements for commercial electric cooking ranges, ovens, hobs, and hob elements
	Part 2-37: Specific requirements for commercial electric doughnut fryers and deep-fat fryers
	Part 3-38: Specific requirements for commercial electric griddles and griddle grills
	Part 2-39: Specific requirements for commercial electric multipurpose cooking pans
	Part 2-40: Specific requirements for electrical heat pumps, air conditioners, and dehumidifiers
	Part 2-41: Specific requirements for pumps
	Part 2-42: Specific requirements for commercial electric forced convection ovens, steam cookers, and steam-convection ovens
	Part 2-43: Specific requirements for clothes dryers and towel rails
	Part 2-44: Specific requirements for irons
	Part 2-45: Specific requirements for portable heating tools
	Part 2-47: Specific requirements for commercial electric boiling pans
	Part 2-48: Specific requirements for commercial electric grillers and toasters
	Part 2-49: Specific requirements for commercial electric appliances for keeping food and crockery warm
	Part 2-50: Specific requirements for commercial electric bain-marie
	Part 2-51: Specific requirements for stationary circulation pumps for heating and service water installations
	Part 2-52: Specific requirements for oral hygiene appliances
	Part 2-53: Specific requirements for sauna heating appliances and infrared cabins
	Part 2-54: Specific requirements for surface-cleaning appliances for household use employing liquids or steam
	Part 2-55: Specific requirements for electrical appliances for use with aquariums and garden ponds
	Part 2-56: Specific requirements for projectors
	Part 2-58: Specific requirements for commercial electric dishwashing machines
	Part 2-59: Specific requirements for insect killers
	Part 2-60: Specific requirements for whirlpool baths and whirlpool spas
	Part 2-61: Specific requirements for thermal storage room heaters
	Part 2-62: Specific requirements for commercial electric rinsing sinks
	Part 2-64: Specific requirements for commercial electric kitchen machines
	Part 2-65: Specific requirements for air-cleaning appliances
	Part 2-66: Specific requirements for waterbed heaters
	Part 2-67: Specific requirements for floor treatment machines for commercial use
	Part 2-69: Particular requirements for wet and dry vacuum cleaners, including power brush, for commercial use
	Part 2-70: Specific requirements for milking machines
	Part 2-71: Specific requirements for electrical heating appliances for breeding and rearing animals
	Part 2-72: Specific requirements for floor treatment machines with or without traction drive for commercial use
	Part 2-73: Specific requirements for fixed immersion heaters
	Part 2-74: Specific requirements for portable immersion heaters
	Part 2-75: Specific requirements for commercial dispensing appliances and vending machines

Standard	*Title of Standard*
IEC 60335-1	Part 2-75: Specific requirements for commercial dispensing appliances and vending machines
	Part 2-76: Specific requirements for electric fence energizers
	Part 2-77: Specific requirements for pedestrian-controlled mains-operated lawnmowers
	Part 2-78: Specific requirements for outdoor barbecues
	Part 2-79: Specific requirements for high-pressure cleaners and steam cleaners
	Part 2-80: Specific requirements for fans
	Part 2-81: Specific requirements for foot warmers and heating mats
	Part 2-82: Specific requirements for amusement machines and personal service machines
	Part 2-83: Specific requirements for heated gullies for roof drainage
	Part 2-84: Specific requirements for toilet appliances
	Part 2-85: Specific requirements for fabric steamers
	Part 2-86: Specific requirements for electric fishing machines
	Part 2-87: Specific requirements for electrical animal stunning equipment
	Part 2-88: Specific requirements for humidifiers intended for use with heating, ventilation, or air-conditioning systems
	Part 2-89: Specific requirements for commercial refrigerating appliances and ice-makers with an incorporated or remote refrigerant unit or motor-compressor
	Part 2-90: Specific requirements for commercial microwave ovens
	Part 2-91: Specific requirements for walk-behind and hand-held lawn trimmers and lawn edge trimmers
	Part 2-92: Specific requirements for pedestrian-controlled mains-operated lawn scarifies and aerators
	Part 2-94: Specific requirements for scissors-type grass shears
	Part 2-95: Specific requirements for drives for vertically moving garage doors for residential use
	Part 2-96: Specific requirements for flexible sheet heating elements for room heating
	Part 2-97: Specific requirements for drives for shutters, awnings, blinds, and similar equipment
	Part 2-98: Specific requirements for humidifiers
	Part 2-99: Specific requirements for commercial electric hoods
	Part 2-100: Specific requirements for hand-held mains-operated garden blowers, vacuums, and blower vacuums
	Part 2-101: Specific requirements for vaporizers
	Part 2-102: Specific requirements for gas, oil, and solid-fuel burning appliances having electrical connections
	Part 2-103: Specific requirements for drives for gates, doors, and windows
	Part 2-104: Specific requirements for appliances to recover and/or recycle refrigerant from air-conditioning and refrigeration equipment
	Part 2-105: Specific requirements for multifunctional shower cabinets
	Part 2-106: Specific requirements for heated carpets and for heating units for room heating installed under removable floor coverings
	Part 2-107: Specific requirements for robotic battery powered electrical lawnmowers
	Part 2-108: Specific requirements for electrolyzers
	Part 2-109: Specific requirements for UV radiation water treatment appliances
	Part 2-110: Specific requirements for commercial microwave appliances with insertion or contacting applicators
	Part 2-111: Specific requirements for electric ondol mattress with a nonflexible heated part
	Part 2-112: Specific requirements for integrated kitchen appliances (in work)

Table 1.8 (continued)

Standard	*Title of Standard*
IEC 60335-1	Part 2-113: Particular requirements for cosmetic and beauty care appliances incorporating lasers and intense light sources.
	Part 2-114: Particular requirements for self-balancing personal transport devices for use with batteries containing alkaline or other nonacid electrolytes
	Part 2-115: Particular requirements for beauty care appliances (in work)
	Part 2-116: Particular requirements for furniture with electrically motorized parts
	Part 2-117: Particular requirements for automatic floor treatment machines for commercial use (in work)
	Part 2-118: Particular requirements for professional ice cream makers
	Part 2-119: Particular requirements for vacuum packaging machines (in work)
IEC 60704	Household and similar electrical appliances–Test code for the determination of airborne acoustical noise–Part 1: General requirements
	Part 2-1: Specific requirements for dry vacuum cleaners
	Part 2-2: Specific requirements for fan heaters
	Part 2-3: Specific requirements for dishwashers
	Part 2-4: Specific requirements for washing machines and spin extractors
	Part 2-5: Specific requirements for electric thermal storage room heaters
	Part 2-6: Specific requirements for tumble dryers
	Part 2-7: Specific requirements for fans
	Part 2-8: Specific requirements for electric shavers, hair clippers, or trimmers
	Part 2-9: Specific requirements for electric hair-care appliances
	Part 2-10: Specific requirements for electric cooking ranges, ovens, grills, microwave ovens, and any combination of these
	Part 2-11: Specific requirements for electrically operated food preparation
	Part 2-13: Specific requirements for range hoods and other cooking fume extractors
	Part 2-14: Specific requirements for refrigerators, frozen-food storage cabinets, and food freezers
	Part 2-15: Specific requirements for household food waste disposers
	Part 2-16: Specific requirements for washer-dryers
	Part 2-17: Specific requirements for dry-cleaning robots
	Part 2-18: Specific requirements for electric water heaters
	Part 3: Procedure for determining and verifying declared noise emission values
IEC 61543	Residual current-operated protective devices (RCDs) for household and similar use–Electromagnetic compatibility
IEC 62301	Household electrical appliances–Measurement of standby power
IEC 63086 (series)	Household and similar electrical air cleaning appliances–Methods for measuring the performance
	Part 1: General requirements
	Part 2-1: Specific requirements for determination of reduction of particles (under development)
	Part 2-6: Specific requirements for fresh-air air cleaners (under development)
UL60335	Safety of Household and Similar Appliances, Part 1: General Requirements

power supplies for laboratory use, transducers, and transmitters, and to test equipment integrated into manufacturing processes and intended for testing manufactured devices. The term measuring instrument is commonly used to describe a measurement system, whether it contains only one or more elements.

Electrical industrial process-control equipment is defined in the IEC 61010-1 as "an equipment which controls one or more output quantities to specific values, with each value determined by manual setting, by local or remote programming, or by one or more input variables." This type of process control is accomplished by observing a parameter, comparing it to some desired value, and initiating a control action to bring the parameter as close as possible to the desired value.

Except for equipment used as monitors during the manufacturing processes, the rest of the measurement activities are done in specialized laboratories or other locations. The IEC 61010-1 standard also covers the equipment used in laboratories and defines this as "equipment which measures, indicates, monitors, inspects or analyses materials, or is used to prepare materials, and in vitro diagnostic (IVD) equipment (including self-test IVD equipment to be used in the home), and inspection equipment to be used to check people or material during transportation." One important category of electrical laboratory equipment is represented by those used for testing manufactured devices. The laboratories need to be accredited by relevant accreditation bodies (see Chapter 7, Section 7.5). This accreditation will depend on the area of activities of that laboratory; for example, according to the ISO/IEC 17025 (for testing and calibration laboratories), or the ISO 15189 (for medical laboratories), or by other regional and national imposed regulations. The measurement equipment shall be calibrated following the requirements imposed by the accreditation body and/or in conformity with the applicable local regulations.

Calibration of the measuring instrument or reference material represents a basic tool for ensuring the traceability of measurement. Calibration determines the performance characteristics of an instrument or reference material. It is achieved using a direct comparison against measurement standards or certified reference materials. A calibration certificate is issued and a sticker is attached to the calibrated instrument. As with the measurements results, the calibration results need to consider the uncertainties that are influencing the evaluation process.

In the EU, the CE Mark regulatory requirements for electrical measurement, control, and laboratory use equipment are specified in LVD 2014/35/EU, EMC Directive (2014/30/EU), Energy Efficiency (2017/1369/EC) (see Chapter 2, Section 2.2), RoHS, REACH, and WEEE Directives (see Chapter 3, Section 3.3). Depending on the specificity of the product, other directives, such as Machinery and Radio, can also apply. IVD laboratory equipment falls under the 2017/746/EU Directive on in vitro diagnostic medical devices (IVDD). LVD 2014/35/EU covers electrical equipment within certain voltage limits (50–1,000V AC supplied or 75–1,500V DC supplied). For equipment with a supply voltage below 50V for AC, or below 75V for DC, GPSD 2001/95/EC applies.

Like other directives, the IVDD contains essential health, safety, and effectiveness requirements that an in vitro diagnostic device must meet, such as physical properties, sterility, calibration, EMC, electrical, mechanical, chemical, and radiation safety.

In the category of medical and machinery equipment, a risk management assessment must be conducted for measurement, control, and laboratory use products.

The purpose of this is to ensure that hazards, such as electric shock and burn, mechanical hazards, effects of mechanical stresses, fluids/fluid pressure, and radiation, spread of fire from the equipment, excessive temperatures; liberated gases, explosion, and implosion to the operator and surrounding area are reduced to a tolerable (low) level. The IEC 61010-1 leaves some flexibility regarding the choice of risk assessment procedures, giving examples of procedures and references to standards that may be used, depending on product application. The steps that need to be followed for risk management are

1. Risk analysis (identifying hazards and estimating risks);
2. Risk evaluation: determining severity and likelihood;
3. Risk reduction: steps taken to reduce risk (by design, by means of protection, inform the user);
4. Reviewing the risks after reduction.

The concept of a single-fault condition, used largely in medical equipment with slight differences, also applies for equipment for measurement, control, and laboratory use.

Specific, or product-specific, requirements (IEC 61010-2-x) are used in conjunction with the IEC 61010-1; they supplement or modify corresponding clauses in the general standard to make it specific to certain type of products. Examples are the IEC 61010-2-030 for equipment having testing or measuring circuits, the IEC 61010-2-101 for IVD medical equipment, and the IEC 61010-2-201 for control equipment.

The EMC requirements for this category of products are included in the IEC 61236 series of standards. Table 1.9 summarizes the list of the IEC 61010 and the IEC 61236 series of standards.

Some existing product standards such as the IEC 61131-2 covering programmable controllers refer to the IEC 61010-2-201 standard for treating the safety requirements. In the United States, the safety requirements for industrial power supplies from the UL 508 (Standard for Industrial Control Equipment) was replaced with the UL 61010-1 and the UL 61010-2-201 requirements; NEC Article 725 specifies the UL 61010-1 and the UL 61010-2-201 as options for assessment of NEC Class 2 power source, because this type of source does not require additional testing according to the UL 1310 (Standard for Class 2 Power Units).

In general, the safety requirements and tests applied to measurement, control, and laboratory use products are similar to these applicable for other electrical and electronic product, but some specific requirements apply to these types of products [11]:

- If an emergency stop is used to prevent a major safety hazard (harm to persons or the environment), there must be an additional type of indication (e.g., sound/alarm, flashing light).
- Information shall be provided on how the operator can tell if the equipment is malfunctioning if a hazard can be caused by misinterpretation.
- Information shall be provided on how to avoid hazards that could not be eliminated (moving parts, radiation, sound, burns from hot surfaces, etc.)
- Equipment for a short-term or intermittent operation that develops significant heat during the startup phase, and that relies on a continued operation

Table 1.9 Standard for Measurement, Control, and Laboratory Use

Standard	*Title of Standard*
IEC 61010 series	Safety requirements for electrical equipment for measurement, control, and laboratory use
	Part 1: General Requirements
	Part 2-010: Part 2: Specific Requirements for Laboratory Equipment for the Heating of Materials
	Part 2-011: Specific Requirements for Refrigerating Equipment
	Part 2-020: Specific Requirements for Laboratory Equipment for Laboratory Centrifuges
	Part 2-030: Specific Requirements for Equipment Having Testing or Measuring Circuits
	Part 2-032: Specific Requirements for Hand-Held and Hand-Manipulated Current Sensors for Electrical Test and Measurement
	Part 2-033: Specific Requirements for Laboratory Equipment for Hand-Held Multimeters and Other Meters, for Domestic and Professional Use, Capable of Measuring Mains Voltage
	Part 2-034: Specific Requirements for Measurement Equipment for Insulation Resistance and Test Equipment for Electric Strength
	Part 2-040: Specific Requirements for Sterilizers and Washer-Disinfectors Used to Treat Medical Materials
	Part 2-051: Specific Requirements for Laboratory Equipment for Mixing and Stirring
	Part 2-061: Specific Requirements for Laboratory Equipment for Laboratory Atomic Spectrometers with Thermal Atomization and Ionization
	Part 2-081: Specific Requirements for Laboratory Equipment for Automatic and Semi-Automatic Laboratory Equipment for Analysis and Other Purposes
	Part 2-091: Specific Requirements for Laboratory Equipment for Cabinet X-Ray Systems
	Part 2-101: Specific Requirements for In Vitro Diagnostic (IVD) Medical Equipment
	Part 2-201: Specific Requirements for Control Equipment
	Part 031: Safety Requirements for Hand-Held Probe Assemblies for Electrical Measurement and Test

to dissipate that heat, shall also be operated for the shortest rated period followed by the shortest rated recovery period.

- The nonhazardous limit for voltage is under 55V AC r.m.s or 140V DC for dry locations and 33V AC r.m.s or 70V DC for wet location.
- If the equipment has a nondetachable power cord, the impedance between the protective conductor plug pin of the mains cord and each accessible part for which protective bonding is specified shall not exceed 0.2 Ω.
- High-integrity resistors and capacitors are required to be rated for twice the working voltage.
- Secondary circuits are circuits where separation from the mains circuits is achieved by a transformer in which the primary windings are separated from the secondary windings by reinforced insulation or a screen connected to the protective conductor terminal. Separation by a relay, an optocoupler, a capacitor, or by protective impedance does not make a circuit "secondary."

Table 1.9 (continued)

Standard	*Title of Standard*
IEC 61326 series	Electrical Equipment for Measurement, Control and Laboratory Use–EMC Requirements
	Part 1: General Requirements
	Part 2-1: Specific Requirements–Test Configurations, Operational Conditions and Performance Criteria for Sensitive Test and Measurement Equipment for EMC Unprotected Applications
	Part 2-2: Specific Requirements–Test Configurations, Operational Conditions and Performance Criteria for Portable Testing, Measuring and Monitoring Equipment Used in Low-Voltage Distribution Systems
	Part 2-3: Specific Requirements–Test Configuration, Operational Conditions and Performance Criteria for Transducers with Integrated or Remote Signal Conditioning
	Part 2-4: Specific Requirements–Test Configurations, Operational Conditions and Performance Criteria for Insulation Monitoring Devices According to IEC 61557-8 and for Equipment for Insulation Fault Location According to IEC 61557-9
	Part 2-5: Specific Requirements–Test Configurations, Operational Conditions and Performance Criteria for Field Devices with Field Bus Interfaces According to IEC 61784-1
	Part 2-6: Specific requirements–In Vitro Diagnostic (IVD) Medical Equipment
	Part 3-1: Immunity Requirements for Safety-Related Systems and for Equipment Intended to Perform Safety-Related Functions (Functional Safety)–General Industrial Applications
	Part 3-2: Immunity Requirements for Safety-Related Systems and for Equipment Intended to Perform Safety-Related Functions (Functional Safety)–Industrial Applications with Specified Electromagnetic Environment

- A floating circuits dielectric test shall be conducted with at least 500V DC or 350V AC.
- Secondary circuits do not have to be low voltage—only accessible parts have to be low voltage. Safety extra-low voltage (SELV) circuits are not defined.
- Products that are not intended to emit radiation cannot exceed 1 μSv/h at 100 mm.
- Conductors located the same two layers molded together shall be separated by a minimum distance of 0.4 mm.
- Human factors and ergonomics shall be considered regarding the interaction between the user, equipment, and their environments.
- No hazards shall arise if adjustments, knobs, or other software-based or hardware-based controls are set in a way not intended, and not described in the instructions.
- Electrical measuring instruments shall have adequate bandwidth to provide accurate readings, taking into account all components (e.g., DC, mains frequency, high frequency, and harmonic content) of the parameter being measured.
- If an r.m.s. value is measured, care shall be taken that the measuring instrument gives a true r.m.s. reading of nonsinusoidal waveforms as well as sinusoidal waveforms.

- Measurements are made with a meter whose input impedance has minimum influence on the measurement.
- The following materials shall not be used as insulation for safety purposes: (a) materials that can easily be damaged (for example, lacquer, enamel, oxides, anodic films), and (b) nonimpregnated hygroscopic materials (e.g., paper, fibers, fibrous materials).

The IEC 61010-1 requires the following markings to appear on the equipment:

a. Manufacturer name and/or trademark;

b. Model designation;

c. Nature of supply;

d. Voltage rating;

e. Supply frequency;

f. Rated power (in VA-apparent power or W-active power) or current input;

g. The type and rating of any operator replaceable fuses (adjacent to each fuseholder);

h. "O" and "I" symbols for ON and OFF positions, adjacent to the mains switch, circuit-breakers, and push-button switch (a lamp alone is not considered to be a satisfactory marking);

i. Accessory mains socket-outlets accepting standard mains plugs shall be marked with the voltage if it is different from the mains supply voltage;

j. All controls and terminals, to indicate their function, settings, and maximum output or input as applicable, as well as the functions of operator controls;

k. The protective conductor terminal for Class I equipment;

l. The functional earth terminal, if applicable;

m. The symbol for Class II equipment protected throughout by double insulation, if applicable;

n. Type of battery, polarity, and mode of insertion, if applicable.

Some specific symbols used in measurement, control, and laboratory use products are summarized in Table 1.10.

See Table 1.9 for a list of the main standards applied to measurement, control, and laboratory use products. Production line tests (routine tests) as recommended in Annex F of the IEC 61010-1 standard should be performed by the manufacturer on every produced item to ensure the right construction of the measurement, control, and laboratory use equipment (see Section 1.1 for more details on routine tests).

1.4 Medical Electrical Equipment

This section is dedicated to medical electrical equipment (MEE), which is a product category to which we give special attention.

Table 1.10 Symbols Applicable to Measurement, Control, and Laboratory Use Products

Symbol	*Meaning*	*Symbol*	*Meaning*
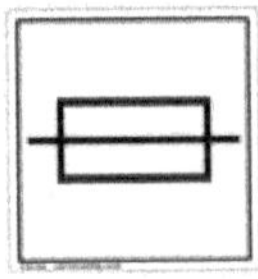	Fuse (rated current may be indicated with the symbol)		Lamp (rated voltage may be indicated with the symbol)
	Hot surfaces	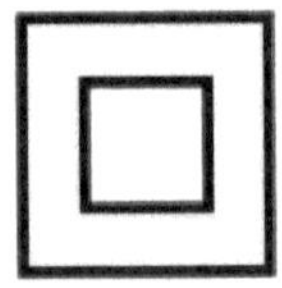	Class II equipment
	"IN" position of a bistable push control	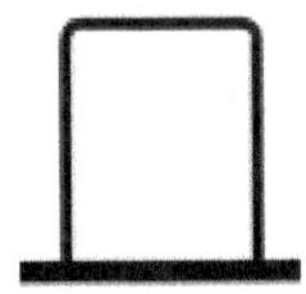	"OUT" position of a bistable push control
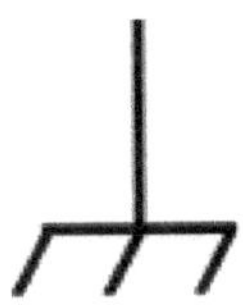	Frame or chassis terminal		Biological hazard
	Ionizing radiation		Nonionizing radiation

The international IEC 60601-1 standard (Medical Electrical Equipment–Part 1: General Requirements for Basic Safety And Essential Performance) [12] defines medical electrical equipment as "electrical equipment having an applied part (defined as part that in normal use–which means 'operation, including routine inspection and adjustments by any operator, and stand-by, according to the instructions for use'; necessarily comes into physical contact with the patient to perform its function) or, assures the transferring of energy to, or from the patient (defined as 'living being–person or animal-undergoing a medical, surgical or dental procedure') or detects such energy transfer to or from the patient" [12].

Such equipment is:

a. Provided with not more than one connection to a specific supply main (defined as "source of electrical energy not forming part of MEE, including battery systems and converter systems in ambulances and the like");

b. Intended by its manufacturer to be used (intended use represent "the use for which a product, process, or service is intended according to the specifications, instructions, and information provided by the manufacturer"):

1. In the diagnosis, treatment, or monitoring of a patient;
2. For compensation or alleviation of disease, injury, or disability.

MEE includes those accessories (defined as "additional part for use with equipment in order to: achieve the intended use, adapt it to some special use, facilitate its use, enhance its performance, or enable its functions to be integrated with those of other equipment") that are necessary to enable this normal use [12].

The implantable parts of active implantable medical devices fall within this definition.

MEE can be used independently or within a medical electrical system (MES), defined [12] as a: "combination, as specified by its manufacturer, of items of equipment, at least one of which is MEE to be interconnected by functional connection (connection, electrical or otherwise, including those intended to transfer signals, data, power, or substances) or by use of a multiple socket-outlet (which means 'one or more socket-outlets intended to be connected to, or integral with, flexible cables, cords or MEE supply mains or equivalent voltage')." A multiple socket-outlet can be a separate item or an integral part of the equipment.

MEE and MES have specific specificities and should be designed and manufactured in such a way that, when used under the conditions for the intended use and specified users, will not compromise the clinical condition or the safety of patients, or the safety and health of users or other persons. The solutions adopted for the design and manufacture of the MEE should conform to safety principles based on the generally accepted state of the art. The essential principles of safety and performance for MEE can be summarized as follows [13]:

- Protection against the harms produced by identified physical hazards (electrical, mechanical, chemical, thermal, radiation, etc.);
- Protection against the harms produced by identified essential performance (informational-labeling and functional-SW, clinical function, etc.);
- Protection against the harms resulting from sterilization, packaging, and microbial contamination;
- Considerations of environment and conditions of use;
- Protection against hazards resulting from implantable MEE;
- Protection against the risks posed to the patient or user by MEE incorporating a substance considered to be a medicinal product/drug or other substances or supplying energy;
- Protection against hazards resulting from diagnostic or measuring function;
- Protection against hazards resulting for treatment or rehabilitation function;
- Fulfilling the clinical evaluation with acceptable results.

When dealing with MEE, two aspects need to be considered: basic safety (prevention of harms due to hazards from physical sources) and essential performance (related to clinical function of the equipment and degradation of this due to internal failures (due to software (SW) or a component), and/or external disturbances: voltage fluctuation, electromagnetic disturbances, environmental influences, etc.). These aspects differ from those for other types of electrical equipment because of the specific relationship between the MEE and the patient, operator, and surroundings, and are separated into means of operator protection (MOOP) and means of patient protection (MOPP). This separation allows in some situations (e.g., MEE with type B applied part) to use in the equipment only one MOPP provided by a power supply nonmedical rated (e.g., approved for information technology equipment). No matter whether MOOP or MOPP is chosen, it is required that the leakage current requirements are met. For the power supply, this means, in normal conditions, 300 μA for the United States and 500 μA for the EU. Three different types of leakage current are present on an MEE: earth, touch, and patient. Patient auxiliary current is also considered (between two or more applied parts). To ensure a traceable simulation of current passing through a human body, measurement circuits have been designed to simulate the average typical electrical characteristics of the human body. These measurement circuits are referred to as body models or measuring device (MD in the IEC 60601-1). Another specificity of MEE is the mandatory management of the risks. The methodology is described in the ISO 14971 standard. The manufacturer is required to analyze the design, identify risks, and use risk control measures to reduce each risk below a tolerable level.

Any risks associated with the use of MEE shall be acceptable when weighed against the benefits. As a result of the risk management process, the related risk acceptability is assessed by estimating both the severity and probability of occurrence of harm resulting from exposure to specific hazards or hazardous situations. Another MEE specificity that can be considered is the use of only a Class II power supply in a home healthcare environment.

The philosophy of the MEE Basic Safety and Essential Performance is described in IEC/TR 60513 [14], which specifies that MEE must not generate unacceptable risks in normal conditions (NCs) and single-fault conditions (SFCs).

One of the main concepts of this philosophy in reference to faults is represented by the single fault [15]. The SFC is defined as the condition in which a single means of protection against a hazard is defective or a single external abnormal condition is present [12]. In a product that is single-fault safe, a lower limit is established on the probability of occurrence of harm from a hazard. If this probability is achieved, then the risk level for the specific hazard is acceptable.

Additionally, the probability of a simultaneous occurrence of two single faults is considered small enough to be neglectable and, in any of the specified single-fault conditions, hazards shall not arise. Where one single fault directly causes subsequent fault conditions, the probability is the same as the one of the first fault, and the product shall remain safe. A single fault should be discovered by an unmistakable and clearly discernible signal (e.g., alarm) that becomes obvious to the operator. Depending on the severity of the failure mode the operator will take corrective action by manual means, or the product will automatically take corrective action. One should not forget that the place where a failure is first detected may not be where it started. Adding verification or validation controls (e.g., alarms on failure) can reduce the probability of a failure being undetected and having a greater effect on the product if a further failure occurs. The single-fault philosophy implies that, in general, the product is expected to have two means of protection against each hazard. This is considered to give a negligible risk, provided that the probabilities of failure of individual systems are low.

When MEE is combined with other electrical equipment to form an MES, additional aspects need to be considered. When laboratory equipment (which falls in the scope of IEC 61010-1) is incorporated by a manufacturer into an MES, this laboratory equipment is treated as an MEE.

Within the EU, MEE falls under 2017/745/EU Medical Devices Regulation (MDR) and under other applicable EU Directives if it consists of specific parts, covered by these Directives (e.g., MEE that has at least one part of which moves, such computed tomography (CT) or radiotherapy gantry, needed to fulfill additionally the Essential Health & Safety Requirements (EHSR) of the 2006/42/EC Machinery Directive; MEE that intentionally emit and/or receive radio waves needed to also fulfill the 2014/53/EU RED Directive, etc.).

For implantable and body-worn MEE, tissue propagation characteristics and specific absorption rate shall be considered. When MEE needs to have primary or secondary radio service classification, this depends on the wireless frequency band selected. When considering commercial off-the-shelf radio frequency (RF) wireless components or systems that conform to industry standards (e.g., IEEE 802.11 standards), MEE manufacturers should take into account that some equipment might not have been adequately tested or qualified to address the needs and risks for use in MEE. This is because such equipment may conform to standards that are not written specifically for MEE. A key factor affecting a wireless MEE's performance is the limited amount of RF spectrum available, which can result in potential competition among wireless technologies for simultaneous access to the same spectrum. If the RF wireless MEE is expected to be used in proximity to other RF wireless in-band sources, is recommended to address such risks through testing for the coexistence of the MEE wireless system in the presence of the number and types of in-band sources expected to be in the proximity.

Actually, a trend exists to harmonize many of the concepts and requirements applicable to audio/video, information, and communication technology equipment specified in the IEC 62368-1 standard to MEE (see Section 1.1).

The IEC 60601 is a series of technical standards for the safety and effectiveness of the medical electrical equipment. The general (primary) standard, known as the "Bible," is IEC 60601-1–Medical Electrical Equipment–Part 1: General Requirements for Basic Safety and Essential Performance, issued in 2005 and amended with Amend.1 in 2012 and Amend.2 in 2020.

The IEC 60601-1 is the harmonized standard for medical electrical equipment recognized by public health authorities in most countries. This standard was globally adopted with specific national deviations related to country-specific requirements. The following countries have national deviations: Austria, Belgium, Brazil, Canada, Czechia, Germany, Finland, France, Great Britain, Hungary, Israel, Italy, Japan, Malaysia, Netherlands, Norway, Poland, Portugal, Russia, Sweden, Switzerland, Singapore, Slovenia, Slovakia, Ukraine, and the United States.

The standard name varies by country: in the EU it is EN 60601-1, in the United States it is ANSI/AAMI ES60601-1, in Japan it is JIS 0601-1, in Canada it is CAN/CSA-C22.2 No. 60601-1, and so on. Within the IEC 60601-1 series are collateral standards that are designated as IEC 60601-1-x. The collaterals are standards that specify additional general requirements applicable to a subgroup of equipment or to a specific characteristic of all equipment not fully addressed in the general standard. The most current issued are:

- IEC 60601-1-2 EMC (the fourth edition from 2014 changed the name from Electrmagnetic Compatibility on Electromagnetic Disturbances);
- IEC 60601-1-3 for radiation protection for diagnostic x-ray systems;
- IEC 60601-1-6 for usability;
- IEC 60601-1-8 for alarms;
- IEC 60601-1-9 relating to environmental design;
- IEC 60601-1-10 for Physiological Closed-loop Controllers;
- IEC 60601-1-11 for home healthcare equipment;
- IEC 60601-1-12 for emergency services environment.

There are also many specific standards designated as IEC 60601-2-X (X representing a specific standard number between 1–90, until now; some under development) that define specific requirements related to specific types of equipment.

A complete list of published IEC standards for medical (MED category) can be found by accessing the IECEE website [16].

The requirements of a specific standard take priority over all part 1 standards (and can have different requirements as in the general and collateral standards).

Based on the cooperation between the IEC and ISO and for prevention of work overlap, part of the specific standards for MEE started to be issued, on some specific applications, as common standards IEC/ISO numbered 80601-2-X.

More information about ISO standards can be found by accessing the ISO website [17].

A database containing the U.S. Food and Drug Administration's (FDA's) recognized consensus standards from medical device standard developers (e.g., AAMI, ANSI, ASME, ASTM, CEN, IEEE, IEC, and ISO) can be searched according to the specialty area (e.g., biocompatibility, cardiovascular, general: risk management, EMC, QMS, software, IVD, materials, radiology, sterility) within the FDA website [18], which allows finding the relevant standards, as shown below:

Sample:
Standard Developer: IES–Illuminating Engineering Society
Specialty: Radiology
Standards:

ANSI/IES RP-27.3-17	Recommended Practice for Photobiological Safety for Lamps–Risk Group Classification and Labeling
IES RP-27.1-2015	Recommended Practice for Photobiological Safety for Lamps and Lamp Systems–General Requirements
ANSI/IES RP-27.2-00/ R17	Recommended Practice for Photobiological Safety for Lamps and Lamp Systems–Measurement Techniques

In addition to technical standards published in the IEC 60601 and the ISO/IEC 80601 series, other standards or regulations dealing with dedicated aspects that apply to MEE are specified in Table 1.11.

Table 1.11 Standards Applicable to Specific Areas of MEE

Quality	ISO 13485, 21 CFR Part 820
Risk management	ISO 14971
Biocompatibility	ISO10993 series
Environment	RoHS, REACH, WEEE EU Directives; IEC 63000
Software	IEC 62304
Storage and shipping	ISO 11607 series, ASTM D4169
Labeling and information	ISO 15223, EN 1041, 21 CFR Part 201
Batteries	IEC 62133, IEC 60086 series, UL 1642, UL 2054, IEC 61960, UN 3480
IP code	IEC 60529
Cleaning and disinfection	AAMI ST98, ASTM E1837, ASTM E2314
Sterilization	ANSI/AAMI ST60, ISO 11135-1, ISO 11137-1, ISO 11737-2, ISO 14937, ISO 14160, ISO 17665-1, ISO 14698-1, ISO 20857
Clinical evaluation	ISO 14155
Bluetooth	ETSI EN 300 328, 47 CFR 15, IEEE 802.15.4
RF Wi-Fi	ETSI EN 302 536, ETSI EN 301 489-29, IEEE 802.11, ANSI C63.10, IEC TR 80001-2-3, IEEE P11073, ISO/TR 21730
Active implants	ANSI/AAMI PC69, ISO 14708 series, ISO 14117, EN 45502
Cybersecurity and privacy	ISO/IEC 27032, HIPAA, NIST SP 800-127, IEC TR 80001-2, AAMI TIR57, UL 2900
Laser	IEC 60825, ANSI Z 136, 21 CFR Part 1040

For example, ISO 13485 establishes the requirements for a quality management system for both the design and manufacturing of MEE. It covers aspects including risk management, design control during product development, and verification and validation systems.

The impact of the environment in which MEE is used represents a challenge of establishing the specific requirements for basic safety and essential performance of MEE. An environmental condition such as temperature, atmospheric pressure, humidity, vibrations, and shocks, can affect the basic safety and essential performance of MEE (refer to Chapter 3).

The fourth edition of the collateral standard IEC 60601-1-2 addresses in a new vision the EMC effects both in the subject MEE and on other equipment located in the vicinity. First, in the title of the standard, the term "compatibility" is replaced with "disturbances," aligning the electromagnetic environment (defined as "the totality of electromagnetic phenomena existing at a given location") with essential performance aspects. The electromagnetic disturbances (defined as "any electromagnetic phenomenon that could degrade the performance of a device, equipment or system: electromagnetic noise, an unwanted signal, or a change in the propagation medium itself") can affect the clinical function of the MEE, taking into account the equipment's intended operating environment. The immunity test levels are specified according to three categories of intended use environments: (1) the professional healthcare facility environment (hospital, intensive care units, dental offices, clinics, etc.), (2) the home healthcare environment (as defined in the IEC

60601-1-11 standard), and (3) special environments (defined as "electromagnetic environment: Specifically high-power medical equipment, that requires emissions limits, immunity test levels or test methods that are different from those specified for the professional healthcare facility environment and the home healthcare environment"). MEE is expected to provide its basic safety and essential performance without interfering with other equipment and systems in the electromagnetic environments in which they are intended to be used.

As an example, laboratory equipment that is intended for use within a professional healthcare facility environment or in a healthcare facility environment will be tested according to the IEC 60601-1-2 and not according to the laboratory equipment EMC standard IEC 61326.

In addition to general testing covering basic safety (visual inspection, ground connection bonding, and continuity, dielectric strength, input current, normal heating, leakage currents, mechanical tests, limitation of voltage, current or energy, durability and legibility of marking, defibrillation-proof applied part protection, creepage distances and air clearances, sound pressure level, identification of the source of ignition, single-fault condition, rechargeable battery overcharge/discharge, mains transformers, etc.) requested by the IEC, ISO, and other applicable standards, MEE needs to be tested for the identified essential performance, electromagnetic disturbances, biocompatibility, software life-cycle, cleaning, disinfection, and sterilization. The risk management file and the usability process also need to be evaluated. When there are applicable specific standards for a specific MEE, additional tests need to be conducted as specified in the specific standards, using dedicated test equipment (e.g., in MEE used in cardiology, radiology, hospital beds, lung ventilators, or anesthetic machine). If the MEE is intended to be used in the sterile state, the sterilization process may be either performed by the manufacturer or by the user. There are four major means employed for sterilizing: heating with steam, ethylene oxide (ETO), radiation, and dry heat. If the MEE consists of radio modules or active implants or needs sterilization, the requirements specified in Table 1.11 also apply.

Guidance on general testing procedures that can be used for MEE is given in IEC/TR 62354 [19].

The labeling (external, internal, and control markings, accompanying documents: instructions for use, technical description, user's and service manuals) and symbols (see Table 1.12) are designated requirements for MEE.

The general information provided in the marking label of an MEE is similar to other categories of electrical equipment and must include, but are not limited to: manufacturer's name and address, model number, serial number, electrical ratings, and class of protection against electrical shock. Safety-critical MEE labeling helps to ensure safety and effectiveness, assisting in the efficient use of the equipment. Symbols are often used on equipment in preference to words to save space and to preclude language differences, allowing easier comprehension. A specific labeling requirement for medical devices refers to the use of a unique device identifier (UDI) that is on the way to becoming a global device identification system. Initially adopted by the FDA, the UDI system now is part of EU regulations covered by MDR ensuring better traceability, more efficient capture of information (including adverse effects) about the product, easier recall of devices, and fights against counterfeiting. UDI is becoming a practical and efficient tool for product identification that

Table 1.12 Specific Symbols Applicable to MEE

Symbol	*Meaning*	*Symbol*	*Meaning*
	Follow instructions for use		Date of manufacturing
	Type B applied part		Defibrillation proof type B applied part
	Type BF applied part		Defibrillation proof type BF applied part
	Type CF applied part		Defibrillation proof type CF applied part
	Category AP equipment		Category APG equipment
	Do not reuse		Biological hazard
	Keep away from rain		Sterile
	Humidity limitation		Atmospheric pressure limitation
	Temperature limitation		Batch code

consists of such information as the date of manufacturing, lot or batch number, serial number, and date of expiration. The FDA UDI and Global Unique Device Identifier Database (GUDID) requirements can be accessed on the FDA website (https://www.fda.gov/medical-devices/unique-device-identification-system-udi-system/global-unique-device-identification-database-gudid).

For Europe, the EUDAMED (https://ec.europa.eu/health/md_eudamed/overview_en) works as a registration system, a collaborative system, a notification system, and a dissemination system (open to the public) IT tool for information about medical devices.

Some of the specific symbols used on MEE are included in Table 1.12 [12, 20]. These symbols may be used on the MEE itself, on its packaging, or in the accompanying documentation.

The accompanying documents should specify the intended use, classifications related to equipment, frequently used functions, explanation of all marked safety signs and symbols, all warning and safety notices, cleaning, disinfection, and sterilization methods, a list of all system messages, error messages, and fault messages, special skills, training and knowledge required of the operator, and environmental restrictions on locations of use, including conditions for transport and storage. In addition, the level of the risks associated with the disposal of waste products, residues, and so on, at the end of the MEE's expected service life shall be specified. In EU RoHS, REACH, and WEEE directives apply and instructions must be provided to prove the implementation.

A manufacturer needs to establish the target markets for MEE and investigate the specific labelling requirements for certification and regulatory approval.

Production line tests (PLT), routine tests, are performed by the manufacturer on every item produced to ensure the right construction of the MEE. These do not need to be identical to the type tests (conducted for type approval and certification of MEE) and can be adapted to manufacturing conditions with less risk for the quality of the insulation or other important characteristics for basic safety and essential performance. These tests are conducted with equipment fully assembled and with settings that would generate the worst-case situation. Suggested basic safety test parameters could be leakage currents, dielectric strength, and the continuity of protective earthing. The equipment shall not be unwired, modified, or disassembled for the test, but snap-on covers and friction-fit knobs may be removed if they would interfere with the tests. The equipment shall not be energized during the tests, but the mains switch shall be in the on-position.

Guidance on general testing procedures that can be used for PLT is given in the IEC/TR 62354 [19]. Specific MEE can have as PLT some tests referring to essential performance (e.g., the output level from a laser MEE, the right work of an alarm function, etc.). Additionally, the IEC 62353 standard Medical Electrical Equipment–Recurrent Test and Test after Repair of MEE defines the minimum requirements of ensuring the in-service electrical safety of MEE can be used. The incorporated tests are beyond those of type testing and are part of the service carried out by the in-hospital clinical engineering department.

The accompanying documents of MEE need to include detailed instructions on preventive inspection, maintenance, calibration, frequency of such maintenance, parts requiring preventive inspection and maintenance, and information on how to conduct safe such maintenance necessary to ensure the continued safe use of the equipment. Special attention refers to instructions ensuring the adequate maintenance of MEE containing rechargeable batteries and instructions for correct replacement of interchangeable or detachable parts specified by the manufacturer as replaceable by service personnel (minimum qualifications of this personnel will be documented in the technical description).

1.5 Luminaires and Lamp Control

The light sources from lighting equipment in addition to daylight are an integral part of our life. Consumers of such equipment include economic, educational, scientific, domestic, commercial, industrial, healthcare, and transportation sectors. Lighting products include lamps and luminaires.

A luminaire is a complete electric light fixture that distributes, filters, or transforms light from one or more lamps. A luminaire has the necessary parts to support and protect lamps, including the mechanism for inserting or holding the lamp(s), wiring, socket, and other protective components. Different types of luminaires include floor, table, wall, pendant, chandelier, spotlight, ceiling, direct, indirect, clear, frosted, and opaque.

A "light source" means an electrically operated product intended to irradiate, or, in the case of a nonincandescent light source, intended to be possibly tuned to emit, light, or both, using incandescence, fluorescence (magnetic induction), halogen, high-intensity discharge (sodium, mercury, metal halide), inorganic light-emitting diodes (LEDs) or organic light-emitting diodes (OLED), or their combinations as lighting technology, and with all of the following optical characteristics [21]:

a. Chromaticity coordinates x and y in the range described by the equations

$$0.270 < x < 0.530$$

$$2.3172\ x^2 + 2{,}3653\ x - 0{,}2199 < y < -\ 2.3172\ x^2 + 2.3653\ x - 0.1595$$

b. A luminous flux <500 lumens per mm^2 of projected light-emitting surface area

c. A luminous flux between 60 and 82,000 lumens;

d. A color rendering index (CRI) > 0

Some specific terms used in lighting equipment are explained next:

- Control gear, which means one or more devices that may or may not be physically integrated with a light source, intended to prepare the mains for the electric format required by one or more specific light sources within boundary conditions set by electric safety and electromagnetic compatibility. It may include transforming the supply and starting voltage, limiting operational and preheating current, preventing cold starting, correcting the power factor and/or reducing radio interference. The term "control gear" does not include power supplies and lighting control parts and nonlighting parts, although such parts may be physically integrated with a control gear or marketed together as a single product.

- Power over Ethernet (PoE) switch means equipment for power supply and data-handling that is installed between the mains and office equipment and/or light sources for data transfer and power supply. PoE is not a control gear.
- Ballast, used in fluorescent lighting systems, that is inserted between the supply and one or more discharge lamps which, utilizing inductance, capacitance, or resistance, single or in combination, serves mainly to limit the current of the lamp(s) to the required value. This is necessary during the start of the fluorescent lamps due to the high voltage involved. It may also include means for transforming from the supply voltage and arrangements that help to provide starting voltage and preheating current, prevent cold starting, reduce the stroboscopic effect, correct the power factor, and suppress radio interference.

Characteristics related to safety, photometry, luminous intensity, light distribution, optical radiation, energy efficiency, and so on, shall be considered when referring to efficient and safe lighting equipment.

The IEC 60598-1 is the standard that specifies general requirements for luminaires, incorporating electric light sources for operation from supply voltages up to 1,000V. The requirements and related tests of this document cover classification, marking, mechanical construction, electrical construction, and photobiological safety [22]. The object of this IEC 60598 Part 1 standard is to provide a set of requirements and tests that are considered to be generally applicable to most types of luminaires and that can be called up as required by the detailed specifications of the IEC 60598-2 series of standards.

The EMC requirements for lighting products are included in the EN 55015 (Emission) and IEC/EN 61547 (Immunity) standards. The EN 55015 requirements apply to the emission (radiated and conducted) of radio frequency disturbances from lighting equipment, the lighting part of multifunction equipment where this lighting part is a primary function, ultraviolet (UV) and infrared (IR) radiation equipment for residential and nonindustrial applications, advertising signs, decorative lighting, and emergency signs.

Within the EU it is mandatory to comply with CE requirements and compliance is largely demonstrated by a self-declaration process. The CE Mark regulatory requirements for lighting products are specified in LVD, EMC Directive (2014/30/EU), ErP (Energy-Related Products) Regulation (2019/2020/EU), RoHS, REACH, WEEE Directives (see Chapter 3, Section 3.3). Depending on the specificity of the product, other directives can also apply, such as machinery and radio. LVD 2014/35/EU covers electrical equipment within certain voltage limits (50–1,000V AC supplied or 75–1,500V DC supplied). For products with a supply voltage below 50V for AC, or below 75V for DC, GPSD 2001/95/EC applies. Energy labeling requirements for light sources are under regulation on energy labelling for light sources (EU) 2019/2015. Using a scale from A (most efficient) to G (least efficient), the new labels will give information on the energy consumption, expressed in kWh per 1,000 hours, and have a quick response (QR) code that links to more information in an online database. Additional information related to energy efficiency labeling can be found in Chapter 2, Section 2.2. Eco-design rules are mandatory for almost all lamps sold in the EU. These regulations set energy efficiency requirements and other factors such as bulb lifetime and warm-up time. The new requirements for light sources and separate control gears are specified in the regulation for eco-design requirements for light sources and separate control gears (EU) 2019/2020. With the new regulation, most halogen lamps and traditional fluorescent tube lighting, which are common in offices, will be phased out beginning in September 2023.

The IEC 62493 standard applies to the assessment of lighting equipment related to human exposure to electromagnetic fields. The assessment consists of the induced internal electric field for frequencies from 20 kHz to 10 MHz and the specific absorption rate (SAR) for frequencies from 100 kHz to 300 MHz around lighting equipment. Additional information related to human exposure to an electromagnetic field can be found in Chapter 5, Section 5.1.4.1.

The United States operates a voluntary approval process like the EU but for American market manufacturers, who are commonly asked to have products tested in a nationally recognized testing laboratory (NRTL).

The Energy Independence and Security Act (EISA) of 2007 [23] laid out changes in lighting legislation that established performance standards and the phase-out of incandescent lightbulbs in order to require the use of more efficient fluorescent lighting. Manufacturers are now required to provide brightness (lumens) and energy-cost information on packaging within a detailed "Lighting Facts" label. A sample for such a label is shown in Figure 1.2 [24]. This applies to all medium-based general-service incandescent, halogen, LED, and compact fluorescent bulbs.

The Consumer Product Safety Act (CPSA) regulates seasonal and decorative lighting such as Christmas light strings. Importers shall comply with the safety requirements provided by the (www.saferproducts.gov) before importing lighting products to the United States. The relevant UL standards applicable in the United States for lighting equipment are presented in Table 1.13.

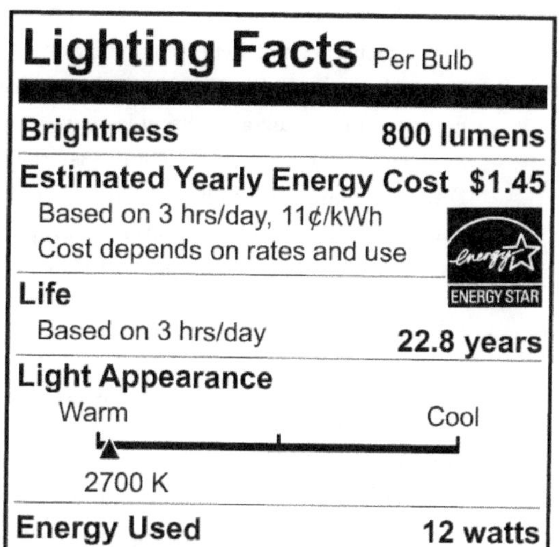

Figure 1.2 Sample of a "Lighting Facts" label.

Table 1.13 Standards for Lighting Equipment

Standard	*Title of Standard*
CSA-C22.2 No. 250.0	Luminaires
EN 1837	Safety of Machinery–Integral Lighting of Machines
EN 1838	Lighting Applications–Emergency Lighting
EN 12464 series	Light and Lighting–Lighting of Work Places
	Part 1: Indoor Work Places
	Part 2: Outdoor Work Places
EN 12665	Light and Lighting–Basic Terms and Criteria for Specifying Lighting Requirements
EN 13032 series	Light and Lighting–Measurement and Presentation of Photometric Data of Lamps and Luminaires
EN 13201 series	Road Lighting
EN 14255 series	Measurement and Assessment of Personal Exposures to Incoherent Optical Radiation
Title 47 CFR Part 15 (FCC Part 15)	Radio Frequency Devices
IEC 60238	Edison Screw Lampholders
IEC 62493	Assessment of Lighting Equipment Related To Human Exposure to Electromagnetic Fields
IEC 61547	Equipment for General Lighting Purposes–EMC Immunity Requirements
EN 55015	Limits and Methods of Measurement of Radio Disturbance Characteristics of Electrical Lighting and Similar Equipment
IEC 60061series	Lamp Caps and Holders Together with Gauges for the Control of Interchangeability and Safety
IEC 60357	Tungsten Halogen Lamps (Non-Vehicle)–Performance Specifications
IEC 60400	Lampholders for Tubular Fluorescent Lamps and Starter-Holders
IEC 60432 series	Incandescent Lamps–Safety Specifications
IEC 60570	Electrical Supply Track Systems for Luminaires
IEC 60598-1	Luminaires–Part 1: General Requirements and Tests
IEC 60598-2 Series	Luminaires–Part 2: Specific Requirements
IEC 60838 series	Miscellaneous Lampholders–Specific Requirements
IEC 61347-1	Lamp Controlgear
	Part 1: General and Safety Requirements
IEC 61347-2 Series	Lamp Controlgear–Part 2: Specific Requirements
IEC 61547	Equipment for General Lighting Purposes–EMC Immunity Requirements
IEC 62031	LED Modules For General Lighting–Safety Specifications
IEC 62035	Discharge Lamps (Excluding Fluorescent Lamps)–Safety Specifications
IEC 62471	Photobiological Safety of Lamps and Lamp Systems
IEC 62493	Assessment of Lighting Equipment Related to Human Exposure to Electromagnetic Fields
IEC 62560	Self-ballasted LED Lamps for General Lighting Services by Voltage > 50 V–Safety Specifications
IEC / TR 62778	Application of IEC 62471 for the Assessment Of Blue Light Hazard to Light Sources and Luminaires
UL 153	Portable Electric Luminaires
UL 588	Seasonal and Holiday Decorative Products
UL 924	Standard for Emergency Lighting and Power Equipment

Table 1.13 (continued)

STandard	*Title of Standard*
UL 1598	Luminaires
UL 2592	Standard for Low Voltage LED Wire
UL 8750	Light Emitting Diode (LED) Equipment for Use in Lighting Products
UL 8753	Field-Replaceable Light Emitting Diode (LED) Light Engines

The OSHA regulation applies to levels related to human exposure to electromagnetic fields. California Proposition 65 is a local regulation referring to restricted substances (similar to EU RoHS and REACH) that regulates over 800 hazardous substances in all consumer products, and importers must carry a warning label if it contains excessive amounts of those restricted substances.

In addition to product safety standards, a luminaire needs to be designed to fulfill any national wiring regulations (e.g., National Electrical Code (NEC) for the United States, British wiring regulation for the United Kingdom).

The aspects related to photometric data for luminaires are based on the consideration of the International Commission on Illumination (CIE). The CIE is an independent, nonprofit organization devoted to worldwide cooperation and the exchange of information on all matters relating to the science and art of light and lighting, color and vision, photobiology, and image technology.

From point of view of protection against electric shock, lighting products are classified as specified in Table 1.14.

Luminaires shall be classified following the IP number system of classification described in IEC 60529.

For LEDs used in general lighting service (GLS), which is white-light sources used to illuminate spaces (see Chapter 5, Section 5.1.4.3).

In general, the safety requirements and tests applied to lighting products are similar to those applicable to other electrical and electronic product, but some specific requirements apply [22]:

- Lighting equipment needs to withstand an endurance test: operate the light source for 1,200 cycles of repeated, continuous switching cycles without interruption. One complete switching cycle consists of 150 minutes of the light source switched ON at full power followed by 30 minutes of the light source switched OFF. The hours of operation recorded (e.g., 3,000 hours) include only the periods of the switching cycle when the light source was switched ON (e.g., the total test time is 3,600 hours).
- If a luminaire with an enclosure of insulating material has provision for earthing internal parts, it is Class I.
- A luminaire intended for use with a flexible cord or cable shall include a protective conductor as part of the flexible cord or cable.
- A Class III luminaire should not be provided with means for protective earthing.
- Where lamp holders are concerned, the United States uses E12, E17, and E26, whereas in Europe uses E14 and E27.

Table 1.14 Classification of Lighting Equipment Depending on the Protection against Electric Shock

Class of Equipment	*Description*
Class 0 Luminaire	Luminaire in which protection against electric shock relies upon basic insulation. This implies that there are no means for the connection of accessible conductive parts, if any, to the protective conductor in the fixed wiring of the installation, reliance in the event of a failure of the basic insulation being placed on the environment.
Class I Luminaire	Luminaire in which protection against electric shock does not rely on basic insulation only, but which includes an additional safety precaution in such a way that means are provided for the connection of accessible conductive parts to the protective (earthing) conductor in the fixed wiring of the installation in such a way that accessible conductive parts cannot become live (exposed to electrical potential and an electrical shock hazard is present) in the event of a failure of the basic insulation.
Class II Luminaire	Luminaire in which protection against electric shock does not rely on basic insulation only, but in which additional safety precautions such as double insulation or reinforced insulation are provided, and there is no provision for protective earthing or reliance upon installation conditions.
Class III Luminaire	Luminaire in which protection against electric shock relies on supply SELV and in which voltages higher than those of SELV are not generated.

- At luminaires the SELV does not need to exceed 50V AC r.m.s.
- Except for type Z attachments, terminations shall be marked to identify live, neutral, and earth in case of connection of the luminaire to the supply mains.
- Leads (tails) used for the connection to extra-low-voltage DC supply shall be color-coded red to indicate its intended connection to the positive terminal, and shall be color-coded black to indicate its intended connection to the negative termination.
- A wiring diagram shall be provided, except where the luminaire is suitable for direct connection to the mains supply.
- Luminaires declared by the manufacturer to be suitable for use outdoors shall not have PVC insulated external wiring.

The flexible cable or cord for external connection to the input or output circuit, fixed to or assembled with the luminaire, can use one of the following methods of attachment:

- *Type X attachment:* Method of attachment of the cable or cord such that it can be easily replaced. The flexible cable or cord may be specially prepared and only available from the manufacturer or their service agent.
- *Type Y attachment:* Method of attachment of the cable or cord such that any replacement can only be made by the manufacturer, their service agent, or a similarly qualified person.
- *Type Z attachment:* Method of attachment of the cable or cord such that it cannot be replaced without breaking or destroying the luminaire.

Lighting and related types of electrical equipment shall be marked by the information provided by the relevant product safety standard. Specific symbols used in lighting products are summarized in Table 1.15.

See Table 1.13 for a list of the main standards applied to lighting products. Compliance for production is the responsibility of the manufacturer and may include routine tests and quality assurance in addition to type testing.

1.6 Industrial Machinery and Semiconductor Manufacturing

The EU Machinery Directive 2006/42/EC (MD) defines a machine as "an assembly of linked parts or components, at least one of which moves...." It covers all hazards that come from machinery, including electrical hazards, and applies to [25]:

- Machinery;
- Interchangeable equipment (a device which, after the putting into service of machinery or of a tractor, is assembled with that machinery or tractor by the operator themself to change its function or attribute a new function, insofar as this equipment is not a tool);
- Safety components (serves to fulfill a safety function and is independently placed on the market);
- Lifting accessories (a component or equipment not attached to the lifting machinery, allowing the load to be held, which is placed between the machinery and the load or on the load itself);

Table 1.15 Specific Symbols Applicable to Lighting Equipment

Symbol	*Meaning*	*Symbol*	*Meaning*
	Class III		Rough service luminaires
	Luminaires suitable for direct mounting on noncombustible surfaces only		Luminaires for use with bowl mirror lamps
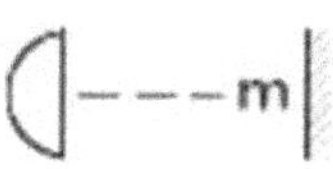	Minimum distance from lighted objects (meters)		Luminaires for use with high-pressure sodium lamps having an internal starting device

- Chains, ropes, and webbing, designed and constructed for lifting purposes as part of lifting machinery or lifting accessories;
- Removable mechanical transmission devices (component for transmitting power between self-propelled machinery or a tractor and another machine by joining them at the first fixed bearing);
- Partly completed machinery (assembly which is almost machinery but which cannot in itself perform a specific application).

The MD has two overall objectives: ensuring a high level of safety and protection for users of machinery and other people exposed to it, and ensuring the free movement of machinery in the European market. An additional objective, protecting the environment, is limited to the machinery used in pesticide applications.

However, according to the Machinery Directive, the safety objectives related to electrical hazards, set out in LVD 2014/35/EU shall apply to the machinery, while obligations concerning conformity assessment of machinery in terms of electrical hazards are governed by the Machinery Directive (Essential Health and Safety Requirement specified in Annex I of the MD).

The above definition covers a large range of machines, from simple hand-held power tools to complete automated industrial production lines, construction machinery, and robots. There are many categories of electrical machines that are exempted from the Machinery Directive and are directed to the Low Voltage Directive or other dedicates EU Directives [25]:

- Safety components intended to be used as spare parts;
- Specific equipment for use in fairgrounds and/or amusement parks;
- Machinery specially designed for nuclear purposes;
- Weapons, including firearms;
- Agricultural and forestry tractors (covered by EU Directive 2003/37/EC), with the exclusion of machinery mounted on these vehicles;
- Motor vehicles and their trailers (covered by EU Directive 70/156/EEC), with the exclusion of machinery mounted on these vehicles;
- Vehicles (covered by EU Directive 2002/24/EC), with the exclusion of machinery mounted on these vehicles;
- Motor vehicles exclusively intended for competition;
- Means of transport by air, on water, and rail networks with the exclusion of machinery mounted on these means of transport;
- Seagoing vessels and mobile offshore units and machinery installed onboard such vessels and/or units;
- Machinery specially designed and constructed for military or police purposes;
- Machinery specially designed and constructed for research purposes for temporary use in laboratories;
- Machinery intended to move performers during artistic performances;
- Mine winding gear;

- Electrical and electronic products falling within the following areas: ordinary office machinery, information technology equipment, household appliances intended for domestic use, audio and video equipment, low-voltage and high-voltage switchgear and control gear, electrical motors, and transformers (covered by EU LVD 2014/35/EU).

The Machinery Directive tries to establish the borderline between its scope and the LVD and as seen above, certain categories of electrical and electronic machinery products are hence excluded from the scope of the MD. In contrast, those household appliances intended specifically for commercial or industrial use are included in the scope of the MD only.

In general, electrical machinery that is not affected by the above-mentioned exclusions is in the scope of the MD. When such machinery has an electrical supply within the voltage limits of the LVD (between 50 and 1,000V for AC or between 75 and 1,500V for DC), it also falls under the scope of LVD. In this case, the safety requirements of the LVD apply to the machinery. In the same manner, a medical device with a "part which moves" (e.g., CT or magnetic resonance imaging (MRI)) needs to fulfill the MDR 2017/745/EU, and also be conformant with the applicable requirements of Annex I from MD. This means that the manufacturer's declaration of conformity for machinery subject to the MD shall not refer to the LVD, but the manufacturer's declaration of conformity for a medical device subject to the MDR shall not refer to the MD. As with the MDR, the EMC Directive, and the RED, the MD also includes conformity assessment procedures involving notified bodies (NB) for certain types of machinery. Aside from the self-assessment procedure, an NB is involved in the conformity assessment procedures according to EC-type examination and full quality assurance.

The fact that for certain product categories the Machinery Directive does not provide a clear definition creates some (incidental) confusion as to when to take the end-use as domestic or industrial (e.g., with laundry machines or 3-D printers), or even which Directive applies. A domestic laundry machine (which is exempted from MD) consists of "parts which move" (for that MD apply), have an electrical motor (which is exempted from MD), and is supplied with electrical energy from mains (for that LVD apply). Which EU Directive applies? Which safety standards need to be considered? These issues can seen as less of a problem with LVD, but in particular with the Machinery Directive, especially with regard to the definitions or lack thereof, even the MD requests a risk assessment according to ISO 12100 Safety of Machinery–General Principles for Design–Risk Assessment and Risk Reduction. But this does not fully clarify the issue, because the hazard identification is only not enough to provide an acceptable level of safety and to clarify the appurtenance to a specific process of conformity. We consider that an HBSE approach (see Section 1.1) will be the right way. For our above example, the intended use and environment use are clear: Washing clothing at home—follow the identification of potential sources of hazardous energy (electrical, mechanical, and thermal), and focus on the transfer mechanism to the human body and environment. In this example, it would seem that the electrical hazard is the most dangerous. Therefore, the need would be to design the laundry machine in a way that provides adequate means of protection for the identified sources of hazardous energies with a focus on electrical hazards. Based on the information provided above, the answers to the questions from our example would be the LVD its and applicable standards.

The concept of machinery safety considers the ability of a machine to execute its intended function throughout its life cycle and whereby risk has been adequately reduced. The standard provides indications on the decisions to be made for the safety of all types of machinery and on the types of documents required to verify the execution of risk assessments.

Application of the ISO 12100 standard alone is not enough to ensure compliance with fundamental requirements of health and safety established (ESHR in Annex I of the MD) by the machinery directive but does, nonetheless, establish an essential framework for the proper application of the said directive.

The risk assessment for machinery follows, in principle, the steps of the similar process for medical devices:

- Risk analysis, which includes determination of the limits of the machine, identification of the hazards (an intrinsic property of the context or object of study not related to external factors and which, due to its properties or characteristics, has the potential to cause harm), and estimation of the risk to determine the probable severity of the harm (loss, damage) and the likelihood of it occurring;
- Risk evaluation, which includes the estimation of the level of the risk [(R) Risk = (P) Probability of harm x (S) Severity of harm], focusing on the risks that need to be adequately reduced. The quantitative determination of the level of risk (R) leads to the implementation of preventive and protective measures to the risk assessment;
- Risk reduction, which includes the implementation of the means to mitigate the hazards or to reduce the severity of the harm and the probability that it will occur.

The risk assessment process requires information related to the description of the machine, applicable standards and regulations, experience of use, and relevant principles of ergonomics.

In the EU the CE Mark regulatory requirements for machinery equipment are specified in the LVD, EMC Directive (2014/30/EU), Energy Efficiency (2017/1369/EC) (see Chapter 2, Section 2.2), RoHS, REACH, WEEE Directives (see Chapter 3, Section 3.3). Apart from the directives mentioned above, other directives that also may affect some equipment falling within the scope of the Machinery Directive include Directive 2000/14/EC on the noise emission of equipment intended for use outdoors, and Directive 2002/44/EC on human body exposure to vibration. For machinery intended for use in a flammable atmosphere, the ATEX Directive applies (see Chapter 6).

In general, the safety requirements and tests applied to machinery products are similar to those applicable for other electrical and electronic products; specific requirements apply to these products [25]:

- Machinery must be supplied with integral lighting suitable for the operations;

- Maintenance areas and internal parts requiring frequent inspection and adjustment must be provided with appropriate lighting;
- Under the intended conditions of use, the discomfort, fatigue, and physical and psychological stress faced by the operator must be reduced to the minimum possible;
- The operating position must avoid any risk due to exhaust gases and/or lack of oxygen;
- Work stations constituting an integral part of the machinery must be designed for the installation of seats;
- A fault in the hardware or the software of the control system does not lead to hazardous situations;
- Reasonably foreseeable (the use of machinery in a way not intended in the instructions for use, but which may result from readily predictable) human behavior human error during operation does not lead to hazardous situations;
- Where fluids are used, prevent risks due to filling, use, recovery, or draining;
- No moving part of the machinery or piece held by the machinery must fall or be ejected;
- For wireless control, an automatic stop must be activated when correct control signals are not received, including loss of communication;
- Control devices must be visible and identifiable, using pictograms where appropriate;
- Machinery must be fitted with indicators as required for safe operation;
- The machinery must not start unexpectedly and must start only by voluntary actuation of a control device provided for this purpose;
- Machinery must be fitted with a control device whereby the machinery can be brought safely to a complete stop and the machinery's stop control must have priority over the start controls;
- Machinery must be fitted with one or more emergency stop devices to enable actual or impending danger to be averted, except for portable hand-held and/or hand-guided machinery, and the emergency stop function must be available and operational at all times, regardless of the operating mode and must be a backup to other safeguarding measures and not a substitute for them;
- The machinery must, where necessary, trigger or permit the triggering of certain safeguard movements;
- The interruption, the reestablishment after an interruption, or the fluctuation in whatever manner of the power supply to the machinery must not lead to dangerous situations;
- If the shape of the machinery itself or its intended installation does not offer sufficient stability, appropriate means of anchorage must be incorporated and indicated in the instructions;
- The instructions must indicate the type and frequency of inspections and maintenance required for safety reasons. They must indicate the parts subject to wear and the criteria for replacement;

- The moving parts of machinery must prevent risks of contact that could lead to accidents or must, where risks persist, be fitted with guards or protective devices;
- Fixed guards must be fixed by systems that can be opened or removed only with tools and the fixing systems must remain attached to the guards or the machinery when the guards are removed;
- Steps must be taken to eliminate any risk of injury arising from contact with or proximity to machinery parts or materials at high or very low temperatures;
- Machinery must be designed and constructed to avoid any risk of fire or overheating posed by the machinery itself or by gases, liquids, dust, vapors, or other substances produced or used by the machinery;
- Machinery must be designed and constructed in such a way that risks resulting from the emission of airborne noise and vibration are reduced to the lowest level, taking account of the availability of means of reducing noise, in particular at the source;
- Undesirable radiation emissions from the machinery must be eliminated or be reduced to levels that do not have adverse effects on persons.

The evaluation of the radiation emitted from machinery products is included in the EN 12198 series of standards. The IEC 61000-6-4 (Emission) and the IEC 61000-6-2 (Immunity) cover the EMC requirements for the industrial environment. For specific machineries EMC requirements are included in the standards in the ISO 13766 series (for construction machinery) and the ISO 14982 (for agricultural and forestry machinery). In the United States, the FCC regulates EMC and spectrum utilization under Title 47 of the Code of Federal Regulations (47CFR). Noise levels for machinery can be measured by following the tests that are included in the EN 1093 series of standards.

In the EU all machinery must be marked visibly, legibly, and indelibly with the following minimum specifics:

- The business name and full address of the manufacturer and, where applicable, their authorized representative;
- Designation of the machinery;
- The CE Marking;
- Designation of series or type;
- Serial number, if any;
- The year of completed construction.

Some specific safety signs used in machinery are summarized in Table 1.16.

Standards play an important role in the regulation and certification of machinery. ISO classifies the technical standards for machinery in the following categories [27]:

- Type-A standards (basic safety standards) giving basic concepts, principles for design, and general aspects that can be applied to machinery.

Table 1.16 Safety Signs Used on Machinery

Symbol	*Meaning*	*Symbol*	*Meaning*
	Rotating parts hazard		Crash hazard
	Moving blade cutter hazard		Moving belt hazard
	Finger trap hazard		Risk of explosion
	Machinery starts automatically hazard		Punch injury hazard
	Belt roller hazard		Foot trap hazard
	Fan hazard		Pressure hazard

From: [26].

- Type-B standards (generic safety standards) dealing with one safety aspect or one type of safeguard that can be used across a wide range of machinery.
  - Type-B1 standards deal with specific safety aspects (e.g., safety distances, surface temperature, noise) and define the methodology of how these can be addressed
- Type-B2 standards provide the performance requirements for the design and construction of specific safeguards (e.g., two-hand control devices, interlocking devices, pressure-sensitive protective devices, guards). Type-B1 and B2 standards can be applied either directly by the designer/manufacturer

or by reference in type-C standards, including, where relevant, means of verification.

- Type-C standards (machine safety standards) dealing with detailed safety requirements for a specific machine or group of machines.

There are in use more than 1,000 standards covering different aspects of machinery. Table 1.17 lists examples of standards applicable to machinery.

Semiconductor Equipment and Materials International (SEMI) is a global industry association with members from all sectors of the semiconductor and related industries [28]. The Environmental Health and Safety (EHS) Committee within SEMI's International Standards Division develops industry standards and guidelines that are globally accepted as ensuring the highest level of safety and quality applicable to all types of semiconductor manufacturing equipment, such as 3D-IC, equipment automation (hardware and software), flat panel displays, facilities, ion implanters, microlithography equipment, wet benches, etchers, mask aligners, saws, spin coaters, clean tracks, chemical dispense systems, bulk gas distribution, gas cabinets, gas source cabinets, metrology systems, test systems, CMP systems, furnaces, subsystems, industrial lasers, and any other equipment used in semiconductor manufacturing. SEMI standards are written documents in the form of specifications, guides, test methods, terminology, practices, and so on. These documents are published in the 16-volume set of SEMI International Standards.

Safety assessment according to SEMI guidelines is a growing global requirement for semiconductor manufacturers. Equipment buyers are demanding third-party assessment according to SEMI guidelines, which have been established to ensure a standard of safety throughout the industry and worldwide. When machinery is assessed according to SEMI standards and guidelines, part of the test results can be used in the EU for CE Mark mandatory requirements included in the European Machinery, Low Voltage, and EMC Directives. Some Semiconductor Industry Association requirements included in SEMI documents are additional to IEC, ISO, and EN standards (e.g., ventilation systems, chemical exposure, noise, environmental impact assessments).

Table 1.18 provides examples of SEMI standards. This list is not meant to be comprehensive; for a complete listing, refer to www.semi.org/standards.

1.7 Electrical Tools

There is a wide range of tools that may be divided into two main groups: hand tools and power tools. Hand tools are operated by the physical strength of the user without external power. They include anything from axes to wrenches.

The power tools are differentiated according to the type of power source that they use: electric, pneumatic, liquid fuel, hydraulic, and powder-actuated.

The advantage of power tools over hand tools is that they do not rely entirely on the physical strength of the operator to perform a job. Power tools are often much faster and, in some cases, more accurate than hand tools.

The term "power tool" describes a tool that is powered by an electrical motor that receives energy from the mains or a battery or batteries. Due to increased use

Table 1.17 Examples of Standards for Machinery

Standard	*Type*	*Title of Standard*
ANSI machine safety standards	A	ANSI B11.0 Safety of Machinery General Requirements–Risk Assessment
	C	ANSI B11.1 to ANSI B11.24 Standards for Specific Machines
ASME B20.1	C	Safety Standards for Conveyors and Related Equipment
EN 1012-1	C	Compressors and Vacuum Pumps. Safety Requirements. Air Compressors
EN 1093 series	B1	Safety of Machinery–Evaluation of the Emission of Airborne Hazardous Substances
EN 1804 series	C	Machines For Underground Mines–Safety Requirements For Hydraulic Powered Roof Supports
EN 12198 series	B1	Safety of machinery–Assessment and Reduction of Risks Arising from Radiation Emitted by Machinery
EN 16524	A	Mechanical Products–Methodology for Reduction of Environmental Impacts in Product Design and Development
IEC 61310 series	B1	Safety of Machinery–Indication, Marking and Actuation
		Part 1: Requirements for Visual, Acoustic and Tactile Signals
		Part 2: Requirements for Marking
		Part 3: Requirements for the Location and Operation of Actuators
IEC 61496 series	B2	Safety of Machinery–Electro-Sensitive Protective Equipment
		Part 1: General Requirements and Tests
		Part 2 To Part 4-3: Specific Requirements for Specific Equipment
IEC 60204 series		Safety of Machinery–Electrical Equipment of Machines
	A	Part 1: General Requirements
	C	Part 11: Requirements for Equipment for Voltages above 1,000V AC or 1,500V DC and Not Exceeding 36 kV
		Part 31: Specific Safety and EMC Requirements for Sewing Machines, Units and Systems
		Part 32: Requirements for Hoisting Machines
		Part 33: Requirements for Semiconductor Fabrication Equipment*
		Part 34: Requirements for Machine Tools†
IEC 61000-6-2	B1	Electromagnetic Compatibility (EMC)–Part 6-2: Generic Standards–Immunity Standard for Industrial Environments
IEC 61000-6-4	B1	Electromagnetic Compatibility (EMC)–Part 6-4: Generic Standards–Emission Standard for Industrial Environments
IEC 62061	A	Safety of Machinery–Functional Safety of Safety-Related Electrical, Electronic and Programmable Electronic Control Systems
IEC/TR 62061-1	A	Guidance on the Application of ISO 13849-1 and IEC 62061 in the Design of Safety-Related Control Systems for Machinery
IEC/TR 62513	B1	Safety of Machinery–Guidelines for the Use of Communication Systems in Safety-Related Applications
ISO 4413	C	Hydraulic Fluid Power–General Rules and Safety Requirements for Systems and Their Components
ISO 4414	C	Pneumatic Fluid Power–General Rules and Safety Requirements for Systems and Their Components
ISO 11161	A	Safety of Machinery–Integrated Manufacturing Systems–Basic Requirements
ISO 11201, 11202, 11204	B1	Acoustics. Noise Emitted by Machinery and Equipment

Table 1.17 (continued)

Standard	*Type*	*Title of Standard*
ISO 11553 series	C	Safety of Machinery–Laser Processing Machines–Part 1: General Safety Requirements
ISO 12100	A	Safety of Machinery–General Principles for Design–Risk Assessment and Risk Reduction
ISO 12499	B2	Industrial Fans–Mechanical Safety of Fans–Guarding
ISO 13732-1	B1	Ergonomics of the Thermal Environment–Methods for the Assessment of Human Responses to Contact with Surfaces–Part 1: Hot Surfaces
ISO 13766 series	B1	Earth-Moving and Building Construction Machinery. Electromagnetic Compatibility (EMC) of Machines with Internal Electrical Power Supply. Part1: General EMC Requirements under Typical Electromagnetic Environmental Conditions
		Part 2: Additional EMC Requirements for Functional Safety
ISO 13849 series	A	Safety of Machinery–Safety-Related Parts of Control Systems
		Part 1: General Principles for Design
		Part 2: Validation
ISO 13850	B2	Safety of Machinery–Emergency Stop Function–Principles for Design
ISO 13851	B2	Safety of Machinery–Two-Hand Control Devices–Principles for Design and Selection
ISO 13854	B1	Safety of Machinery–Minimum Gaps to Avoid Crushing of Parts of the Human Body
ISO 13855	B1	Safety of Machinery–Positioning of Safeguards with Respect to the Approach Speeds of Parts of the Human Body
ISO 13856 series	B2	Safety of Machinery–Pressure-Sensitive Protective Devices–Part 1: General Principles for Design and Testing of Pressure-Sensitive Mats and Pressure-Sensitive Floors
ISO 13857	B1	Safety Distances to Prevent Hazard Zones Being Reached by Upper and Lower Limbs
ISO 14118	B2	Safety of Machinery–Prevention of Unexpected Start-up
ISO 14119	B2	Safety of Machinery–Interlocking Devices Associated with Guards–Principles for Design and Selection
ISO 14120	B2	Safety of Machinery–Guards–General Requirements for the Design and Construction of Fixed and Movable Guards
ISO 14122 series	B1	Safety of Machinery–Permanent Means of Access to Machinery–Part 1: Choice of Fixed Means and General Requirements of Access
ISO 14123 series	A	Safety of Machinery–Reduction of Risks to Health Resulting from Hazardous Substances Emitted by Machinery
		Part 1: Principles and Specifications for Machinery Manufacturers
ISO 14159	A	Safety of Machinery–Hygiene Requirements for the Design of Machinery
ISO 14982	B1	Agricultural and Forestry Machinery–Electromagnetic Compatibility–Test Methods and Acceptance Criteria
ISO 19296	C	Mining–Mobile Machines Working Underground–Machine Safety
ISO 19353	A	Safety of Machinery–Fire Prevention and Fire Protection
ISO 20430	C	Plastics and Rubber Machines. Injection Moulding Machines. Safety Requirements
ISO 20607	A	Safety of Machinery–Instruction Handbook–General Drafting Principles
ISO 21469	A	Safety of Machinery–Lubricants with Incidental Product Contact–Hygiene Requirements

Table 1.17 (continued)

Standards	*Type*	*Title of Standards*
OSHA machine safety standards	A	29 CFR 1910 General Industry Regulations
		29 CFR 1910.211 Definitions
	C	29 CFR 1910.212 to 29 CFR 1910.219 Standards for Specific Machines
	A	29 CFR 1910.333 Electrical Safety
	C	29 CFR 1926.300 General Requirements
		29 CFR 1926.301 Hand Tools
		29 CFR 1926.302 Power-Operated Hand Tools
		29 CFR 1926.555 Conveyors
		OSHA 1910 Subpart D Walking–Working Surfaces
		OSHA 1910 Subpart O Machinery and Machine Guarding
		OSHA 1910 Subpart P Hand and Portable Tools
		OSHA 1910 Subpart S Electrical
		OSHA 1910.6 Incorporation by Reference–Includes all ANSI, NFPA Standards
		OSHA 1910.24 Fixed Industrial Stairs
		OSHA 1910.27 Fixed Ladders
		OSHA 1910.147 Control of Hazardous Energy Lockout/Tagout
UL 2011	A	Outline of Investigation for Machinery

The above ISO standards were adopted as EN standards.
*Refer also to Table 1.18.
†Refer also to Section 1.7.

of the batteries as a power source for tools, an alternate name was designated for this type of tool: cordless tool. Currently, there are three major rechargeable battery technologies in use for batteries dedicated to cordless power tools: lithium-ion (Li-ion), nickel-cadmium (NiCd), and nickel-metal hydride (NiMH).

Depending on the application area the power tool can be classified as either professional or consumer tools. These tools can be portable or stationary (fixed).

Professional power tools are designed to withstand heavier workloads and as a result have more available power. They also have motors able to withstand use for long periods without burning up the motor. Professional tools feature motors with a resin coating on the motor wire to protect from grit and dust, ball bearings to reduce vibration, and gears made of heat-treated wrought steel. They also have tough plastic housing to withstand impact and to function as a superior electrical insulator. Consumer power tools are for domestic private use, are low-cost, and do not include all the features of professional power tools.

Electrical power tools are subdivided according to the power level on

- Low-power hand/small bench tools with the ratings 2-4A, 120 VAC/230 VAC, or 18 VDC or less from a battery. Example of such tools are light-duty drills, palm sanders, power screwdrivers, trim saws, laminate trimmers, soldering irons and guns, heat guns, hot-melt glue guns, and sewing machines.
- Medium-power portable tools with the ratings 4–10A, 120 VAC/230 VAC, or 18V–24 VDC from battery, less than 1/4–1/2 horsepower (HP). Example

of such tools are heavy-duty drills, hammer drills, rotary hammers, jig saws, laser cutters/engravers, thermal foam cutters, impact drivers, angle drills, disc sanders.

- Powerful portable and small benchtop tools with the rating 10–15A, 120 VAC/230 VAC, or 24V–36 VDC from a battery, over 1/2 HP. Examples of such tools are circular saws, worm drive circular saws, sidewinder circular saws, scroll saws, belt sanders, framing nailers, geared drills, reciprocating saws, chop/miter saws, routers, plate (biscuit) joiners, mini-lathes, welders, angle grinders, and small printing presses.
- Large stationary industrial power tools, manual and CNC (computer numerical control)–controlled. Examples of such tools are full-sized milling machines, full-sized metal lathes, radial arm saws, large drill presses, large band saws, tile saws, surface grinders, large jointer/planers, shaper/molders, power shears, and 3D-printers.

Table 1.18 Examples of SEMI Standards

Standard	*Title of Standard*
IEEE 62659	IEC/IEEE International Standard–Nanomanufacturing–Large Scale Manufacturing for Nanoelectronics
IEST-RP-CC012.2	Considerations in Cleanroom Design
ISO 14644-1	Cleanrooms and Controlled Environments, Part 1: Classification of Air Cleanliness
SEMI S1	Safety Guideline for Equipment Safety Labels
SEMI S2	Environmental, Health, and Safety Guideline for Semiconductor Manufacturing Equipment
SEMI S7-0310	Safety Guideline for Evaluation of Personnel and Evaluating Company Qualifications
SEMI S8	Safety Guideline for Ergonomics Engineering of Semiconductor Manufacturing Equipment
SEMI S10	Safety Guideline for Risk Assessment and Risk Evaluation Process
SEMI S13	Environmental, Health and Safety Guideline for Documents Provided to the Equipment User for Use with Manufacturing Equipment
SEMI S14	Safety Guidelines for Fire Risk Assessment and Mitigation for Semiconductor Manufacturing Equipment
SEMI S16	Guide for Semiconductor Manufacturing Equipment Design for Reduction of Environmental Impact at End of Life
SEMI S19	Safety Guideline for Training of Manufacturing Equipment Installation, Maintenance and Service Personnel
SEMI S22*	Safety Guideline for the Electrical Design of Semiconductor Manufacturing Equipment
SEMI S23	Guide for Conservation of Energy, Utilities and Materials Used by Semiconductor Manufacturing Equipment
SEMI E6	Guide for Semiconductor Equipment Installation Documentation
SEMI E33	Specification for Semiconductor Manufacturing Facility Electromagnetic Compatibility
SEMI E129	Guide to Assess and Control Electrostatic Charge in A Semiconductor Manufacturing Facility
SEMI E169	Guide for Equipment Information System Security

*Nearly identical to IEC 60204-33.

The electrical power tools need to comply with compulsory requirements contained in national legislation and within the relevant codes and directives.

Such regulations require manufacturers, importers, product suppliers, and other entities to deliver only safe products to the marketplace; these products should be delivered with instructions for their safe use in the field.

In the United States the documents 29 CFR Part 1926, 29 CFR 1910 Subpart P, and the OSHA 3080 specifiy the regulations for safe work with power tools. In addition to standards set forth by OSHA, ANSI, NEC, and other organizations, the U.S. Army Corps of Engineers (USACE) has expanded their Safety and Health Requirements Manual, the EM 385-1-1 (2014 edition), specifying health and safety rules to help keep jobsites safe and efficient. Section 13 of the EM 385-1-1 refers to power tools [29].

In the EU the CE Mark regulatory requirements for household products are specified in MD 2006/42/EC or GPSD 2001/95/EC, EMC Directive (2014/30/EU), RoHS, REACH, and WEEE Directives (See Chapter 3, Section 3.3).

In many countries, additional requirements are specified by the national health authorities, the national authorities responsible for the protection of labor and similar authorities.

In general, fulfilling such regulations is proved by quality or safety marks/identifiers present in the product labeling. The association of these marks/identifiers with electrical power tools in the marketplace indicates a level of quality and/or safety that has been verified by an independent body or agency. Purchasers of equipment need to be aware of possible unauthorized use of these marks/identifiers. This means that a purchaser does not have to rely on the claims of a manufacturer/supplier, including self-declarations. Only objective evidence based on certificates or test reports issued by accredited organisms represent credible arguments. It is important to distinguish between quality marks/identifiers and those relating to the safety features of the product based on a dedicated standard.

Due to the specificity of electrical power tools, during their use the following injuries can result:

- Major shocks that may cause a fatality;
- Electric flash burns;
- Minor shocks that may lead to injury from the tool itself or resulting falls from ladders or platforms;
- Eye injuries from flying chips and cuttings;
- Gashes, cuts, and puncture wounds.

Typical causes of electrical power tool accidents include the following:

- Using the wrong tool for the job;
- Tools falling from overhead;
- Sharp tools carried in pockets;
- Excessive vibration;
- Failure to support or clamp work into position;
- Using damaged electrical cords or end connectors;

- Failure to use ground fault circuit interrupters (GFCIs), especially outdoors and in wet locations.

The following general practices should be followed when operating electric tools:

- Operate electric tools within their design limitations;
- Use gloves and appropriate safety footwear;
- Store electric tools in a dry place when not in use;
- Do not use electric tools in damp or wet locations unless they are approved for that purpose;
- Keep the work areas well lit;
- Ensure that cords from electric tools do not present a tripping hazard;
- Recommend ear plugs, as necessary (many power tools can be noisy);
- Use safety glasses and a dust mask as additional personal protective equipment.

To protect the user from shock and burns, electric tools must have a three-wire cord with ground and be plugged into a grounded receptacle, be double insulated, or be powered by a low-voltage isolation transformer. Three-wire cords contain two current-carrying conductors and a grounding conductor. Any time an adapter is used to accommodate a two-hole receptacle, the adapter wire must be attached to a known ground. The third prong must never be removed from the plug.

Grounding of electric power tools and the use of GFCIs will protect the worker under most circumstances. GFCIs detect current leaking to the ground from an electric tool or cord and shut off the power before injury or damage can occur.

Double-insulated tools provide reliable protection against electrical shock without third-wire grounding. On these tools, an internal layer of protective insulation completely isolates the external housing of the tool. This type of protection is permanently marked with a double insulation symbol as shown in Table 1.19.

From the point of view of protection against electric shock, electrical power tools are classified as specified in Table 1.20.

An important aspect that needs to be considered is represented by periodic maintenance, which ensures the continuous level of safety needed for an electrical power tool. Table 1.21 summarizes the time periods of appliable maintenance depending on the operating conditions and based on manufacturer recommendations, visual inspection and, if applicable, on specific tests such as earth continuity, dielectric withstand, or leakage current.

In general, the safety requirements and tests applied to electrical power tools are similar to those applicable for other electrical and electronic products; specific requirements apply to these products [30]:

- Tools that are expected to be connected to more than two supply conductors shall be provided with a connection diagram, affixed to the tool, unless the correct mode of connection is obvious.

Table 1.19 Specific Symbols Applicable to Electrical Tools

Symbol	*Meaning*	*Symbol*	*Meaning*
	Double Insulated	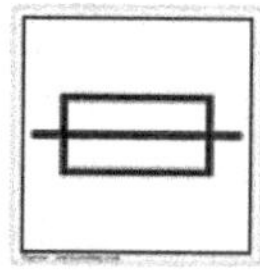	Fuse (rated current may be indicated with the symbol)
	Protective Earthing		Special tool
	Circular saw	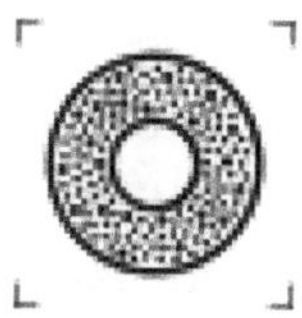	Grinding/abrasive wheel

Table 1.20 Classification of Electrical Power Tools Depending on Protection Against Electric Shock

Class of Equipment	*Description*
Class I tool	Tool in which protection against electric shock does not rely on basic, double, or reinforced insulation only, but which includes an additional safety precaution in that conductive accessible parts are connected to the protective earthing conductor in the fixed wiring of the installation in such a way that conductive accessible parts cannot become live (exposed to electrical potential and an electrical shock hazard is present) in the event of a failure of the basic insulation. Also considered as class I tools are tools with double insulation and/or reinforced insulation throughout having an earthing terminal or earthing contact.
Class II tool	Tool in which protection against electric shock does not rely on basic insulation only, but in which additional safety precautions, such as double insulation or reinforced insulation, are provided, there is no provision for protective earthing or reliance upon installation conditions.
Class III tool	Tool in which protection against electric shock relies on supply at safety extra-low voltage, and in which voltages higher than those of safety extra-low voltages are not generated.
Class II construction	Part of a tool for which protection against electric shock relies upon double insulation or reinforced insulation.
Class III construction	Part of a tool for which protection against electric shock relies upon safety extra-low voltage, and in which voltages higher than those of safety extra-low voltages are not generated.

- The supply cord of an electrical power tool may have one of the X, Y, or Z types of attachment as for a household product (see Section 1.2).
- Except for type Z attachments, terminals shall be indicated as follows:
  - Terminals intended exclusively for the neutral conductor shall be indicated by the letter N.

Table 1.21 Period of Maintenance of Electrical Tools Depending on the Operating Conditions

Operating Conditions	*Example of Location*	*Recommended Maintenance (Testing) Period*
Heavy use	Areas with conductive dust	Weekly
	Outdoor	3 months
Normal use	Indoor general engineering	6 months
Light use	Domestic, office	1 year

- Protective earthing terminals shall be indicated by the symbol as in Table 1.19. Power tool plugs must match the outlet. Never modify the plug in any way. Do not use any adapter plugs with earthed (grounded) power tools.

- For power tools with type Z attachment, the supply cord of this power tool cannot be replaced, and the power tool shall be scrapped.
- Labels need to withstand a specific heating test.
- Overload protection devices shall not operate under normal starting conditions.
- The electrical power tool shall withstand an endurance test.
- Tools shall be so constructed that accidental changing of the setting of control devices is unlikely to occur.
- It is not to be expected that two independent parts will become loose or fall out of position at the same time.
- For electrical connections, spring washers are not considered to be adequate for preventing the loosening of the parts.
- Insulating materials in which heating conductors are embedded serves as basic insulation, and shall not be used as reinforced insulation.
- Switches and reset buttons on a non-self-resetting control shall be so located that accidental operation is unlikely to occur.
- Parts separated by protection impedance shall comply with the requirements for double insulation or reinforced insulation.
- When sleeving is used as supplementary insulation on internal wiring, it shall be retained in position by positive means. A sleeve is considered to be fixed by positive means if it can only be removed by breaking or cutting, or if it is clamped at both ends.
- Mains switches shall have adequate breaking capacity, and shall be switching for 50,000 cycles of operation.
- Hand-held, powered circular saws with blade diameter greater than 5 cm (2”), power drills, tappers, fastener drivers, grinders with wheels, disc sanders (with discs greater than 5 cm (2”) in diameter), belt sanders, reciprocating, saber, scroll, and jig saws (with blade shanks greater than 0.6 cm = ¼”) must be equipped with constant pressure switch (when a let-go of the switch occurs, the tool stops) and as a result, it will shut off the power.

- A positive on-off control (which means a switch that must push to turn the tool on and then push again to turn it off) is allowed on platen sanders, grinders with wheels, disc sanders (discs 5 cm (2") diameter or less), routers, planers, laminate trimmers, nibblers, shears, and saber, scroll, and jig saws (blade snakes 0.6 cm (1/4") or less).
- All hand-held powered drills must be equipped with a momentary contact off switch. Momentary switches require continuous compression. They will switch on when the user compresses the switch and will remain on only for as long as there is pressure on the switch. Once the pressure is removed, they will switch off. They may have a lock-on provided that they can be turned off by a single motion of the finger that turned them on.
- Guards are required for all moving parts, including the point of operation where the work is performed, at all power transmission components, and at all other moving parts of the electrical tool.
- Make sure guards and shields are in place and vents are clear of debris before turning on a tool.
- Use only the battery and the battery charger specified by the manufacturer for the tool.
- Do not let the battery completely discharge before charging it again. Do not let it go lower than 20% before recharging.
- Do not charge the battery at temperatures below 0°C (32°F) or above 41°C (105°F). Extreme temperatures will shorten battery lifespan.
- Better batteries have a fuel gauge that allows seeing if it is time to charge the battery.

The selection of the type of bits used in drilling power tools represents an important aspect of improving the efficiency of the tool. Some practical recommendations are [31]:

- Steel bits are best suited for drilling softwoods, which may dull quickly.
- High-speed steel drill bits will stay sharp longer than regular steel bits. They are more efficient and versatile than steel bits. Some high-speed steel bits have a titanium coating that allows them to stay sharp even longer.
- Carbide-tipped means these bits will stay sharper longer that high-speed steel or titanium-coated bits.
- Cobalt bits are best for drilling into metal.

Electrical Power Tools shall be marked with:

- Rated voltage(s) or rated voltage range(s), in volts. Tools for star-delta connection shall be clearly marked with the two rated voltages (e.g., 230Δ/400 Y).
- The symbol for the nature of the supply, unless the rated frequency is marked.
- The rated input, in watts, or rated current, in amperes. The rated input or current to be marked on the tool is the total maximum input or current that

can be on the circuit at the same time. If a tool has alternative components that can be selected by a control device, the rated input or rated current is that corresponding to the highest loading possible.

- The name, trademark, identification mark, or company name of the manufacturer or any other person responsible for placing the tool on the market.
- The designation of series or type.
- The symbol for class II construction, for class II tools only.
- The IP number according to the degree of protection against ingress of water other than IPX0. The first numeral of the IP number need not be marked on the tool.
- The following statement: "Warning–To reduce the risk of injury, the user must read and understand instruction manual."
- The date of manufacture.

Specific symbols used in electrical power tools are summarized in Table 1.19.

The safety standards for specific types of electrical power tools are primarily covered by the IEC 60745 series [36] and the IEC 62841 series [32]. For EMC the EN 55014-1 (Emissions), EN 55014-2 (Immunity), IEC 61000-3-2, and IEC 61000-3-2 (Harmonics) standards apply.

See Table 1.22 for a list of the main standards applied to electrical power tools.

1.8 Alarm Systems

Today, almost every dwelling is equipped with some sort of alarm system that will warn about smoke or an intruder. What exactly is an "alarm system"?

An alarm system is defined as per the IEC Standard 60839-1-1 as "an electrical installation designed to detect and signal the presence of an abnormal condition indicating the presence of a hazard."

Due to the diversity of the standards and the involved organizations, there are other definitions of alarm systems, but all have the same objective: to detect and indicate the presence of a hazard.

As of today, the New Standard IEC62368-1 represents the main standard used for the evaluation of alarm systems from the point of product safety based on HBS. Along with it, specific performance requirements are presented for each market depending on the local standards and local rules, regulations, organizations (insurance companies, police, government offices, etc.).

The alarm systems industry is governed by specific standards for alarm systems that impose general requirements for the design, installation, commissioning, operation, and maintenance of manual and/or automatic alarm systems. Alarm systems perform an important safety function, which is the protection of persons, property, and the environment.

Currently around the world there are hundreds of standards covering alarm systems from a safety perspective and also in terms of providing reliable performance within a prescribed environment.

Table 1.22 Standards Applied for Electrical Power Tools

Standard	*Title of Standard*
29 CFR 1910 Subpart P	Hand and Portable Tools
29 CFR 1926.302 to 29 CFR 1929. 304	Specific Standards for Specific Tools
ANSI B11.21	Machine Tools Using Lasers
EN 847 (series)	Tools for Woodworking–Safety Requirements
EN 55014-1	Electromagnetic compatibility. Requirements for Household Appliances, Electric Tools and Similar Apparatus. Emission.
EN 55014-2	Electromagnetic Compatibility. Requirements for Household Appliances, Electric Tools and Similar Apparatus. Immunity.
IEC 61000-3-2	Electromagnetic Compatibility (EMC)–Part 3-2: Limits–Limits for Harmonic Current Emissions (Equipment Input Current ≤16A per Phase)
IEC 61000-3-3	Electromagnetic Compatibility (EMC)–Part 3-3: Limits–Limitation of Voltage Changes, Voltage Fluctuations and Flicker in Public Low-Voltage Supply Systems, for Equipment with Rated Current 16A per Phase and Not Subject to Conditional Connection
IEC 61029 series	Safety of Transportable Motor-Operated Electric Tools
	Part 1: General Requirements
	Part 2-1 to Part 2-8: Specific Standards for Specific Tools
IEC 60745 series	Hand-held Motor-Operated Electric Tools–Safety
	Part 1: General Requirements
	Part 2-12 to Part 2-23 Specific Standards for Specific Tools
IEC 62841 series	Electric Motor-Operated Hand-Held Tools, Transportable Tools and Lawn and Garden Machinery–Safety
	Part 1: General Requirements
	Part 2-1 to Part 2-21: Specific Standards for Specific Tools
	Part 3-1 to Part 3-14: Specific Standards for Other Specific Tools
	Part 4-1 to Part 4-4: Specific Standards for Other Specific Tools
UL 745-1	Portable Electric Tools
UL 987	Standard for Stationary and Fixed Electric Tools
UL 2565	Standard for Industrial Metalworking and Woodworking Machine Tools

The most relevant market, the United States, uses a large number of standards, which, along with safety aspects, combine performance requirements as prerequisites of acceptance of equipment into their markets. The United States has dozens of UL Standards covering equipment and even imposing supplementary requirements at the state level regarding the performance and safety of alarm systems.

Why are so many standards necessary for alarm systems? It is a legitimate question and the answer is due to many reasons:

- There are many categories of alarm systems.
- There are different places and settings in which the alarm systems are used.
- There are many types of organizations involved in the alarm manufacturing business and each has the right to impose specific requirements, based on their side of the business or with their local specific requirements (govern-

ment, police, insurance, local organizations, alarm systems organizations, associations, etc.).

- By its nature, an alarm system comprises many components, such as a central control equipment, which has its own power supply; a communication bus, which will connect the detectors; visual and/or audible indicating equipment; devices that are actuated by the detectors; communication/signaling device/devices; a program input device that provides the brain of the alarm system (very often embedded within the controller); and the execution element (or elements), the device/devices that are activated by the control equipment following the programming and external conditions.

There are standards for each type of the above-specified components. Each component must perform safely and reliably. All perform a safety function.

The main scope of these standards is to ensure a high level of safety for the users of the alarm systems, to ensure that the systems are reliable and perform as intended, to reduce the incidents of false alarms, and finally—despite the competitive market in which all suppliers of alarm systems participate—to ensure that systems are compatible as necessary.

The multitude of standards for alarm systems are providing a grading system for alarm/transmission systems, the involved equipment, the systems using dedicated paths, wired networks, and wire-free systems that are used to ensure that an appropriate security system is applied.

Next we expand our discussion by explaining the main categories of alarm systems.

1.8.1 Video Surveillance Systems

Video surveillance systems are operational systems with closed-circuit TV (CCTV) that are able to monitor customers, staff, and the premises where they are installed. Cameras can act as a deterrent to would-be criminals or record sufficient footage to capture or prosecute actual criminals. Since these systems record all activities, they can also protect companies in the long term with video evidence in the event of a liability suit.

1.8.2 Fire Alarm Systems

Fire alarm systems detect smoke and alert occupants of a potential problem. Generally speaking, these can work in tandem with automated fire suppression systems and security alarm systems to prevent the total loss of a business in the event of a fire.

1.8.3 Security Alarm Systems

Security alarm systems are the most popular alarm systems and are dedicated to deter and protect people and businesses against theft and general intruders. With a wide range of detectors, including glass break detectors, motion detectors, door and window sensors, flood sensors, and carbon monoxide detectors, these systems can monitor and prevent break-ins and unwanted damages. When security alarm systems are externally monitored, they add the benefit of 24/7 coverage and involve monitoring centers, which provide restricted-access locations for their equipment.

These alarm systems currently use sensors and/or self-contained battery-powered equipment. Sensors are available in both hardwired or wireless configurations for protection against unauthorized access, to detect hazards based on a specific function, including various gases such as carbon monoxide and other toxic/or combustible gases. Under the stated application conditions, the sensors will produce an alarm and in some cases will be able to initiate executive actions whenever the levels exceed a preset alarm volume ratio.

Self-contained battery-powered equipment is provided with an internal battery that supplies the necessary amount of energy for a predefined duration of operation to detectors that are working in a continuous operation, equipment that is continuously powered with continuous or intermittent automatic sensing, such as motion detectors, wireless keypads, and remote controls.

1.8.4 Alarm Transmission Systems

Alarm transmission systems (ATS) are designated to transmit alarm signals in a reliable, resilient, and secure manner. An alarm transmission system may use any type of transmission network: telecommunications, radio, cellular, or satellite.

1.8.5 Remote Monitoring Systems

Remote monitoring systems provide the ability to remotely monitor information and data supplied by security systems connected to intelligent alarm keypads and to get instant notifications of any issues that appear on the premises that are monitored and protected by the system.

1.8.6 Access Control Systems

Electronic access control systems are able to control any number of access points. These systems are widely used around the world in some residential and mainly commercial environments.

Such systems (as with almost all the alarm systems) include in their configurations a few basic mandatory functions as follows. The number of the mandatory functions to be performed explains the large number of performance Standards that are used within the alarm systems; each function is characterized individually and shall be conformant with the minimal requirements set within the corresponding standard.

Processing: The function by which the system compares the changes occurring within the system with preset rules to produce predefined actions.

Communication: Represents the function of transmission of signals between components of the access control system to ensure the application of preset rules.

Configuration (programming): The action sets the processing rules.

Access point interface: Through this function the system.

- Releases and secures access according to preset rules.

- Continuously reports the open/closed status of the portal, the status of releasing/securing status of locking devices, releasing/securing of the portal according to preset rules without recognition.

Recognition: With this function the system recognizes authorized users requesting access.

Annunciation: With this function the system alerts, displays, and/or provides a log of the functionalities as follows:

- Alert: The annunciation subfunctionality related to the activation of an indicator to prompt human assessment.
- Display: The annunciation subfunctionality related to the visual and/or audible presentation of changes occurring within the system.
- Logging: The annunciation subfunctionality related to the logging and retrieving of changes occurring within the system.

Duress signaling: The function of providing a silent warning by system users of ongoing coercive access request conditions.

Interfacing with other systems: The function by which the system shares the functionalities and/or changes occurring within systems.

Self-protection: With this function the system provides the prevention, detection, and/or reporting of deliberate and accidental tampering and/or interfering with the system operation.

Power supply: The function of supplying power to the access control system.

User interface: The function by which the user requests access (e.g., keypad or card reader) and receives an indication of the access status.

Additional functions may be included in the electronic access control system providing that they do not interfere with the correct operation of the mandatory functions.

By performing all of the above functions, an access control system, and an alarm system, generally speaking, performs the overall safety function of ensuring the safety and security of staff and of the premises in a building or location where it is installed and operated.

Specifics

The alarm system represents a domain which within the last couple of years has changed at a very fast pace: the technology of the sensors, the major shift regarding the function of transmission of data, the protocols used. The increasing threats of cyberattacks that affect security and life-safety-critical products, along with the desire to circumvent the negative implications associated with cybersecurity, has changed almost everything within the alarm system industry. We predict that in just a few years, telecommunications lines will not be used as a means of transmission even by the most basic alarm systems.

In terms of safety, the electrical safety compliance analysis of the main hardware components—the alarm controllers and the detectors using sophisticated technologies—is based on the IEC 62368-1 Standard as well as the standard that covers the degree of protection for enclosures (IEC 60529) and the standard that is used to determine the degree of protection at impact (IEC 62262).

Alarm systems have many specifications, as detailed next.

Indicators and Alarms

Visual indicators shall be fitted and colored as follows:

- Power supply indicators shall be colored green;
- Alarm indicators shall be colored red;
- Where fitted, the visual fault indication shall be yellow.

If a sensor "end-of-life" visual indication is fitted, this shall be clearly different from all other visual indications.

The visual indicators shall be labeled to show their function.

The visual indicators shall be visible when the equipment is installed in its operating position according to the manufacturer's instructions.

Fault Signals

The alarm system shall provide a fault signal in the event of loss of continuity or a short circuit to a sensor. The fault signal shall be clearly identified and differentiated from a gas alarm.

Marking

The equipment shall carry durable label(s) with the following information:

- The manufacturer's or supplier's name, trademark, or other means of identification;
- The name of the equipment and the type of gas to be detected (e.g., "propane gas detector") and the model number;
- The number of the European Standard;
- The type of equipment;
- The serial number or manufacturing date code of the equipment;
- For mains-powered equipment, the electricity supply voltage, frequency, and maximum power consumption;
- For battery-powered equipment, the type and size of replacement batteries;
- Recommendations on the replacement procedures and lifetime of the equipment;
- The expected lifetime of the sensor, if different from the lifetime of the equipment;
- When the equipment has a replaceable sensor, the equipment shall be de-energized before the sensor is replaced;

- When the sensor is replaceable, the equipment shall carry the next replacement date of the sensor;
- It is important to know that the Marking as it is requested by the IEC 62368-1 Standard shall be complemented with all of the applicable requirements stated within the specific standard for the alarm system.

1.8.7 Cautions

All gas detection equipment shall carry a caution, on a label attached to the equipment, such as:

> CAUTION: READ THESE INSTRUCTIONS CAREFULLY BEFORE OPERATING OR SERVICING.

Regarding Installation Instructions

The equipment shall be provided with installation instructions. The installation guides shall give complete, clear, and accurate instructions for the installation, safe and proper operation, and shall provide instructions regarding maintenance.

Table 1.23 list a few of the applicable standards for alarm systems.

References

[1] Wang, Z., Y. H. Hsu, and R. Nute, "Hazard-Based Safety Standard Comparison between ECMA 287, IEC 60950-1, IEC 60065." *IEEE Symposium on Product Safety Engineering*, 2005.

[2] IEC 60950-1, Information Technology Equipment–Safety–Part 1: General Requirements," Geneva, 2005–2013.

[3] IEC 62368-1, Audio/Video, Information and Communication Technology Equipment–Part 1: Safety Requirements, Geneva, 2018.

[4] ECMA TR106, Guidance and Comparison between 60950-1 and 62368-1, Geneva, 2013.

[5] Astrodyne TDI, https://www.astrodynetdi.com/blog/iec-en-ul-62368-1-compliance-standards.

[6] Loznen, S., C. Bolintineanu, and J. Swart, *Electrical Product Compliance and Safety Engineering, Vol. I*, Norwood, MA: Artech House Publishing, 2017.

[7] IEC 62911, Audio, Video and Information Technology Equipment–Routine Electrical Safety Testing in Production," Geneva, 2016.

[8] 15 US Code 2052, United States Code Title 15–Commerce and Trade Chapter 47–Consumer Product Safety-Sec. 2052–Definitions, 2011, https://www.govinfo.gov/app/details/USCODE-2011-title15/USCODE-2011-title15-chap47-sec2052.

[9] Marucheck, A., et al., "Product Safety and Security in the Global Supply Chain: Issues, Challenges and Research Opportunities," *Journal of Operations Management*, Vol. 29, 2011.

[10] IEC 60335-1 Household and Similar Electrical Appliances–Safety - Part 1: General Requirements, Geneva, 2020.

[11] IEC 61010-1 Safety Requirements for Electrical Equipment for Measurement, Control, and Laboratory Use, Geneva, 2010–2019.

[12] IEC 60601-1–International Electrotechnical Commission, Medical Electrical Equipment–Part 1: General Requirements for Basic Safety and Essential Performance, Geneva, 2005–2020.

[13] IMDRF GRRP WG(PD1)/N47–International Medical Devices Regulators Forum, Essential Principles of Safety and Performance of Medical Devices and IVD Medical Devices, 2018.

[14] IEC/TR 60513–International Electrotechnical Commission, Fundamental Aspects of Safety Standards for Medical Electrical Equipment, Geneva, 1994.

[15] Mellish, R. G., "The Single Fault Philosophy: How It Fits with Risk Management," *ACOS Workshop VI, Safety of Electromedical Equipment—An Integrated Approach through IEC Standards*, Toronto, May 6–7, 1998.

[16] IECEE–List of Medical Standards (MED), https://www.iecee.org/dyn/www/f?p=106:48:0.

[17] ISO-List of Standards, http://www.iso.org/iso/home/standards.htm.

[18] FDA-Recognized Consensus Standards, http://www.accessdata.fda.gov/scripts/cdrh/cfdocs/cfStandards/search.cfm.

[19] IEC/TR 62354–International Electrotechnical Commission, General Testing Procedures for Medical Electrical Equipment, Geneva, 2014.

[20] ISO 15223-1–Medical Devices–Symbols to Be Used with Medical Device Labels, Labelling and Information to Be Supplied–Part 1: General Requirements, Geneva, 2016.

[21] Regulation (EU) 2019/2020, Eco-Design Requirements for Light Sources and Separate Control Gears, Brussels, 2019.

[22] IEC 60598-1, Luminaires–Part 1: General Requirements and Tests, Geneva, 2020.

[23] US Energy Independence and Security Act (EISA), 2007, https://www.govinfo.gov/content/pkg/PLAW-110publ140/pdf/PLAW-110publ140.pdf.

[24] Philips, https://www.usa.lighting.philips.com/support/support/legislation/labels.

[25] EU MD 2006/42/EC, Machinery Directive, Brussels, 2006.

[26] Label Source, https://www.labelsource.co.uk/label/safety-signs/hazard-warning-safety-signs/general-hazard-warning-safety-signs.

[27] G. T. Engineering, https://www.gt-engineering.it/en/Insights/e-norme-armonizzate-di-tipo-a-b-c-1-2-3.

[28] SEMI, https://www.semi.org/en/Standards/P_000787.

[29] USACE, Safety and Health Requirements Manual EM 385-1-1, 2014, https://www.usace.army.mil/Portals/2/docs/Safety/EM%20385-1-1,%202014%20Sections/EM%20385-1-1%202014%20Section%2013.pdf.

[30] IEC 60745, Hand-Held Motor-Operated Electric Tools–Safety, Series, Geneva, 2003–2012.

[31] Study Guide: Power Tools, Yournhpa, 2016, https://yournhpa.org/BasicTraining/Power_Tools/Power_Tools_Study_Guide.pdf.

[32] IEC 62841, Electric Motor-Operated Hand-Held Tools, Transportable Tools and Lawn and Garden Machinery–Safety, Geneva, 2012–2020.

Selected Bibliography

Bolintineanu, C., and S. Loznen, "Product Safety and Third Party Certification," *The Electronic Packaging Handbook*, G. R. Blackwell (ed.), Boca Raton, FL: CRC Press, 2000.

CIE (International Commission on Illumination), https://cie.co.at/about-cie.

FDA, 21 CFR Parts 808 and 812, Medical Devices; Current Good Manufacturing Practice (CGMP); Final Rule, Federal Register, October 7, 1996.

Lighting Europe, https://www.lightingeurope.org.

Table 1.23 List of Standards Applicable to Alarm Systems

Standard	*Title of Standard*
EN 50130 series	Alarm Systems.
	-Part 4: Electromagnetic Compatibility. Product Family Standard: Immunity Requirements for Components of Fire, Intruder, Hold-Up, CCTV, Access Control and Social Alarm Systems
	-Part 5: Environmental Test Methods
EN 50131 series	Alarm Systems–Intrusion and Hold-Up Systems
	–Part 1: System Requirements
	-Part 2-2 to Part 2-11: Specific Standards for Specific Parts
EN 50136 series	Alarm Systems. Alarm Transmission Systems and Equipment.
	-Part 1: General Requirements for Alarm Transmission Systems
	-Part 2 to Part 4: Specific Standards for Specific Parts
EN50194-1	Electrical Apparatus for the Detection of Combustible Gases in Domestic Premises
	Part 1: Test Methods and Performance Requirements
EN 50518-1	Monitoring and Alarm Receiving Center–Part 1: Location and Construction Requirements
EN 60079-29-1	Explosive Atmospheres–Part 29-1: Gas Detectors–Performance Requirements of Detectors for Flammable Gases
IEC 10118 series	Information Technology–Security Techniques–Hash-Functions
IEC 18033 series	Information Technology–Security Techniques–Encryption Algorithms
IEC 60839-5 series	Alarm and Electronic Security Systems
	Part 5-1: Alarm Transmission Systems–General Requirements
	Part 5-2: Alarm Transmission Systems–Requirements for Supervised Premises Transceiver (SPT)
	Part 5-3: Alarm Transmission Systems–Requirements for Receiving Center Transceiver (RCT)
IEC 60839-7 series	Alarm Systems
	Part 7-1 to Part 7-20: Message Formats and Protocols for Serial Data Interfaces in Alarm Transmission Systems
IEC 60839-10-1	Alarm Systems–Part 10: Alarm Systems for Road Vehicles–Section 1: Passenger Cars
IEC60839-11 series	Alarm and Electronic Security Systems
	Part 11-1 To Part 11-5: Specific Standards for Specific Parts
IEC 62599 series	Alarm Systems
	Part 1: Environmental Test Methods
	Part 2: Electromagnetic Compatibility–Immunity Requirements for Components of Fire And Security Alarm Systems
IEC 62642 series	Alarm Systems–Intrusion and Hold-Up Systems
	Part 1: System Requirements
	Part 2-2 to Part 2-7: Specific Standards for Specific Parts
	Part 3: Control and Indicating Equipment
	Part 4: Warning Devices
	Part 5-3: Interconnections–Requirements for Equipment Using Radio Frequency Techniques
	Part 6: Power Supplies
	Part 7: Application Guidelines
	Part 8: Security Fog Device/Systems

CHAPTER 2

Energy Management

Energy management aspects have a strong impact on electrical product safety engineering. Some of these aspects were presented in the first volume of this book, such as safe current and voltage limits, power distribution systems, and batteries and power source; reference is made to Chapters 8 and 9 of the first volume. This chapter will explore other aspects of energy management.

Within the last few years, the energy sector has undergone a significant transformation that affects all of us. This transformation has been driven by the realization that having an affordable and fully reliable supply of electricity is a key factor for the evolution and prosperity of our society. Throughout this period, the power grid itself has undergone many and major technological changes and has become ubiquitous and indispensable to us. More changes will continue at an even faster pace that will introduce new elements into to this sector.

In this chapter, we show how product safety engineering is an inseparable tool in the development process of energy management.

2.1 Smart Grid

The linking of new technologies in power engineering and information technologies to solve the challenges in the energy sector is referred as the smart grid (an intelligent energy supply system) [1]. The smart grid comprises the networking and control of intelligent generators, storage facilities, renewable sources, and loads and network operating equipment in power transmission and distribution networks with the aid of information technology. The objective is to ensure a sustainable and environmentally responsible power supply by means of transparent, energy-efficient, safe, and reliable system operation by increasing the number of and expanding the functionality of the present schemes that include private household and commerce products. The IEC 61970 and ISO 50001 series of standards deals with the application program interfaces for energy management systems (EMS).

Within the last few years, the IEC and other standards developers identified over 120 standards as relevant to the smart grid. These standards deal with the main aspects related to the smart grid, starting with the architecture of it (IEEE 2030), continuing with energy management (IEC 61970 series), communication networks and systems for power utility automation (IEC 61850), distribution management (IEC 61968), security (IEC 62351, NISTIR 7628), meter reading related aspects–load control (IEC 62056), and functional safety (IEC 61508).

2.2 Energy Efficiency

Energy efficiency is about tackling energy losses and using less energy to get the same job done. Losses occur in processes of energy transformation, transmission, and distribution as well as in the final uses of energy. These aspects have an important impact on product compliance and safety. Many products are designed with deficiencies affecting the quality of distributed power. Due to this, many worldwide regulations were issued that called for the use of only energy verified products. Some examples of products regulated for energy efficiency are lighting products, heating and cooling equipment, electrical motors, household appliances, automatic dispensers, and nonprofessional arc welding equipment. These products have great potential for energy saving.

Energy efficiency labels are informative labels affixed to manufactured products indicating their energy performance and efficiency in a way that allows for comparison between similar products or endorses the products use [2].

Energy efficiency standards and labels (EE S&Ls) are sets of procedures and regulations that prescribe the minimum energy performance of manufactured products and the informative labels on these indicating products energy performance. EE S&Ls are key mechanisms to promote energy efficiency, especially in relation to household appliances, lighting products, automobiles and other mass-produced consumer and commercial energy-using equipment. The Energy Efficiency, Energy Performance & Energy Consumption (E3) Program is also used, which is a collection of coordinated subprograms that implement standards and labels for appliances. However, the E3 Program only provides a statement of test results (STR) and is not yet an established certification process.

Organizations that want to put in place an energy management system (EMS) and use their energy more efficiently and effectively should use the ISO 50001 [3]. The ISO 50001:2018, Energy management systems–Requirements with Guidance for Use, is a strategic tool toward EMS implementation.

The ISO 50001 does not require an independent auditor for certification. To certify or not is a decision to be taken by the organization, unless imposed by regulation. ISO technical committee ISO/TC 301, Energy Management and Energy Savings, have developed a number of other related standards (Table 2.1) to complete the energy management and energy savings family.

Many standards are used for evaluation of product performances referring to energy efficiency. Some examples are provided in Table 2.2.

With regard to product safety, all appliance products, both stationary and portable, manufactured or marketed in the specific countries should be submitted to an appropriate independent laboratory for testing and approval in conformance with the product safety standards, local electrical codes, and energy performance standards and procedures followed by those laboratories, depending on the specific countries. We will focus further on the energy efficiency aspects related to these evaluations.

Table 2.1 List of Standards Related to Energy Management and Energy Savings

Standard	*Title of the Standard*
ISO/IEC 13273-1	Energy efficiency and renewable energy sources – Common international terminology – Part 1: Energy efficiency; – Part 2: Renewable energy sources;
ISO 17741	General technical rules for measurement, calculation and verification of energy savings of projects;
ISO 17742	Energy efficiency and savings calculation for countries, regions and cities;
ISO 17743	Energy savings – Definition of a methodological framework applicable to calculation and reporting on energy savings;
ISO 50002	Energy audits – Requirements with guidance for use;
ISO 50003	Energy management systems – Requirements for bodies providing audit and certification of energy management systems;
ISO 50004	Energy management systems – Guidance for the implementation, maintenance and improvement of an energy management system;
ISO 50006	Energy management systems – Measuring energy performance using energy baselines (EnB) and energy performance indicators (EnPI) – General principles and guidance;
ISO 50007	Energy services – Guidelines for the assessment and improvement of the energy service to users;
ISO 50008	Energy management and energy savings - Building energy data management for energy performance – Guidance for a systemic data exchange approach;
ISO 50009	Energy management systems – Guidance for implementing a common energy management system in multiple organizations
ISO 50015	Energy management systems – Measurement and verification of energy performance of organizations – General principles and guidance;
ISO 50021	Energy management and energy savings – General guidelines for selecting energy savings evaluators;
ISO 50044	Energy saving projects (EnSPs) – Economics and financial evaluation of energy saving projects;
ISO 50045	Technical guidelines for evaluation of energy savings of thermal power plants;
ISO 50046	General quantification methods for predicted energy savings (PrES);
ISO 50047	Energy savings – Determination of energy savings in organizations;
ISO 50049	Calculation methods for energy efficiency and energy consumption variations at country, region and city levels: relation to energy savings and other factors;

In the United States, the Energy Star label is a joint electronic energy efficiency certification project launched by the U.S. Department of Energy (DOE) and the Environmental Protection Agency (EPA). The products that are eligible for Energy Star qualification are included in Table 2.3.

The EPA requires all Energy Star products to be third-party certified (3PC). Product performance data must be certified by an EPA-recognized certification body prior to labeling. Tests need to be conducted in recognized laboratories with demonstrated capabilities.

Table 2.2 List of Standards for Product Performances Referring to Energy Efficiency

Standard	*Title of the Standard*
ISO 5151	Non-ducted air conditioners and heat pumps – Testing and rating for performance;
ISO/IEC 10561	Information technology - Office equipment - Printing devices - Method for measuring throughput - Class 1 and Class 2 printers;
ISO 16358-1	Air-cooled air conditioners and air-to-air heat pumps – Testing and calculating methods for seasonal performance factors – Part 1: Cooling seasonal performance factor;
ISO 16358-2	Air-cooled air conditioners and air-to-air heat pumps – Testing and calculating methods for seasonal performance factors – Part 2: Heating seasonal performance factor;
IEC 60107 (series)	Methods of measurement on receivers for television broadcast transmissions
	- Part 1: General considerations - Measurements at radio and video frequencies;
	- Part 2 to Part 7: Particular standards for specific TV product;
IEC 60379	Methods for measuring the performance of electric storage water-heaters for household purposes;
IEC 60456	Clothes washing machines for household use - Methods for measuring the performance;
IEC 60969	Self-ballasted compact fluorescent lamps for general lighting services - Performance requirements;
IEC 62301	Household electrical appliances - Measurement of standby power;
IEC 62552-1	Household refrigerating appliances - Characteristics and test methods - Part 1: General requirements;
IEC 62087 (series)	Audio, video, and related equipment - Determination of power consumption
	- Part 1: General;
	- Part 2 to Part 7: Particular standards for specific Audio, video product;
IEC 62623	Desktop and notebook computers – Measurement of energy consumption;
EN 14511-1	Air conditioners, liquid chilling packages and heat pumps for space heating and cooling and process chillers, with electrically driven compressors. Terms and definitions;
EN 50229	Electric clothes washer-dryers for household use. Methods of measuring the performance;
EN 50285	Energy efficiency of electric lamps for household use. Measurement methods;
EN 60350-2	Household electric cooking appliances. Hobs. Methods for measuring performance;
EN 60436	Electric dishwashers for household use. Methods for measuring the performance;
EN 60456	Clothes washing machines for household use. Methods for measuring the performance;
EN 61121	Tumble dryers for household use. Methods for measuring the performance;
ANSI/AHAM DH-1[1]	Dehumidifiers- procedure for measuring the capacity and energy input;
ANSI/UL 858	Standard for Household Electric Ranges;
AS/NZS 4665.2[2]	Performance of external power supplies Part 2: Minimum energy performance standard (MEPS) requirements;

Table 2.2 (continued)

Standard	*Title of Standard*
GB 21456[3]	Minimum allowable values of the energy efficiency and energy efficiency grades for household induction cookers;
GB/T 24908	Self-ballasted LED lamps for general lighting services-Performance requirements;
GB 30255	Minimum allowable values of energy efficiency and energy efficiency grades of LED products for indoor lighting;
GOST 51380[4]	General requirements and the methods for the verification of energy efficiency indicators for energy consuming products;
GOST 51388	Provision of Information for consumers about energy efficiency of products for household application;
GOST 51565	Energy classes for refrigerators and freezers;
JIS C 9603[5]	Ventilating Fans;
JIS C 9606	Electric washing machines;
JIS S 2103	Gas Cooking Appliances for Domestic Use;
10 CFR 430 Subpart B[1]	Test Method for Calculating the Energy Efficiency of Single Voltage External AC-DC and AC-AC Power Supplies: 2004.

Notes:
1. USA
2. Australia
3. Guobiao standards or GB standards are the Chinese national standards issued by the Standardization Administration of China (SAC), the Chinese National Committee of the ISO and IEC.
4. Russia
5. Japan

The EPA provides an online search tool that can locate recognized evaluation laboratories, certification bodies, locations, and product types for the Energy Star program [4].

In the EU, Directive 2012/27/EU on energy efficiency establishes the legal framework for the subject. In addition, Regulation (EU) 2017/1369 sets a framework for energy labeling specifying the requested documentation and criteria applicable for energy efficiency labeling. This regulation is the tool through which the consumer can recognize the best performing products. The energy label shows the level (A~G) of energy efficiency, A being the most efficient [5].

Depending on the product, the energy labels will display not only electricity consumption, but also other energy and nonenergy information, with intuitive pictograms. Figure 2.1 shows the European Union energy efficiency label for lamps (lighting sources).

Since January 1, 2019, suppliers (manufacturers, importers, or authorized representatives) need to register their appliances, which require an energy label in the European Product Database for Energy Labelling (EPREL) before selling them on the European market [6].

The energy labeling requirements for individual product groups established under the EU's Energy Labelling Framework Regulation (2017/1369) mandate for some product categories' registration and labeling (see Table 2.3) [5].

Many other countries around the world have issued their own labeling for energy efficiency [7]. Some examples include Canada (Energy Efficiency Regulations,

Table 2.3 Products Eligible to Earn the Energy Star Label and Request Registration and Labeling per EU Regulation 2017/1369

Product Category	*Products*	*Eligible for Energy Star*	*Request Registration and Labeling per EU Regulation (2017/1369)*
Appliances	Air purifiers (cleaners)	Yes	Yes
	Clothes dryers	Yes	Yes
	Clothes washers	Yes	Yes
	Dehumidifiers	Yes	Only energy efficiency
	Dishwashers	Yes	Yes
	Freezers	Yes	Yes
	Refrigerators	Yes	Yes
Building products	Residential windows, doors, and skylights	Yes	
	Roof products	Yes	
	Seal and insulate	Yes	
	Storm windows	Yes	
Commercial food service equipment	Coffee brewers	Yes	Only energy efficiency
	Dishwashers	Yes	Only energy efficiency
	Fryers	Yes	Yes
	Griddles	Yes	Only energy efficiency
	Hot food holding cabinets	Yes	Only energy efficiency
	Ice makers	Yes	Only energy efficiency
	Ovens	Yes	Yes
	Refrigerators and freezers	Yes	Yes
	Steam cookers	Yes	Only energy efficiency
Data center equipment	Data center storage	Yes	Only energy efficiency
	Enterprise servers	Yes	Only energy efficiency
	Large network	Yes	Only energy efficiency
	Small network	Yes	Only energy efficiency
	Uninterruptible power supplies	Yes	Only energy efficiency
Electronics	Audio/video	Yes	Only energy efficiency
	Digital media player	Yes	Only energy efficiency
	Set-top boxes	Yes	Only energy efficiency
	Signage displays	Yes	
	Slates and tablets	Yes	
	Telephones	Yes	
	Televisions	Yes	Yes
	External power supplies	Yes	Only energy efficiency

SOR/2016-311, 2016); China (Code of Practice on Energy Labelling of Products 2018 [8]); India (BEE Star Rating Program, 2019); Japan (Ministry of Economy, Trade and Industry (METI) Top Runner Program, 2010); and Romania (Law on Energy Efficient Utilization (Law no. 199/2000, modified and completed by Law 56/2006)); Turkey (Law on the Preparation and Implementation of the Technical Regulations on Products) (Law No: 4703).

Table 2.3 (continued)

Product Category	*Products*	*Eligible for Energy Star*	*Request Registration and Labeling per EU Regulation (2017/1369)*
Heating and cooling	Air-source heat pumps	Yes	Only energy efficiency
	Central air conditioner	Yes	Yes
	Boilers	Yes	Yes
	Ductless heating and cooling	Yes	Only energy efficiency
	Furnaces	Yes	Yes
	Geothermal heat pumps	Yes	Only energy efficiency
	Light commercial heating and cooling	Yes	Only energy efficiency
	Room air conditioner	Yes	Yes
	Smart thermostats	Yes	
	Ventilation fans	Yes	Only energy efficiency
Lighting*	Ceiling fans	Yes	Yes
	Decorative light strings	Yes	Yes
	Lightbulbs	Yes	Yes
	Light fixtures	Yes	Yes
Office equipment	Computers	Yes	Only Energy Efficiency
	Imaging equipment	Yes	Only energy efficiency
	Monitors	Yes	Only Energy Efficiency
Water heaters	Water heaters	Yes	Yes
	Water pumps	Yes	Only energy efficiency
	Storage water heaters	Yes	Yes
	Solar water heaters	Yes	Yes
	Home gas water heaters	Yes	Yes
Other	Electric vehicle chargers	Yes	
	Pool pumps	Yes	Only energy efficiency
	Smart home energy management systems	Yes	
	Vending machines	Yes	Yes
	Water coolers	Yes	Yes

*Emergency lighting and lamps designed for very specific uses, for instance in theaters, or lamps sold in very small quantities per year (less than 200), are excluded from these EU regulations.

2.2.1 Power Quality

Power quality is a very important issue when evaluating energy efficiency. This is due to aspects caused by distortion of the nominal waveform characteristics of a current and the voltage of the power supplied to the consumer by the power distribution system. This distortion has several negative effects (e.g., overheating of the components of the distribution system, mechanical oscillations in generators and motors, increased noise) and impaired performance of power-line systems involving the unpredictable behavior of protection systems, voltage fluctuations, and generating radio-frequency interference.

These negative effects on power quality have their origin in a few factors, which include:

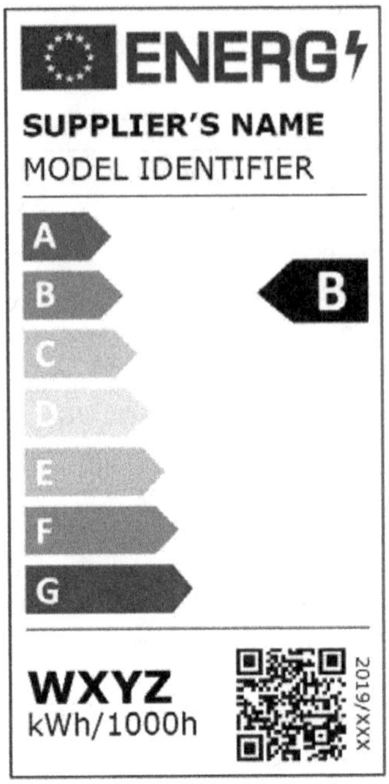

Figure 2.1 Example of EU energy efficiency label for lamps (lighting sources).

- Power factor;
- Voltage regulation (sag and surges);
- Frequency regulation;
- Voltage and phase imbalance;
- Harmonic distortion.

The main cause of all negative phenomena is the harmonics on the power system due to nonlinear loads on the products connected to a public distribution system. Due to harmonics, two current or voltage waveforms with the same root mean square (RMS) value can have different peak values that generate different DC output of the product power supply, and the product performance may be affected.

2.2.2 Power Factor

The AC power supplied through the power mains (apparent power, denoted by capital letter S measured in VA units) will produce the required amount of active power (denoted by capital letter P measured in W units) for real needs. The apparent power S is formed from active power P and reactive power (denoted by capital letter Q measured in VAR units). The measure of the fraction of current in phase with the voltage contributing to the active power is the power factor (PF) that can be between 0 and 1. The PF is represented by the active power (P) divided to apparent power (S). When the PF is close to 1 the energy of the mains is used more effectively with less loss of power through reactive power. In an ideal situation, when both current and voltage waveforms are pure sinusoidal, the value of the apparent power is equal with active power. In reality the current draw by power supplies is nonsinusoidal due to either inductive or capacitive lags, such that the current waveform does not follow the voltage waveform and the PF is less than 1. When separate measurements are made of voltage and current, the current (I) × voltage (U) is not the AC power since the amperemeter and voltmeter read an average (RMS value of the voltage or current) without referring to the phase shift that may be present. If there is a difference in phase between the voltage and current waveforms, the peak current will not be in coincidence with the voltage peak and as such the PF will be less than 1. Other causes that generate a reduced PF are

- Waveform distortion caused by capacitor input filter circuit following rectifiers;
- When the load current is drawn over only a part of each mains cycle;
- The high current peaks cause some clipping distortion on the peaks of the voltage sinusoid.

By reducing the inductive and capacitive lags, the current and voltage waveforms end up being in phase, causing PF to become equal with 1 and the circuit become quite purely resistive (with nonreactive influences as in a DC circuit). Reducing reactive currents and increasing the power factor of a line is an effective way of reducing energy waste.

The action to increase the PF is done by a power factor correction circuit (in general mandatory in a power supply with more than 50W output power), which additionally minimizes the input current distortion. The power factor correction (PFC) circuit that reshapes the input nonsinusoidal AC waveforms (Uin and Iin), can be active or passive. The amount of output power (Pout) available after using a PFC can be estimated as

$$\text{Pout} = \text{Uin (RMS)} \times \text{Iin (RMS)} \times \text{PF}$$

When the PF is close to 1, the current harmonics will be close to zero since all the power is contained in the fundamental frequency. The total RMS value of the current waveform is the sum of the RMS values of all the harmonics, so apparent power, which is the product of RMS voltage and RMS current, is greater than active (real) power. The higher the harmonic content of the current, the lower the PF becomes.

High concentration of low power factor loads (consumers of the power from mains) caused problems and led to the introduction of regulations to limit the harmonic content of current taken from public mains supplies. All harmonic sources are referred to as nonlinear loads because they draw nonsinusoidal currents when a sinusoidal voltage is applied. Although power factor correction is not the correct wording, it has become synonymous for harmonic line current reduction [9].

Table 2.4 summarizes the standards covering the requirements for harmonics of power supply systems.

There are four different classes of equipment in the IEC 61000-3-2 that have different limit values (see Table 2.5).

Table 2.4 Standards Covering the Requirements for Harmonics of Power Supply Systems

Standard	*Name of Standard*
EN 50160	Voltage Characteristics of Electricity Supplied by Public Distribution Systems
IEC 61000-2-2	Compatibility Levels for Low-Frequency Conducted Disturbances and Signaling in Public Low-Voltage Power Supply Systems
IEC 61000-2-4	Compatibility Levels in Industrial Plants for Low-Frequency Conducted Disturbances
IEC 61000-2-12	Compatibility Levels for Low-Frequency Conducted Disturbances and Signaling in Public Medium-Voltage Power Supply Systems
IEC 61000-3-2	Limits for Harmonic Current Emissions for Equipment with Input Current Not Exceeding 16A per Phase
IEC 61000-3-3	Limitation of Voltage Changes, Voltage Fluctuations and Flicker in Public Low-Voltage Supply Systems, for Equipment with Rated Current ≤16A per Phase and Not Subject to Conditional Connection
IEC 61000-3-4	Limitations of Emissions of Harmonic Currents in LV Power Supply Systems for Equipment Rated > 16A
IEC/TR 61000-3-6	Assessment of Emission Limits for the Connection of Distorting Installations to MV, HV and EHV Power Systems
IEC 61000-3-12	Limits for Harmonic Currents Produced by Equipment Connected to the Public Low-Voltage Systems with Input Currents > 16A and ≤ 75A per Phase
IEC 61000-4-7	Testing and Measurement Techniques–General Guide on Harmonics and Inter Harmonics Measurements and Instrumentation, for Power Supply Systems and Equipment Connected Thereto
IEC 62301	Household Electrical Appliances–Measurement of Standby Power
IEEE 519	IEEE Recommended Practice and Requirements for Harmonic Control in Electric Power Systems
IEEE 3002.8	IEEE Recommended Practice for Conducting Harmonic Studies and Analysis of Industrial and Commercial Power Systems

Table 2.5 Classification of Products According to the IEC 61000-3-2 Standard

Class	*Type of Equipment*
A	Balanced three-phase equipment
	Household appliances excluding equipment identified as class D
	Tools, excluding portable tools
	Dimmers for incandescent lamps
	Audio equipment
	Other equipment that is not classified as B, C, or D
B	Portable tools
	Arc welding equipment that is not professional equipment
C	Lighting equipment
D	Personal computers, personal computer monitors
	Radio, TV receivers with input power P ≤ 600W
	Refrigerators and freezers having one or more variable-speed drives (VDS) to control compressor motor(s)

There are no limits for symmetrical controlled heating elements with input power P ≤ 200W, or independent dimming devices for incandescent lamps with a rated power of less than or equal to 1 kW.

2.2.3 Stability of the Power Source

Characteristics of the electrical power source used to power the product under test can influence the results of the tests. Results such as the temperature on electrical heat generating parts are affected by the voltage applied, by the frequency, and by harmonic distortion of the power source. Issues related to the harmonic distortion affect also include the leakage currents of the product.

The specifications in standards of these characteristics are made with the understanding that the specified characteristics are maintained as stated throughout the testing, but in the real world, a power source that meets these ideal specifications does not exist. Some standards recognize this and include tolerances for the power source specifications. Some bodies, such as the IECEE-CTL, recommend default power source stability requirements to be followed when the test standard does not contain tolerances for the power source to be used [11]. These power source stability requirements define the characteristics of real-world power sources that can be used in the testing laboratory so that laboratories can obtain consistent, uniform, and repeatable results. To evaluate the power source stability, it is required to conduct measurements of voltage, frequency, and total harmonic distortion (THD) of the laboratory's power source. The aspects related to source capacity, such as short-circuit current testing, abnormal testing, and switching testing, are not addressed when the stability of power source is evaluated. The power source stability requirements apply to power sources used for testing products connected to ordinary branch circuits in residence and business environments (e.g., 120V, 15 and 20A; 240V, 15A circuits in North America, or 230V, 10 and 15A branch circuits in Europe). Laboratory power source characteristic suitability measurements shall be performed upon initial installation, modification, and repair of the laboratory power source(s).

According to the IECEE-CTL, the recommended power source used for testing shall meet the following requirements:

- Voltage stability: +/–3%;
- Frequency stability: +/–2%;
- THD: max. 5%.

The IEC 61000-3-2 specifies different requirements for the testing of a supply source:

- Voltage stability: +/–2%;
- Frequency stability: +/–0.5%;
- Phase angle: +/–1.5° between the fundamental voltage of each phase.

Characteristics of electrical power sources representing electrical mains connections used in the testing laboratory shall be measured at the point where the tests are performed. Typically, this point is considered the receptacle or wiring terminals of the test station where the test setup is connected.

Before the start of the tests, the voltage, frequency, and harmonic distortion of the power source shall be measured with an open circuit over a period of 1 hour. The voltage shall be adjusted to one of the nominal voltages used for testing.

Afterward, the power source shall be loaded to rated maximum normal resistive load (continuous duty) for a period of 1 hour without readjustment of the power source during which time the voltage, frequency, and harmonic distortion shall be measured. The power source shall comply with the requirements throughout the duration of the test.

2.3 Stored Energy Systems

Recent development of modern electricity networks have begun to implement electrical energy storage (EES) technologies to provide balancing auxiliary services, useful to grid integration of large-scale renewable energy systems. The issue of energy storage for electricity overlaps with the wider matter of general energy storage that is represented by fuel such as coal, gas, hydro, and nuclear [12].

Electricity storage technologies, either in a distributed manner or as bulk storage interfacing to the energy system, will play an important role in the future. By storing electrical energy in a cheaper and more consistent way, technology will improve the efficiency of the existing methods and of the renewable methods of the generation of electrical energy. The parameters that allow the characterizations of an EES are energy capacity, power capacity, responsiveness to discharge/charging, and discharge/charge duty cycle. One of the problems of the storage systems is how to obtain a good load match between the supplier of the electrical energy and the storage system.

The IEC 62933 series of standards refer to system aspects on EES systems and on the interaction between EES systems and electric power systems (EPS) rather than energy storage devices. This standard series considers all storage technologies as long as they are capable of storing and discharging electrical energy.

There are a wide variety of electrical energy storage technologies used currently and many others in different stages of development.

In today's technologies, capacitors (using potential electrical energy) and batteries (using chemical reaction) are frequently used for the storage of electrical energy. In addition to these technologies, superconducting magnetic energy storage (SMES) and flywheel energy storage (FES) systems are also considered. The power capacity for an electrical energy storage system is dictated by the application and is dependent on the power quality, power system stability, and load leveling.

A simple capacitor is represented by two metal plates separated by a short distance. Capacitors are devices for storing electrical charge, but are not used for long-term storage because the charge leaks away. Capacitors may be dangerous, even after the equipment has been de-energized, because they can accumulate a dangerous residual charge without an external source. Capacitors may also be used to store large amounts of electrical energy, but an internal failure of one capacitor in a bank could result in an explosion when all other capacitors in the bank discharge into the fault. High-voltage cables should be treated as capacitors because they have capacitance and thus can store electrical energy.

The UL 810A Electrochemical Capacitors standard covers electrochemical capacitors for use in equipment such as electronic products, uninterruptible power supplies, emergency lighting, engine starting, and power equipment. These capacitors, also known as energy storage capacitors, electric double layer capacitors, ultracapacitors, double layer capacitors, or supercapacitors, consist of either individual capacitors or multiple series and/or parallel connected capacitors with or without associated circuitry.

Stationary batteries configurations such as lithium-ion, nickel-cadmium, sodium-sulfur (NaS), sodium nickel chloride, high-capacity lead-acid (valve regulated), NiMH, or flow (vanadium redox and zinc-bromine) store electrical energy in the chemical structures and can be used as stand-alone systems or hybrid (comprised of multiple power sources) systems.

The electrical shock, fire, explosion, arc flashes, and exposure to dangerous substances are hazards with a high level of unacceptable risks developed by batteries in fault situations. Means to mitigate the hazards is a main goal of designers of such systems.

Performance and reliability testing for stationary batteries includes capacity, charge/discharge cycling, overcharge abilities, environmental and altitude simulation, combined temperature cycling, and vibration testing. The life cycle of a battery depends on the following attributes: depth of discharge, cycling range of charging and discharging in each cycle, operating temperature, and rate of charging/discharging. The battery storage end of life is typically 80% for automotive and approximately 65% for stationary batteries. As batteries age, they not only lose capacity, they also have increasing internal resistance—which leads to decreased rate capabilities.

Next we provide a brief list of the standards applicable to batteries when they are used as electricity energy storage systems (Table 2.6).

Superconducting magnetic energy storage (SMES) is a novel electromagnetic technology that stores electrical energy from the grid with near-zero loss of energy in a magnetic field created by the flow of direct current in a superconducting coil, which has been cryogenically cooled to a temperature below its superconducting critical temperature. The energy stored in an SMES system is discharged by connecting an AC power convertor to the conductive coil. SMES is a grid-enabling device that stores and discharges large quantities of power almost instantaneously but has very low energy densities.

The Flywheel energy storage (FES) system is an electromechanical system based on large rotors that store energy kinetically. This energy is stored as kinetic energy by causing a disk or rotor to spin on its axis. During discharging; that is, when power is required, the FES takes advantage of the rotor's inertia and the stored kinetic energy is converted to electricity. Modern FES systems consist of a rim attached to a shaft (rotating mass), which is supported by bearings and is connected to a motor/generator. FES systems are characterized by high power density, relatively low maintenance needs, high cycling rate, deep discharges, and high self-discharge rate. FES systems are used mainly to ensure short-duration power quality and to provide a reliable option for UPS applications [14].

2.4 DC/DC Conversion

DC/DC converters are one of the components of the power supplies. They are considered components because they are integrated within the end-use equipment.

Table 2.6 List of Standards and Guides for Batteries Used as Electricity Energy Storage Systems

Standard	*Title of Standard*
ANSI/UL 9540	Energy Storage Systems and Equipment-Safety Standard
UL 1973	Batteries for Use in Light Electric Rail (LER) Applications and Stationary Applications
UL 2271	Standard for Batteries for Use in Light Electric Vehicles (LEV) Applications
UL 2580	Standard for Batteries for Use in Electric Vehicles
IEC 62485 (series)	Safety Requirements for Secondary Batteries and Battery Installations Part 1: General Safety Information Part 2: Stationary Batteries Part 3: Traction Batteries Part 4: Valve-Regulated Lead-Acid Batteries for Use in Portable Appliances
IEC 62619	Secondary Cells and Batteries Containing Alkaline or Other Non-Acid Electrolytes–Safety Requirements for Large Format Secondary Lithium Cells and Batteries for Stationary and Motive Applications
IEC 62620	Secondary Lithium Cells and Batteries for Use in industrial applications
IEC 62840 (series)	Electric Vehicle Battery Swap System Part 1: General and guidance Part 2: Safety requirements
IEC 62932 (series)	Flow battery energy systems for stationary applications Part 1: Terminology and general aspects Part 2-1: Performance general requirements and test methods Part 2-2: Safety requirements
IEEE 1375	Guide for Protection of Stationary Batteries
IEEE 1679.1	Guide for Characterization and Evaluation of Lithium-Based Batteries in Stationary Application
JIS C 8715 (series)	Secondary Lithium Cells and Batteries for Use in Industrial Applications Part 1: Tests and Requirements of Performances Part 2: Tests and Requirements of Safety

Each type of DC/DC converter offers a method to increase or decrease the voltage from a DC voltage supply (e.g., battery), with the purpose of saving space, instead of using multiple batteries to accomplish the same thing, and at the same time, these circuits also regulate the output voltage.

A very simple definition of these devices is that it represents the equipment/circuits that converts one DC voltage into another DC voltage.

The operating voltage of various electronic components may vary over a wide range, thus making it a requirement to supply a voltage for each component; a buck converter outputs a lower voltage than the original input voltage, while a boost converter will offer a higher voltage at its output.

DC/DC converters are high-frequency power conversion circuits that use high-frequency switching and inductors, transformers, and capacitors to smooth out switching noise into regulated DC voltages. Closed feedback loops maintain constant voltage output even when changing input voltages and output currents. Their disadvantages are noise and complexity.

As component power supplies, the DC/DC converter modules are regarded from a regulatory point of view that by themselves, they are not able to entirely meet the safety standard's requirements. Generally speaking, almost each DC/DC converter comes on the market along with a series of conditions of acceptability. Then we consider that it may become a supplementary responsibility for the manufacturer of the end-use equipment to ensure that the end product fulfills all regulatory approvals to in turn ensure that the final equipment will be compliant with the applicable standards equipment. Within the first volume of this book, we explained that including one or more recognized components within a piece of equipment may not necessarily result in a fully compliant end-use piece of equipment. This statement should be taken in consideration when we discuss power supplies, DC-DC converters, and any other involved components (e.g., radio frequency interferences (RFI) filters, transformers). EMC compliance considerations should also be taken into account.

The recognized component scheme that Underwriters Laboratories and other testing houses operate assures that any qualified compliant component with this scheme will be safe to use within end-use equipment, providing that "installation takes place within a controlled environment—that is, a factory rather than a field site." At the same time, utmost attention must be paid to the fine print called "Conditions of Acceptability." These conditions are in fact a statement document that includes a series of clear requirements about how the installation must be performed, and how to use a component within end products in order to remain compliant with the applicable standard for the end equipment. The manufacturer of the equipment who adheres to these requirements must be able to offer objective evidence that the component meets all of the conditions imposed by its recognition.

DC/DC converters are electronic circuits used to change DC electrical power from one voltage level to another by modifying the input energy into a different impedance level. A DC/DC converter receives power from an input DC source and converts this into regulated output power for delivery to a load. Some of the input power dissipates as heat within the converter. No energy is generated inside the converter. The delivered output power divided to input power designate the efficiency of the DC/DC converter, which for some models is higher than 90%. DC/DC converters can operate from different input sources: power modules, wall suppliers, or batteries.

In a DC/DC converter the switch-mode regulation is used (as in switching mode power supplies) to allow high levels of efficiency and power density. A switching regulator chops the filtered DC input voltage into a high-frequency square wave and a pulse width modulation (PWM) circuit controls the average DC output voltage by setting the duty cycle of the chopper.

From an insulation point of view, some DC/DC converters are nonisolated and other provide a total dielectric isolation between input and output circuits. The nonisolating type of converter is generally used where the voltage needs to be stepped up or down by a relatively small ratio (e.g., less than 4:1) [15]. Actually, based on incomplete statistics, more than 500 prototype topologies of DC/DC converters exist. All existing DC/ DC converters were designed to meet the requirements of certain applications. [16]. Table 2.7 presents the main topologies of DC/DC converters.

The isolation voltage between input and output of the DC/DC converter can be between 500 to 4500 VDC (depending on applications: medical equipment, systems used in noisy environments, and machine control systems require a higher

Table 2.7 Main Topologies of DC/DC Converters

Isolation between IN-OUT	*Topology*	*Function*	*Advantages*	*Disadvantages*	*Comment*
NO	Buck	Regulated output DC voltage lower than the DC input unregulated voltage	High efficiency; low ripple;	Overvoltage if switch shorts	One supply line must be common to both IN and OUT
	Boost	Regulated output DC voltage higher than the DC input unregulated voltage	High efficiency; low input ripple	High switch peak current; high output ripple	One supply line must be common to both IN and OUT
	Buck-boost	The output voltage is of opposite polarity to the input and can be higher, equal, or lower than input	Output waveforms with sharp edges	High output ripple	One supply line must be common to both IN and OUT
	Cuk	The output voltage is of opposite polarity to the input and can be higher or lower than input	Very high efficiency; less ripple; less switching losses		Based on capacitive energy transfer
	Charge-pump	Used for low and moderate load current (output currents less than 300 mA and output voltage less than 6V)	Inductor not used; high efficiency; cost-effective; less noise	Internal regulation not present	Capacitor used for storage and transfer of the energy
	Resonant	Used for high power (motor drivers, UPS)	Without PWM	Very complex	Process power in a sinusoidal form
YES	Fly-back (boost derived)	Used for low power up to 50W	Good efficiency; cheaper	Higher levels of output noise and ripple; higher stress levels on output components	The coupled inductor provides energy storage and isolation
	Push-pull (buck derived)	Very robust; used from low to medium power	Good regulation at low input voltage; easily adaptable to multiple outputs	Moderate efficiency; low output voltage; expensive	Use a center tapped transformer and two switches that are driven ON and OFF alternatively
	Forward (buck derived)	Used in medium and high power 150-200;	Lower ripple and noise; better transient response; high reliability	Expensive; high transistor voltage	Use a transformer and an output inductor that store the energy
	Full bridge	More efficient in high power more than 400W; high input voltage	Low output ripple	Needs four power devices; control more complicated; expensive	Four transistors are connected in a bridge configuration to drive a transformer primary
	Half bridge	Used in medium and high power 100–400W; high input voltage	Very robust; low output ripple	Control more complicated; expensive; poor transient response	Two transistors are connected in a bridge configuration with the other two replaced by capacitors

isolation level), with a resistance of insulation of min. 100 MW and an isolation capacitance less than 100 pF. These values are determined by the incorporated transformer (ferrite core for the high frequency) characteristics. Many DC/DC converters are provided with output current limitation and short-circuit protection.

The overvoltage protection (OVP) is usually provided by a Zener diode that senses the rise in converter output voltage level and PWM is shut down. Another OVP is the overvoltage crowbar, when an SRC (connected between the DC output and common) shorts and clamps the output to ground.

Usually the DC/DC converter requests an external fuse on +IN for protection against electrical shock. The fuse should be sized large enough so that will not blow out under steady-state conditions. In general, this fuse is sized at 150%–200% of the maximum steady-state input current at maximum load and minimum DC input voltage. Additionally, the transient or start-up current of the DC/DC converter should be considered for the fuse selection. Slow blow (SB or T) fuses are the best option because they have larger inrush current (the peak instantaneous input current draw at turn-on of the circuit) capability. It is important to know that for a medical application the DC/DC converter may be turned ON/OFF many times per day, in contrast to radio equipment with very few ON/OFF cycles during its life cycle. The best way to select the correct fuse is to test the subject DC/DC conversion circuit in the particular application.

Actually, a DC/DC converter module (DCM) provides isolation, regulation, fault protection, and monitoring in a single module.

Heat is removed from the converter though the flat metal baseplate (which in general is grounded and electrically isolated from all heat-generating parts). Heat is transferred by the following mechanisms [16]:

- Radiation (electromagnetic transfer of heat between locations at different temperatures);
- Conduction (transfer of heat through a solid medium);
- Convection (transfer of heat through the air).

In terms of the approval process, the standards applicable for power supplies are also applicable for DC/DC converters; the applicable end-use equipment standard and all of the conditions of applicability (if stated) will be complementary and will dictate the final compliance.

The best way to optimize the use of a DC/DC converter in an application is to fully characterize the converter and intended load early in the design cycle.

Energy efficiency and power density are always important considerations in any power system and will impact the design of the DC/DC converters. We will not discuss the advantages and/or disadvantages of using DC/DC converters; as of today, they are widely used in many applications, including electric vehicles, medical electrical equipment, the railway and aerospace industries, military equipment, and/or to interface with power grid connection equipment.

2.5 DC/AC Inverters

A DC/AC inverter is a type of circuit designed to convert a low DC voltage (e.g., 12V) from a battery or a DC source to 120V or 230V AC used to supply power on a load. This module consists primarily of an oscillator circuit followed by alternate switching elements working at a frequency of 50 or 60 Hz (depending on the desired output frequency), and a step-up transformer that steps up the voltage to

120V or 230V AC (inverters without a transformer also exist, but here users can be exposed to a hazardous situation when dangerous DC fault current is generated).

As topologies, single-phase inverters are full-bridge inverters, the half-bridge inverters, and push-pull inverters.

Another classification of the inverters refers to line-commutated inverters (which use an external source to trigger switching elements and synchronize their output) and self-commutated inverters (which control the switching elements and regulate their waveform output with internal software and controls).

According to the mode of operation the inverters can be

- *Stand-alone:* Operates from DC batteries and supplies AC power independent of utility grid.
- *Interactive:* Operates from other DC sources, such as PV arrays, and supplies AC power in parallel with the utility grid. The optimal input DC power must be 1.1–1.25 multiplied with the maximum AC power output.
- *Multimode:* Operates in both utility stand-alone and interactive modes in case of utility failure. For inverters identified as interactive, the standard addresses power quality, synchronization of power back into the grid, and anti-islanding protection.

Low-frequency inverter designs use a half-bridge or push-pull inverter circuit, and the resulting AC output is stepped up to higher voltages through a transformer.

High-frequency inverters use DC/DC converters and smaller transformers, resulting in highly efficient and lightweight designs. Table 2.8 presents the types of inverters depending on application and power.

Typically, the specifications for all types of inverters include:

- AC output power;
- AC output voltage;
- DC input voltage;
- Maximum input current;
- Output power factor;
- Operating frequency;

Table 2.8 Main Application Types of DC/AC Inverters

Inverter Type	*Power*	*Application*
Module-level	200–300W	Residential and small commercial
String	2–12 kW	Residential and small commercial
Central	30–500 kW	Public and private properties interconnected to the grid at service voltages less than 600 VAC
Utility-scale	500–1,000 kW	Up to 500-kW public and private properties interconnected to the grid at service voltages less than 600 VAC. Up to 1,000 kW for service voltage up to 38 kV.
Bimodal	2–19 kW	Provide backup AC power to critical loads when the utility grid is not energized

- Power conversion efficiency;
- Maximum ambient temperature;
- Installation environment.

Inverters less than 1 kW may use a 12V battery, while large inverters use a nominal DC bus voltage of 24V, 48V, or higher. Some inverters need to be installed and operated with an external transformer and/or overcurrent protection (input or output) sized at 125% of the inverter's output current rating. Most inverters incorporate monitoring and communications functions to record and display system operating parameters, fault conditions, and performance information. Table 2.9 consists of a short list of standards applicable to AC/DC inverters.

2.6 Uninterruptible Power Systems

Power distribution systems, both public and private, theoretically supply electrical equipment with a sinusoidal voltage of fixed amplitude and frequency (e.g., 400V RMS, 50 Hz, on low-voltage systems).

In real-life conditions, however, utilities indicate the degree of fluctuation around the rated values. Standards IEC 60038 and EN 50160 defines the normal fluctuations in the low voltage supply on distribution systems as follows:

- Voltage +10% to –14% or –15% (average RMS values over 10-minute intervals), of which 95% must be in the +10% range each week;
- Frequency +4% to –6% over 1 year with ±1% for 99.5% of the time (synchronous connections in an interconnected system).

For efficient use of energy by consumers (ITE, medical, household, lamps, etc.) and to protect data against loss and corruption, when the AC input (utility power) supply falls, is affected by disturbances or is out of the above allowed fluctuation, a continuity of energy supply on loads must be ensured. For this purpose, uninterruptible power systems (UPS) equipment is generally used.

Utility power can be disturbed or even cut by:

- Atmospheric phenomena affecting overhead lines or buried cables:
  - Lightning, which can produce a sudden voltage surge in the system;

Table 2.9 List of Standards for DC/AC Inverters

Standard	*Title of Standard*
IEC 62109 (series)	Safety of Power Converters for Use in Photovoltaic Power Systems
	Part 1: General Requirements
	Part 2: Particular Requirements for Inverters
UL 1741	Inverters, Converters, Controllers, and Interconnection System Equipment for Use with Distributed Energy Resources
IEEE 1547	Interconnecting Distributed Resources with Electric Power Systems

    - Frost, which can accumulate on overhead lines and cause them to break.
- Accidents:
    - A branch falling on a line, which may produce a short circuit or break the line;
    - Cutting of a cable, for example during trench digging or other construction work;
    - A fault on the utility power system.
- Phase unbalance;
- Switching of protection or control devices in the utility power system, for load shedding or maintenance purposes.

Some equipment can also disturb the utility power system, as in the following examples:

- Industrial equipment:
    - Motors, which can cause voltage drops due to inrush currents when starting;
    - Equipment, such as arc furnaces and welding machines, which can cause voltage drops and high-frequency interference.
- Power electronics equipment (switch-mode power supplies, variable speed drives, electronic ballasts, etc.), which often cause harmonics;
- Building facilities, such as lifts, which provoke inrush currents or fluorescent lighting, which causes harmonics [20].

A UPS represents a combination of converters, switches, and energy storage means (e.g., batteries) constituting a power system for maintaining continuity of load power in case of input power failure or deviation from allowable variations [21]. The UPS converts and controls direct current (DC) energy to alternating current (AC) energy. It uses a conventional battery of 12V rating as the input source and by the action of an inverter circuitry, it produces an alternating voltage that is sent to the load.

The UPS is the central component of any well-designed power protection system and over many years, has become an integral part of high-quality power distribution to the customer. The UPS provides backup power when utility power fails, either long enough for critical equipment to slowly shut down so that no data is lost, or long enough to keep required loads operational until a generator comes online and conditions incoming power, and surges don't damage sensitive electronic parts of the system. The IEC 62040 series of standards specify the configuration, safety, EMC, performance, and test requirements for UPS. In the United States, the applicable standard is the UL 1778: Standard for Safety Uninterruptible Power Systems.

The UPSs have the following topologies [22]:

- A single-conversion system is a configuration which, when the AC input supply falls, the UPS utilizes an inverter to draw current from the internal battery disconnecting the AC input supply and staying on battery power until the AC input returns to normal value. There are two types of such systems: standby UPSs (the most basic type of single-conversion UPS; generally, the best option for smaller loads) and line-interactive UPSs (preferable in locations with access to relatively trouble-free AC input power).

- Double-conversion systems consist of an input rectifier which converts AC power into DC and feeds it to an output inverter that processes the power back to AC before sending it on the load. This double-conversion process provides a total separation of the critical loads from the input power, ensuring that the loads receive only clean, reliable electricity. When the AC input power fails, the input rectifier shuts off and the output inverter begins drawing power from the internal battery instead (as in a single-conversion system).
- Multimode systems combine features of both single- and double-conversion topologies providing substantial improvements in both efficiency and reliability. When AC input power fails, the system automatically switches to double-conversion mode, completely isolating the loads from the input AC source. If AC input power fails outside the tolerances of the double-conversion rectifier, the UPS uses the internal battery to keep supported loads up and running (see Figure 2.2) [22]. When the input AC come back, the UPS switches to double-conversion mode until input power stabilizes.

A UPS consists of the following parts:

- *Rectifier/charger:* Circuit that converts AC voltage to DC voltage and charges the battery;
- *Inverter:* Circuit that converts DC voltage to AC voltage regulated and filtered in terms of voltage and/or frequency;

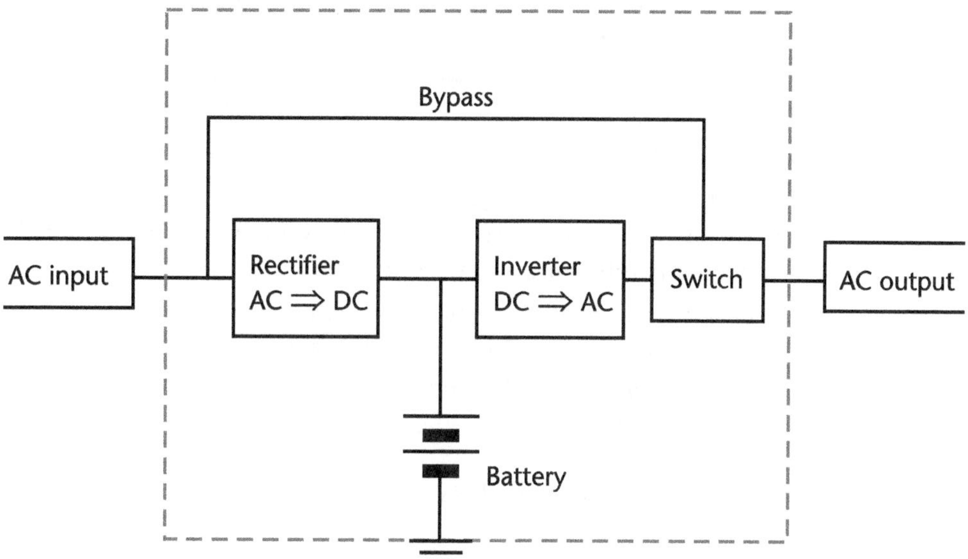

Figure 2.2 Block diagram of a UPS.

- *Bypass:* Circuit that directly outputs AC voltage in the case of UPS overload or fault;
- *Switch:* Circuit t swhatitches between inverter output and bypass output;
- Battery (generally lead-acid type) enabling energy to be stored and instantly recovered as required over a 5- to 30-minute period.

The UPS needs to be monitored. Some audible and visual alarms should be provided when:

- Input power supply fails (voltage and frequency);
- Ground fault;
- Operation of battery backup;
- Battery is being discharged.

For a UPS selection the following needs to be considered [23]:

- *Topology:* Single-conversion UPSs are more efficient than double-conversion devices but offer less protection. Multimode UPSs, although they may be more expensive than either single- or double-conversion systems, are the best choice to achieve an optimal result of both efficiency and protection.
- *Rating:* 20% above the sum of all loads in VA, connected to UPS, to estimate the requested rating for the needed UPS.
- *Installation:* The UPS must be installed in an enclosure made of one or more floor-mounted, free-standing, self-supporting steel cabinets, or wall mounted cabinets. The enclosure must have a minimum degree of protection of IP31.
- *Battery runtime:* Typical UPS battery provides 5 to 30 minutes of backup power.
- *Crest factor:* CF = Ipk/Irms; for nonlinear loads CF>3; Irms–effective load current measured with an oscilloscope; Ipk–peak value of the load current(true RMS voltage).
- *Efficiency:* η = Po/Pi; ratio between active output power and active input power of the UPS.
- *Noise:* 52 dBA in an office, 60 dBA in a computer room, 65/75 dBA in an electrical equipment room.
- *Rectifier unit:* Must restart automatically upon restoration of the mains power supply following a power interruption.
- *Output voltage:* Must be sinusoidal with a relative harmonic content not exceeding 5% for linear and nonlinear loads.
- *Output current limiting:* When the output voltage falls below 50% of nominal rated output voltage, that the UPS must shut down within 5 seconds.

The basic functionality of an UPS can be described as follows:

- Ensure that there is no break in the power supply at any point in time unless major faults like fuse opening occurs;

- Provide the battery with an adequate charge to maintain the optimum conversion rate to AC;
- Ensure overcharge protection to prevent battery damage;
- All forms of surges and undesired waveforms that may be generated by an inverted source voltage must be filtered and well adapted to the output level;
- Must maintain stability when the battery safe voltage is exceeded;
- Provide an overload protection.

Special attention needs to paid to AC input wiring ampacity to not be damaged when feedback from the inverter to the mains utility occurs. This can represent a high risk of fire. If the load is nonlinear (usually the case for single phase power supplies) and connected to the output of a three-phase UPS between phases and neutral, then it is likely that the neutral conductor will have to carry third harmonic currents, which are additive, for all the loads connected.

In this situation, the output neutral conductor should be increased in size in accordance with national wiring rules or IEC 60364-5-53 clause 532.2.1. Many UPS types use the input supply neutral to reference the output neutral. When providing a means of supply isolation or input supply change-over circuits to the UPS, care needs to be taken to ensure that the input supply neutral reference is not disconnected while the UPS is in service. This also applies to installations where the bypass supply is separate from the normal input supply to the UPS and only one supply neutral is connected to the UPS for both supplies [20].

Batteries installed remotely from the UPS itself should be provided with protective devices suitably rated for operation on DC as close to the terminals as possible. A means of isolation should also be fitted to enable maintenance of the battery.

It is a well-known fact that the battery is the most vulnerable part of a UPS. In fact, battery failure is a leading cause of load loss. If the battery consists of more than one battery string in parallel, then each battery string should have a means of isolation. Ambient temperature-lead-acid battery design life is reduced by half for every 10 degrees rise above the design reference temperature of 25°C. For an optimum service life it is required to install the battery in a temperature-controlled environment. Proper ventilation must also be provided to reduce any potential dispersed explosive mixtures of hydrogen and oxygen below hazardous levels. Ventilation must be calculated per EN 50272-2 Prescriptions for Safety of Batteries and Installations. The valve-regulated batteries (VRLA), best known as sealed lead batteries with internal gas recombination, can be installed into sites without particular recommendations for safety as the air flow needed for these batteries is very small. Due to the fact that water cannot be added to VRLA batteries, recombination of water is critical to their life and health, and any factor that increases the rate of evaporation or water loss—such as temperature or heat from the charging current—reduces the life of the battery. The standard lifespan for VRLA batteries is 3 to 5 years. The IEEE defines end of useful life for a UPS battery as being the point when it can no longer supply 80% of its rated capacity in ampere-hours. If batteries sit unused for a long time with no charging regimen, their life will decrease. Due to the self-discharge characteristics of lead-acid batteries, it is necessary that they be charged after every 6 to 10 months of storage; otherwise, a permanent loss of capacity will occur between 18 and 30 months. To prolong shelf life without charging, batteries need to be stored at 10°C (50°F) or less.

2.7 Fuel Cells

A typical fuel cell is an electrochemical device that converts hydrogen-rich gases or hydrocarbons and oxygen from air into electrical power and heat. Current technologies are:

- The proton exchange membrane fuel cell (PEMFC);
- The phosphoric-acid fuel cell (PAFC);
- The solid oxide fuel cell (SOFC);
- The molten carbide fuel cell (MOFC).

The stationary fuel cells are used as electric power sources for residential and commercial and backup voltage sources for uninterruptible power supplies for critical computer and telephone applications, portable battery replacement power sources for cell phones and other electronic equipment, and battery replacement power sources for industrial trucks. Additionally, fuel cells are beginning to be used by the utility companies as a source of supplemental power during periods of peak demand. A basic structure of a fuel cell power system consists of either a reformer to extract hydrogen from fuel or can be supplied with a direct source of hydrogen, a fuel cell, and power conditioning circuitry, which may include an inverter [23].

The IEC 62282 series of standards refer to requirements for the safety and performance of fuel cell modules. These standards specify the conditions that can create hazards to persons and cause damage outside the fuel cell modules.

In United States and Canada, applicable codes and standards are NFPA 853, NEC Articles 692 and 705, CSA FC-1, UL 2262, UL 2265A, UL 2265C, ANSI/UL 2267, and UL 1741[23].

2.8 Photovoltaic and Solar Energy

Among renewable energy technologies, solar photovoltaic (PV) electricity (solar cells as a converter of solar radiation to electricity) seems to be one of the most effective.

The principal points of interest on the installation and use of distributed or localized alternative energy generation are safety, power quality, harmonic distortion, and islanding (when the distributed generation equipment continues to feed power to the grid when the utility source has been disconnected, resulting in sourcing an "island" or part of the grid).

There are four well-recognized photovoltaic technologies:

- Thin crystalline Si solar cells;
- Organic solar cells;

- High-efficiency photovoltaic stacks for terrestrial concentrators;
- Stand-alone germanium (Ge) low-bandgap cells for thermophotovoltaics application, which convert the radiation from heat sources that are at a lower temperature compared to the sun (Figure 2.3).

The DC voltage produced by the solar cells is connected to the input of a DC/AC inverter and after this to a step-up transformer that provides the requested AC voltage and frequency to consumers (see Section 2.5). The installation requirements for interconnection with electrical power production sources are covered by NEC Article 705.

The list of standards from Table 2.10 refers to standards covering the systems for PV conversion of solar energy and to all elements in the entire photovoltaic energy system. The concept of a photovoltaic energy system includes the entire field from light input to a solar cell to the interface with the electrical systems to which energy is supplied.

The IECRE–Renewable Energy System, the IEC system for certification to standards relating to equipment for use in renewable energy application, refer to PV systems in the document IECRE 04, which defines the rules and procedures for conformity assessment and certification of PV power plants with respect to standards and technical requirements for photovoltaic equipment, as well as installation and operation of the system. It is intended to facilitate mutual recognition by participants (reciprocal acceptance) of inspection results and certificates issued by other participants for obtaining certification at a national or international level. The IECRE Operational Document (OD) 407 defines the requirements for the data to be reported in conjunction with issuance of electronic certificates at milestones in the PV system life cycle and specifies that performance data for all PV systems certified under the IECRE PV Certification Scheme shall be reported [25].

Figure 2.3 Photovoltaic panel.

Table 2.10 List of Standards for Photovoltaic Conversion

Standard	*Title of Standard*
IEC 60364-7-712	Low voltage electrical installations Part 7-712: Requirements for special installations or locations - Solar photovoltaic (PV) power supply systems
IEC 60904 (series)	Photovoltaic devices –
	Part 1: Measurement of photovoltaic current-voltage characteristics
	Part 1-1: Measurement of current-voltage characteristics of multi-junction photovoltaic (PV) devices
	Part 2: Requirements for photovoltaic reference devices
	Part 3: Measurement principles for terrestrial photovoltaic (PV) solar devices with reference spectral irradiance data
	Part 4: Photovoltaic reference devices - Procedures for establishing calibration traceability
	Part 5: Determination of the equivalent cell temperature (ECT) of photovoltaic (PV) devices by the open-circuit voltage method
	Part 7: Computation of the spectral mismatch correction for measurements of photovoltaic devices
	Part 8: Measurement of spectral responsivity of a photovoltaic (PV) device
	Part 8-1: Measurement of spectral responsivity of multi-junction photovoltaic (PV) devices
	Part 9: Classification of solar simulator characteristics
	Part 10: Methods of linear dependence and linearity measurements
	Part 13: Electroluminescence of photovoltaic modules
	Part 14: Guidelines for production line measurements of single-junction PV module maximum power output and reporting at standard test conditions
IEC 61215 (series)	Terrestrial photovoltaic (PV) modules—Design qualification and type approval
	Part 1: Test requirements
	Part 1-1: Special requirements for testing of crystalline silicon photovoltaic (PV) modules
	Part 1-2: Special requirements for testing of thin-film Cadmium Telluride (CdTe) based photovoltaic (PV) modules
	Part 1-3: Special requirements for testing of thin-film amorphous silicon based photovoltaic (PV) modules
	Part 1-4: Special requirements for testing of thin-film Cu(In,GA)(S,Se)2 based photovoltaic (PV) modules
	Part 2: Test procedures
IEC 61701	Photovoltaic (PV) modules - Salt mist corrosion testing
IEC 61724 (series)	Photovoltaic system performance Part 1: Monitoring
	Part 2: Capacity evaluation method
	Part 3: Energy evaluation method
IEC 61730 (series)	Photovoltaic (PV) module safety qualification
	Part 1: Requirements for construction
	Part 2: Requirements for testing
IEC 62108	Concentrator photovoltaic (CPV) modules and assemblies - Design qualification and type approval
IEC 62109 (series)	Safety of power converters for use in photovoltaic power systems
	Part 1: General requirements
	Part 2: Particular requirements for inverters
	Part 3: Particular requirements for electronic devices in combination with photovoltaic elements

Table 2.10 (continued)

Standard	*Title of Standard*
IEC 62716	Photovoltaic (PV) modules - Ammonia corrosion testing
IEC 62759-1	Photovoltaic (PV) modules - Transportation testing
	Part 1: Transportation and shipping of module package units
IEC 62788 (series)	Measurement procedures for materials used in photovoltaic modules
	Part 1-7: Encapsulants - Test procedure of optical durability
	Part 2: Polymeric materials - Frontsheets and backsheets
	Part 5-1: Edge seals - Suggested test methods for use with edge seal materials
IEC 62790	Junction boxes for photovoltaic modules - Safety requirements and tests
IEC 62817	Photovoltaic systems - Design qualification of solar trackers
IEC 62892	Extended thermal cycling of PV modules - Test procedure
IEC 62941	Terrestrial photovoltaic (PV) modules - Quality system for PV module manufacturing
UL 1703	Standard for Flat-Plate Photovoltaic Modules and Panels
UL 61730	Standard for Photovoltaic (PV)ModuleSafety Qualification

The test program for determination of performance, safety, durability, and reliability of the PV modules consists of the following tests:

- Performance at nominal operating cell temperature and low irradiance;
- Isolation test;
- Long time outdoor test;
- Hot-spot test;
- Thermal cycling;
- Humidity freeze;
- Mechanical testing;
- Twist test;
- Hailstone and ultraviolet test.

To conduct the tests above, expensive equipment such as climatic chambers and sun simulators is required. In addition, well-trained technical experts must perform the tests in order to ensure correctness and reproductible results.

Because PV is a young area of business, the annual degree of innovation is rather high. Due to this, test results from previous years are usually no longer relevant to the improved product of today. However, consistent quality for a PV product is possible if the manufacturer repeats certain tests at regular time intervals. In terms of safety and performance characteristics, these tests have to be performed prior to shipment to the customer. After a fixed time period established by regulatory bodies, a complete type approval has to be repeated. Demonstration of a test report of a singular evaluation is not enough.

In the EU, solar thermal collectors are covered by the EN 12975 standard and solar domestic hot water systems by the EN 12976 standard. These standards are used in the EU for product testing and certification for the Solar Keymark quality mark applicable to boilers and collectors.

2.9 Wind Turbines

The main objective of a wind turbine (giant propellers on high towers) is to generate electricity without producing harmful emissions that pollute the environment. Among renewable technologies, wind power is the most successful one and can continuously generate up to 1 megawatt of power depending on wind velocity. The main trends in wind-turbine development is increased size and rating for offshore installations; variable-speed operation, and use of direct-drive generators. The IEC 61400 series of standards refer to wind energy generation systems, design, marking, lighting, performance, safety, EMC, noise characteristics, site suitability, wind turbines, and wind power plants.

In United States, the UL 1640, UL 6141, and UL 6142 are the American safety standards developed specifically for wind turbines. These UL standards are simplifying the approval process for wind-turbine electrical systems and cables through local Authorities Having Jurisdiction (AHJ) inspectors.

A typical wind turbine system is a complex array of systems and components that capture as much energy as possible from the wind and efficiently convert it into electricity. Such a system consists of the rotor (blades and hub), generator, speed increaser (gearbox), conversion system, drive train, support structure, controls, power collection, power distribution system, protection systems, and tower. For units connected to a utility grid, 50 or 60 Hz, the generators can be synchronous or induction connected directly to the grid, or a variable-frequency alternator or direct current generator connected indirectly to the grid through an inverter. The process begins with the wind impacting the turbine's blades, which can be longer of 30–100 meters and rotate a shaft at the top of a tower more than 100 meters above the ground, where wind activity is most intense (see Figure 2.4). The wind

Figure 2.4 Photo of a wind turbine.

turbine blades are made from lightweight, strong composite materials, and are designed to efficiently capture wind energy, survive gusty winds, and use aerodynamic controls or brakes to control their speed. The wind energy is then transferred from the blades to the flexible rotor hub, where the blades are connected together. The rotating rotor hub turns a direct drive, low-speed generator, and a power train designated to produce electrical energy at variable speeds. This arrangement makes generators more efficient and easier to control, and eliminates or reduces the size of the gearboxes (one of the most expensive components in the array). The wind turbine power control system acts as a breaking mechanism and controls the speed of the turbine in varying wind conditions, optimizes the electrical power produced, and minimizes fatigue damage to the wind turbine system components. The power control system also distributes the electrical energy in a number of forms, including three-phase AC. Safety is a major concern to manufacturers of wind turbine systems because of the high voltage generated by the large towers and blade turbines.

These systems are evaluated for risk of fire and shock, including safety-related control system electrical performance and utility grid-interconnect performance for utility interactive models.

As a requirement for wind turbines, the approval process starts by performing a preliminary and detailed design evaluation to verify compliance with the basic design requirements from IEC 61400-1. By incorporating qualification test results of wind turbine components into a detailed dynamics analysis, the results of the tests and analyses are used to prepare a loads document that defines all design loads and verifies predicted loads. At the final design stage all analyses, design drawings, specifications, manufacturing, operation, maintenance, and installation manuals must be available. This documentation gives a complete description of the design, and instructions to build, install, and operate the wind turbine.

Wind turbines are shut down for faults such as loss of load, vibration, loss of phase, current, or voltage anomalies. Each of these safety features could save the unit, but the most important feature is a method of controlling the rotor when there is a loss of load (fault on the utility grid) during high winds (overspeed control). If the unit is not shut down within a few seconds, it will reach such high-power levels that it cannot be shut down and will self-destruct. The large torque excursions as well as the emergency application of mechanical brakes may damage the gearbox. Faults result in power spikes, large current, and voltage drops.

In the United States, the National Renewable Energy Laboratory (NREL), supported by the DOE, is the body that verifies the design, operational characteristics, installation, and use of wind turbines.

The IECRE–Renewable Energy System, the IEC system for certification to standards relating to equipment for use in renewable energy application, have accredited 29 bodies that ensure testing Renewable Energy Testing Laboratory (RETLs) and certification of wind energy systems [26]. The IECRE System includes conformity assessment of any particular material, product (services, software, hardware, or processed materials), installation, process, system, person, or body covered by IEC 61400 series standards and IECRE OD-551-11 to -16 (for tests) and IECRE 02 (for certification) [27] as proposed by the Renewable Energy Management Committee (REMC) and approved by the IEC Conformity Assessment Board (CAB). The IECRE System may also provide for the assessment and certification of competence of persons and bodies working in or conducting work affecting IECRE Sectors.

References

[1] Energy Independence and Security Act of 2007, https://www.govinfo.gov/content/pkg/PLAW-110publ140/html/PLAW-110publ140.htm.

[2] Energy Charter, "Policies that Work: Introducing Energy Efficiency Standards and Labels for Appliances and Equipment," Brussels, 2009.

[3] ISO 50001, Energy Management Systems, Geneva, 2018.

[4] Environmental Protection Agency, https://www.energystar.gov/index.cfm?fuseaction=recognized_bodies_list.show_RCB_search_form.

[5] European Commission, Regulation 2017/1369 Setting a Framework for Energy Labelling, Brussels, 2017.

[6] European Commission, European Product Registry for Energy Labelling (EPREL), ver.2.80, Brussels, 2018.

[7] European Commission, New Energy Efficiency Labels Explained, Fact Sheet MEMO/19/1596, Brussels, 2019.

[8] Code of Practice on Energy Labelling of Products, Electrical and Mechanical Services Department, Hong Kong, 2018, http://www.energylabel.emsd.gov.hk.

[9] European Power Supplies Manufacturers' Association (EPSMA), *PFC Harmonic Current Emissions–Guide to EN61000-3-2:2014*, Oxfordshire, UK, 2018.

[10] IEC 61000-3-2, Electromagnetic Compatibility (EMC)–Part 3-2: Limits–Limits for Harmonic Current Emissions (Equipment Input Current ≤16 A per Phase), Geneva, 2018.

[11] IECEE, *Procedure for Measuring Laboratory Power Source Characteristics*, CTL-OP 110, Geneva, 2007.

[12] IEC, *Electrical Energy Storage*, White Paper, Geneva, 2011.

[13] UL 1973, Batteries for Use in Light Electric Rail (LER) Applications and Stationary Applications, 2018.

[14] Kavadias, K. A., "Wind Energy," in Comprehensive Renewable Energy, 2012.

[15] Jaycar Electronics, Reference Data Sheet: DCDCCONV, 2001.

[16] Luo, F. L., and H. Ye, Advanced DC/DC Converters, Boca Raton, FL: CRC Press, 2004.

[17] Vicor, Design Guide & Applications Manual, 2007.

[18] UL 1471, Inverters, Converters, Controllers and Interconnection System Equipment for Use with Distributed Energy Resources, 2010.

[19] IEEE 1547, IEEE Standard for Interconnection and Interoperability of Distributed Energy Resources with Associated Electric Power Systems Interfaces, 2018.

[20] CEMEP, UPS European Guide, 2008.

[21] IEC 62040-1, Uninterruptible Power Systems (UPS)–Part 1: Safety Requirements, Geneva, 2017.

[22] Eaton, *UPS Basics*, White Paper WP153005EN, 2011; http://www.eaton.com/pq/whitepapers.

[23] Eaton, *UPS Handbook*, 2012.

[24] UL, Alternative Energy Equipment and Systems–Marking and Application Guide, 2016.

[25] IECRE, OD-407 PV SYSTEM Performance Data Reporting Requirements," Geneva, 2017.

[26] IECRE, https://www.iecre.org/members/testlabs/.

[27] IECRE, IEC System for Certification to Standards Relating to Equipment for Use in Renewable Energy Applications (IECRE System)-IECRE 02 Rules of Procedure, Geneva, 2020.

Selected Bibliography

Compendium of Energy Efficiency Policies of APEC Economies, http://aperc.ieej.or.jp/file/2012/12/28/Compendium_2011.pdf.

FM Global Property Loss Prevention Data Sheets, https://www.fmglobal.com/research-and-resources/fm-global-data-sheets.

Hargreaves, N., G. Taylor, and A. Carter, "Smart Grid Interoperability Use Cases for Extending Electricity Storage Modeling within the IEC Common Information Model," in 2012 47th International Universities Power Engineering Conference (UPEC), 2012, pp. 1–6.

IEC Standards for Smart Grid, https://www.iec.ch/smartgrid/standards/.

Survey of Market Compliance Mechanisms for Energy Efficiency Programs in APEC Economies, http://publications.apec.org/publication-detail.php?pub_id=1285.

CHAPTER 3

Environmental Aspects

When referring to the compliance and safety of electrical products, the environmental aspects must be treated as an essential element with implications on design, performances, manufacturing, installation, operation, transport, storage, and proper disposal of waste products. In order to realize this, we need to fully understand how the environment influences functionality and safety, considering the environmental stresses on electrical products throughout all phases of their service life.

On the other hand, we must also pay attention to minimize the environmental problems that may result during the manufacturing, use of the product, and disposal of it at the end of its working life.

The first part of this chapter presents the features of electrical products affected by environment and the simulation of the environmental stresses in-use, storage, and transportation. The second part of the chapter focuses on the environmental impact of electrical products and what needs to be done for protection of the environment.

3.1 Environmental Influences on Electrical Products

An electrical product should be designed to survive and operate during its lifetime, without expected hazards, in permissible environmental conditions as declared by the manufacturer. If these conditions are different from the normal use conditions, the performance of the product and the qualification test results can be adversely influenced.

Environmental conditions are climatic, mechanical, chemical, radiation, or biological conditions, external to an electrical product, to which it is subjected in-use, storage, and transportation during its life cycle.

In a practical way, the technical description of the product must provide the main data related to the permissible environmental conditions of use, storage, and transportation.

Depending on the type of product, the manufacturer needs to specify the range of environmental conditions over which the product shall operate safely. As an example, the values listed below are recommended as normal use conditions for medical electrical equipment [1]:

(a) Ambient temperature range of + 10° C to + 40° C;

(b) Relative humidity range of 30% to 95%; condensing or noncondensing;
(c) Atmospheric pressure range of 70.0 kPa to 106.0 kPa;
(d) Temperature of the water at the inlet of the equipment, water-cooled not higher than 25°C (where applicable).

Another example, for outdoor equipment:

(a) Operating temperature: −25°C + 40°C;
(b) Relative air humidity: 50% to 80%;
(c) Atmospheric pressure: 800 hPa to 1,060 hPa;
(d) Storage temperature: –25°C +80°C;
(e) Protection degree: IP66, IP67, IP69K, IK07;
(f) Environmental categories during operation according to IEC 60721: 4K6, 4C3, 4S2, and 4M6.

If not specified by the manufacturer, the range of ambient temperature for outdoor information technology equipment is requested by the IEC 60950-22 standard to be considered as between –33°C to +40°C [2].

The same environmental conditions shall be maintained on the locations in which the compliance tests (performance and basic safety) are conducted.

For electrical products intended for operation in special environments; for example: extremes of temperature, excessive dust accumulation, moisture, vibration, increased humidity, flammable gases, corrosive gases or vapors, explosive atmospheres, or to be used in vehicles, on board ships or aircraft, in tropical countries, or at altitudes greater than 2,000m, adequate permissible environmental conditions and applicable restrictions shall be indicated.

The instruction for use needs to specify any special skills, such as training and knowledge required of the intended user and any other restrictions on locations or environments in which the electrical product can be used.

In real life the environmental conditions could differ from those declared by the manufacturer and the product is still expected to remain safe. To ensure this, the customer needs to perform periodic inspection and maintenance as prescribed by the manufacturer. These activities are expected to prevent any deterioration of the safety level and also detect signs of the beginnings of any deterioration related to aging or different faults.

The IEC 60721 series of standards specifies information on environmental parameters and a number of their severities within the range of conditions met by electrical products, specifying six environmental conditions classes, as follows:

K: Climatic environmental conditions
Z: Special climatic conditions
B: Biological environmental conditions
C: Chemically active substances
S: Mechanically active substances
M: Mechanical environmental conditions

This classification allows a number of possible combinations of environments in which electrical products are used, stored, or transported. It represents the real situation concerning worldwide environmental conditions due to local specificity (see Table 3.1) [3, 4].

The first numeral is specific for in-use (indoor, outdoor, mobile, or stationary), storage, or transport locations and applicability of the product. The letter refers to the environmental conditions. The last numeral refers to the severity category of the requirement.

Referring to weather protection on a specific location, IEC 60721 standard designates three categories (C1: totally, C2: partially, and C3: non-weather-protected). In a totally weather-protected location, the direct weather influences are totally excluded, whereas in a partially weather-protected location the direct weather influences are not completely excluded. A non-weather-protected location is a location at which equipment is not protected from direct weather influences.

For a specific product, reference should be made to the total set of classes as identified, and as such the tests at which the product needs to comply will be specified.

Some product standards (e.g., IEC 60601-1-11, IEC 61800-5-1) specify clearly which environmental tests need to be conducted, but other standards do not specify this. In such a situation based on the intended use, location of in-use, storage, or transportation, and the applicable class characteristics of the product, the manufacturer can draft the environmental test profile themselves and as such, establish which environmental sequence of tests need to be conducted and with which parameters, to qualify the suitability of the product for specific conditions.

The IEC 60721-3 (all parts) may be used as a means to establish expected requirements for use, storage, and transportation. Based on these parameters it will be determined what tests needs to be done. On each of the determined tests the location-dependent severities expected to produce the effects of a real environment can be added. The information on environmental testing procedures and severities of tests are contained in the IEC 60068 series of standards. The IEC TR 60721-4-7 covers the correlation and transformation of the conditions given in the IEC 60721-3 (7K1 to 7K7; 7M1 to 7M3) to the environmental tests defined in the IEC 60068-2 standard.

Table 3.1 IEC 60721 Environmental Classes Specification for Worldwide Use*

	In-Use				
	In Buildings (Indoor)		*Non-Weather-Protected (Outdoor)*		
Environmental Conditions	*Mobile*	*Stationary*	*Stationary*	*Storage*	*Transport*
Climatic	7K1	3K3	4K2	1K3	2K4
Biological	7B2	3B2	4B1	1B2	2B2
Chemically active substances	7C2	3C2	4C2	1C2	2C2
Mechanically active substances	7S1	3S2	4S2	1S2	2S2
Mechanical	7M2	3M3	4M3	1M2	2M2

*Except in extreme conditions such as in Alaska or Sahara.

Each environmental condition class is determined by the values of the environmental parameters of climatic conditions, biological conditions, chemical contaminants, mechanically active substances, and mechanical conditions [3, 4]. Table 3.2 contains examples of each type of these conditions.

Products conforming to a particular class shall function at all conditions within the boundary of the class climatogram (envelope of climatic conditions in X-Relative Humidity and in Y- Air Temperature defined by the characteristic severities) [5].

As an example, for a product designated to operate stationary in an indoor partially weather-protected location should be considered in the following environmental conditions classes: 3K3, 3B2, 3C3, 3S2 and 3M3; for storage: 1K3, 1B2, 1C2, 1S2, 1M2; and for transportation: 2K4, 2B2, 2C2, 2S2, 2M2. These are the classes (with moderate severities) accepted in the majority of countries. If some high severity applies, the classes will be changed by increasing the second numeral (e.g., rough transport conditions, class 2M3 should be applied). The above classification means that the product during use, storage, and transportation may be exposed to solar radiation and to heat radiation. It may also be exposed to movements of the surrounding air due to drafts in buildings. Humidity is not regulated. It is also not subjected to condensed water, precipitation, water from sources other than rain or icing, not situated in proximity to sources of sand or dust. The permissible mechanical environmental condition for these classes does not include noticeable shock, impact, and vibration.

The ETSI EN 300 019 series of standards [6] specifies test methods and severities for verification of the required resilience of telecommunication equipment according to the relevant environmental class.

The design of the electrical product, selection of components, and type of insulating materials should take into account the working environment and environmental stresses that may occur during the anticipated lifetime of the product. The insulation and other safety-related parts or parameters of an electrical product should be less affected by the environmental conditions during in-use, installation, storage, and transport.

Note that each climatogram has an area of points with a high probability of occurrence due to statistical distributions of temperature and humidity. This implies

Table 3.2 Examples of Environmental Parameters

Condition	*Examples of Environmental Parameters*
Climatic	Air temperature, relative humidity, absolute humidity, rain intensity, rate of change of temperature, solar radiation, heat radiation, low air pressure, change of air pressure, movement of surrounding air, condensation, precipitations carried by wind, water other than rain and condensation, formation of ice and frost, wetness.
Mechanical	Vibration sinusoidal, vibration random, shock, impact, free fall
Chemical contaminants	Sulfur dioxide (SO_2), hydrogen sulfide (H_2S), chlorine (Cl), hydrochloric acid (HCl), hydrofluoric acid (HF), ammonia (NH_3), ozone (O_3), nitrogen dioxide (NO_2), sea salt
Biological	Flora, fauna
Mechanically active substances	Sand, dust

that the environment can remain for long periods in that area, but shall not remain near the boundary of the climatogram for long periods because it will stress the product.

Some features of electrical products can be affected by environment. In particular, insulation coordination [7] and mechanical strength are the most influenced.

Insulation coordination, defined as mutual correlation of insulation characteristics of electrical equipment (creepage distances, air clearances, distance through insulation, coatings, encapsulation, etc.) taking into account the expected microenvironment (nearby environment) and other influencing stresses [1], is one of the important electrical product features affected by environment.

Insulation coordination implies the selection of the electric insulation characteristics of the equipment, with regard to its application and in relation to its surroundings, and can be achieved if the design of the equipment is based on the stresses to which the product is likely to be subjected during its anticipated lifetime regarding voltage and microenvironmental conditions.

The behavior of air to withstand a maximum peak voltage value between two parts at different voltages is in relation to air pollution. The effect of pollution on insulation, especially provided by air clearances and creepage distances, is critical because these spacings are dimensioned taking in consideration the four pollution degrees (PD) established in direct relation to the microenvironment.

Means may be provided to reduce pollution at the insulation under consideration by the effective use of enclosures with a specific IP numeral, encapsulation, potting, molding, coating, or hermetic sealing. Such means to reduce pollution may not be effective when the equipment in normal use generates pollutants itself.

Small clearances can be bridged completely by solid particles, dust, and water, and therefore minimum clearances are specified where pollution may be present in the microenvironment.

It must be understood that pollution at the surface of a material is generally not conductive, but the humidity is an influencing parameter, inducing the pollution to become conductive. Pollution caused by contaminated water, soot, metal, or carbon dust is inherently conductive. A higher conductivity allows circulation of a tracking current at the surface of the materials either between live parts or between live parts and earth. There are some materials, such as ceramics and glasses, which do not track because the scintillation cannot break the chemical bonds at the surface of the material.

In the presence of humidity, a surface-related phenomenon called water adsorption may trap water at the surface of the insulation materials leading to a higher risk of flashover. This risk along the creepage distance at the surface of the insulating material increases with the humidity level (HL) and with the ability of the insulating material to trap water.

The spacing, especially the air clearance, and electric strength test voltages can be affected by the atmospheric pressure at which the product is in-use. Unless otherwise declared by the manufacturer an electrical product (e.g., ITE, video, audio) is rated to operate at an altitude less than 2,000m. Where the product is intended to be operated in a higher altitude, the estimated values of the air clearance for normal environmental conditions and the value of the electric strength test voltages shall be multiplied with a factor depending on the air pressure at the higher altitude for use, as specified in Table 3.3 [1, 2].

Table 3.3 Multiplication Factors for Altitudes Up to 5,000m

Operating Altitude (a) (m)	*Barometric Pressure (kPa)*	*Multiplication Factor for Air Clearance on Mean of Operator Protection*	*Multiplication Factor for Air Clearance on Mean of Patient Protection*	*Multiplication Factor for Electric Strength Test Voltages*	
				Air clearance ≥1 mm to <10 mm	*Air clearance ≥10 mm to <100 mm*
< a ≤ 2000	80	1.00	1.00	1.00	1.00
2000 < a ≤ 3000	70	1.14	1.00	1.07	1.10
3000 < a ≤ 4000	62	1.29	1.14	1.15	1.20
4000 < a ≤ 5000	54	1.48	1.29	1.24	1.33

The creepage distances are not subject to the multiplication factors but shall always be at least as large as the resulting value for the air clearance. A vacuum chamber can be used to simulate higher altitudes.

Related on the usage of an electrical equipment at a higher rated altitude, the user needs to pay special attention to the correlation between the rated altitude at which a product will operate and the rated altitude of parts (e.g., power supply) of this product, which was approved for another altitude. If a product will operate in a location at, for example, a 4,000 m altitude, it needs to be designed with a high value of air clearance, and cannot use a power supply that was tested and approved for a smaller rated altitude.

Altitude testing (e.g., low-pressure testing) makes use of a vacuum chamber to simulate the effects of high-altitude conditions. The pressure inside the altitude chamber can be reduced to correspond to the air pressure at a specific altitude. Products can be energized and placed inside the altitude chamber and tested to determine if they will still function during and after exposure to a given duration at a specified altitude. Components sealed with internal fluid such as batteries or capacitors may fail or leak during altitude testing because an internal pressure results at rising altitudes as the external pressure is reduced. The lower pressure at higher altitudes can reduce the cooling of components, which can lead to possible failures. For this type of testing, it is necessary to have power and signal wires that can be fed into the altitude chamber without causing vacuum leaks. Altitude testing is normally conducted at room temperature. It is also possible to perform altitude testing at hot and cold temperatures. This will increase the test time as additional time will be required to heat up or cool down the altitude chamber.

The mechanical strength of any means of protection shall be so designed or protected that it is not likely to be impaired by environmental stresses including deposition of dirt or by dust resulting from wear of parts within the product. A few aspects that need to be considered are:

- Mechanical hazards arising from temperature, static, dynamic, vibration, impact, and pressure loading and other movement conditions shall be taken into consideration in support systems;
- The selection of tensile safety factors should to take account any impairing effects of wear and corrosion due to environmental effects;

- An atmosphere containing a high amount of halogen gases (chlorine and fluorine) can corrode aluminum;
- Nonmetallic parts of an outdoor enclosure shall be sufficiently resistant to ultraviolet (UV) radiation degradation;
- Polymeric materials can be affected by low temperatures, and can also become embrittled due to the effect of UV radiation.

3.2 Simulation of Environmental Stresses In-Use and During Storage and Transportation

An electrical product during its life cycle may be exposed to a single extreme environmental condition or simultaneously to a number of extreme environmental conditions (temperature and humidity, solar radiation and temperature, vibration and temperature change, etc.). Combinations of environmental conditions may increase the stress effect on the product, in two ways:

1. By the effects of short-term extreme environmental conditions that may occur at any time in a product's life and may directly cause malfunction or destroy the product;
2. By the effect of long-term subjection to nonextreme environmental stresses which may slowly degrade the product and eventually cause malfunction or destruction of the product.

A product may be unaffected by an extreme environmental condition when it is new, but fail when it is subjected to the same condition after being in operation for a long period of time due to the effects of aging.

Considering the above reasons, a product should be designed to survive and operate in different environments and needs to have the ability to withstand the environmental stresses associated with in-use, storage, and transportation during its operational life. To evaluate this ability, environmental tests are conducted for the environmental stresses that will be presented next.

The main IEC standards series for environmental tests is the IEC 60068, in which these tests are designated by uppercase letters [8]:

A: Cold
B: Dry heat
C: Damp heat (steady state)
D: Damp heat (cyclic)
E: Impact (for example shock and bump)
F: Vibration
G: Acceleration (steady state)
J: Mold growth
K: Corrosive atmospheres (e.g., salt mist)
L: Dust and sand

M: Air pressure (high or low)
N: Change of temperature
R: Water (e.g., rain, dripping water)
S: Radiation (e.g., solar, but excluding electromagnetic).

The letter Z is used to denote combined tests and composite tests as follows: Z is followed by a solidus and a group of uppercase letters relating to the combined or composite stresses; for example, Test Z/AM: Combined cold and low air pressure tests.

Each family of tests may be subdivided and identified by the addition of a second letter in lowercase (e.g., Bd Dry heat; Fc vibration sinusoidal).

Any test may be designated as "primarily intended for components" or "primarily intended for equipment." In the following we will refer to tests primarily intended for equipment.

The following environmental tests are intended to evaluate the ability of an electrical product to withstand environmental stresses. Components or parts will be referred to as a "specimen." A product with or without its packaging will be referred to as a "test specimen." Other methods of environmental tests specific to the individual type of product may be included in the relevant specifications. The environmental classes may be used as a basis for the selection of tests and test severities with respect to the consequence of failure. To provide tests appropriate to the different intensities of an environmental condition, some of the test procedures have several degrees of severity. These different degrees of severity are obtained by varying the time, temperature, air pressure, or some other determining factor separately or in combination [8]. The order (sequence) in which the environmental stresses are applied may affect the results of the evaluation and should be determined on the basis of information about the conditions of in-use (see also IEC 60068-2-61 Test Z/ABDM). When this information is not available, it is recommended that a sequence giving the most significant effect be used. An example of a general sequence of tests, suitable for most types of products, follows [8]:

1. A Cold
2. B Dry heat
3. N Rapid change of temperature
4. F Vibration
5. E Impact
6. G Acceleration
7. M Air pressure
8. Db Damp heat (12 + 12 h cycle)
9. C Damp heat (steady state)
10. K Corrosion
11. L Dust and sand

Note that "steady state" means a state of a condition (e.g., temperature, humidity) that does not change over time.

A detailed presentation of environmental engineering and laboratory tests is done in the following U.S. Department of Defense documents: MIL-STD 810 Rev.

H, MIL-STD 2164, and MIL-STD 1165. These standards are for military applications, but the theoretical and methodology aspects specified can serve as a reference for a better understanding of environmental tests in general.

The standard range of atmospheric conditions for carrying out measurements and tests is as follows [8]:

- Temperature: 15°C to 35°C;
- Humidity: 25% to 75% relative humidity (RH);
- Air pressure: 86 kPa to 106 kPa (860 mbar to 1 060 mbar).

Note that RH represents the ratio of the partial pressure of water vapor in moist air at a given temperature to the partial pressure of water vapor in saturated air at the same temperature. Related to RH is the dew point, which is the temperature to which air must be cooled to become saturated with water vapor. When cooled further, airborne water vapor will condense to form liquid water. When air cools to its dew point through contact with a surface that is colder than the air, water will condense on the surface.

Before each test, the product is evaluated from functional and safety points of view as specified by the manufacturer.

During the test, temperature, humidity, and acceleration, need to be monitored continuously.

After each test, the product is allowed to stabilize at ambient conditions (temperature 15°C to 35°C, relative humidity 25% to 75%, atmospheric pressure 86 kPa to 106 kPa, for a minimum of 1 hour) and evaluated for functionality and safety based on the following acceptance criteria, to determine if the environmental test has affected the specification:

- No degradation of any safety-relevant component;
- No sign of component overheating;
- No live parts have become accessible;
- No cracks in the enclosure and not damaged or loose in isolation;
- Pass a routine dielectric AC or DC voltage test (some specific product standards do not request this verification);
- Pass protective earth continuity test;
- No hazardous behavior when the product is operated following the test.

Each nonconformity will be assessed based on the risk management methodology and after fixing the deficiency the specific test should be repeated.

Table 3.4 summarizes the main environment tests as described in the IEC 60068-2 series of standards for products.

The IEC 60068-3-4, IEC 60068-3-5, IEC 60068-3-6, IEC 60068-3-7 standards provide the information for supporting documentation and relevant specifications to select appropriate tests, test severities, and testing conditions for: dump heat test, performance of temperature chambers, performance of temperature/ humidity chambers, and measurements in temperature chambers for tests A (cold) and B (dry heat).

Table 3.4 Main Environment Tests from the IEC 60068-2 Series of Standards

Type of Test	*Test Letter*	*Standard*	*Scope of Test*	*Test Details*		*Evaluation after Test*
				Energized	*Packaged*	
Cold	A	IEC 60068-2-1	Ability to withstand low temperature[1]	Yes or no Temperature shall be monitored	Yes or no for storage	Functionality and safety
Dry heat	B	IEC 60068-2-2	Ability to withstand high temperature[1]	Yes or no Temperature shall be monitored	Yes or No for storage	Functionality and safety
Damp heat (steady state)	Cab	IEC 60068-2-78	Effect of high humidity at constant temperature, without condensation[2, 3]	No Temperature and humidity shall be monitored	No	Allow to dry and tested for functionality and safety
Dump heat, cyclic (12h + 12h cycle)	Db	IEC 60068-2-30	Effect of high humidity combined with cyclic temperature changes, with condensation	No Temperature and humidity shall be monitored	Yes or no	Allow to dry and tested for functionality and safety
Composite temperature/humidity cyclic test	Z/AD	IEC 60068-2-38	Effect of combined high temperature/humidity and cold[3]	No Temperature and humidity shall be monitored	No	Allow to dry and tested for functionality and safety
Combined temperature and humidity with low air pressure	Z/DM	IEC 60068-2-39	Effect of combined temperature and humidity with low air pressure	No Temperature and humidity shall be monitored	No	Allow to dry and tested for functionality and safety
Shock	Ea	IEC 60068-2-27	Ability to withstand nonrepetitive or repetitive shocks[4]	No; yes, for functional shock	Yes or no	Functionality and examined for any internal damage
Shock	Ec	IEC 60068-2-31	Effects of rough handling[5]	No	Yes or no	Functionality and examined for any internal damage
Shock	Ee	IEC 60068-2-55	Ability to withstand specified severities of bounces[6]	No	Yes or No	Functionality and examined for any internal damage
Vibration sinusoidal[9]	Fc	IEC 60068-2-6	Ability to withstand specified severities of sinusoidal vibration[7]	No	Yes10 or no	Functionality and examined for any internal damage

Vibration sinusoidal[9]	Fc	IEC 60068-2-6	Ability to withstand specified severities of sinusoidal vibration[7]	No Acceleration shall be monitored	Yes[10] or no	Functionality and examined for any internal damage
Vibration random[9])	Fh	IEC 60068-2-64	Ability to withstand specified severities of broadband random vibration[8]	No	Yes[10] or no	Functionality and examined for any internal damage
Vibration transient[9]	Ff	60068-2-57	Ability to withstand specified severities of transient vibration	No Acceleration shall be monitored	Yes[10] or no	Functionality and examined for any internal damage
Combined climatic (temperature/ humidity) and dynamic (vibration/ shock)	Z/DEF	IEC 60068-2-53	To investigate to what extent the product is affected by combined climatic and dynamic tests	No Temperature, humidity, and acceleration shall be monitored	Yes or no	Functionality, safety, and examined for any internal damage
Vibration–acoustically induced method	Fg	IEC 60068-2-65	Ability of a product to withstand vibration caused by a specified sound-pressure level	No	Yes or no	Functionality, safety, and examined for any internal damage
Vibration–mixed mode	Fi	IEC 60068-2-80	Adequacy of the product to resist the specified mixed-mode excitation[11]	No Acceleration shall be monitored	Yes or no	Functionality and examined for any internal damage
Acceleration (steady state)	Ga	IEC 60068-2-7	Ability of structures and performances to withstand the forces produces by acceleration	No Acceleration shall be monitored	Yes or No	Functionality and examined for any internal damage
Salt mist (steady state)	Ka	IEC 60068-2-11	Compares resistance to deterioration between products of similar construction	No	No	Functionally
Salt mist (cyclic)	Kb	IEC 60068-2-52	Ability to withstand a salt-laden atmosphere[12]	No The salt fog condensate rate is monitored	No	Functionally
Flowing mixed gas corrosion	Ke	IEC 60068-2-60	Determines the corrosive influence on contacts and connections of a product	No	No	Functionally and safety

Table 3.4 (continued)

Type of Test	*Test Letter*	*Standard*	*Scope of Test*	*Test Details Energized*	*Packaged*	*Evaluation after Test*
Dust and sand	L	IEC 60068-2-68	Ability to withstand the effects of dust and sand suspended in air	No	No	Functionally and safety
Air pressure	M	IEC 60068-2-13	Ability of a product to be stored, transported, or used under low air pressure	Yes	No	Functionally and safety
Rapid change of temperature	N	IEC 60068-2-14	Ability to withstand the stresses of alternate exposure to hot and cold temperatures	Yes or no	No	Functionally and safety
Water (rain, dripping water)	R	IEC 60068-2-18	Ability of enclosures, covers, and seals to maintain the product working after and under falling water drops, impacting water, immersion, or high-pressure water impact	Yes	No	Functionally and safety
Solar radiation	S	IEC 60068-2-5	Ability to withstand simulated solar radiation conditions at the surface of the earth	No	No	Functionally

1. These tests refer, in particular, to the limits of the rated operating and storage temperature.
2. May be utilized to check the resistance of a product to some forms of corrosion attack.
3. Parts sensitive to humidity, normally used in controlled environments, and which do not influence safety need not be subjected to this test.
4. To reveal mechanical weakness and/or degradation in performances, or accumulated damage or degradation due to transport or operational environment. Shock testing consists of the following shock impulses: half sine, haversine, sawtooth, and trapezoid. The half-sine waveform simulates impact with a rebound while a sawtooth waveform simulates impact with no rebound.
5. Not simulate the effects of shock applied to installed products.
6. When transported as loose cargo on wheeled vehicles.
7. Generated by rotating, pulsating, or oscillating forces. Sine sweep vibration is useful for identifying resonances by comparing response vibrations of the product to the vibrations on the shaker table and sine dwell vibration is performed at resonances found in the product under test.
8. Used to identify accumulated stress effects and the resulting mechanical weakness and degradation in the performance and structure.
9. Using IEC 60068-3-8 the dynamic vibration environment of the product can be estimated, and based on that, the appropriate test method is selected.
10. For testing packaged products, the standards may be used in conjunction with IEC 60068-2-47.
11. Reveal the accumulated effects of stress induced by random vibration, mixed with sine, and the resulting mechanical weakness and degradation of performances.
12. Evaluate how salt can degrade the performance of parts manufactured using metallic and/or nonmetallic materials.

The IEC 60068-2-47 Mounting of Specimens for Vibration, Impact and Similar Dynamic Tests standard provides methods for mounting products, whether packaged or unpackaged, for the series of dynamic tests from IEC 60068-2: shock impact (Test E), vibration (Test F) and acceleration, and steady-state (Test G).

Table 3.5 presents examples of environmental temperature and humidity tests for which the parameters for tests are established based on the classes from IEC 60721-4-7 pending on the locations and conditions of product utilization (tests 1 and 3 are for normal operation conditions of home healthcare products, tests 2 and 4 are for normal operation conditions of adjustable speed power drive systems, tests 5 and 6 are for storage and transport of home healthcare products).

Table 3.6 presents examples of environmental mechanical tests for which the parameters for tests are established based on the classes from IEC 60721-4-7 pending on the locations and conditions of utilization of the product (test 1 is for adjustable speed power drive systems, tests 2 and 3 are for nontransportable home

Table 3.5 Example of Temperature and Humidity Tests

No.	*Type of Test*	*Class as in IEC TR 60721-4-7*	*Test Parameters (Based on Classes from IEC TR 60721-4-7)*		*Standard for Testing*
1	Dry heat (steady state)	7K1	Range of rated temperature operating conditions	+ 5°C to + 40°C	IEC 60068-2-2
			Duration of test at each limit temperature from range	Minimum 6 h for each extreme temperature	
2	Dry heat (steady state)	7K1	Temperature operating conditions	+ 40°C ± 2°C	IEC 60068-2-2
			Duration of test	16 ± 1 h	
3	Damp heat (steady state)	7K1	Range of rated temperature operating conditions	+ 5°C to + 40°C	IEC 60068-2-78
			Humidity	≤15% RH noncondensing for + 5°C	
				90% RH noncondensing for + 40°C	
			Atmospheric pressure	700 hPa to 1,060 hPa	
			Duration of exposure	Minimum 6 h for each extreme temperature	
4	Damp heat (steady state)	7K1	Temperature	+ 40°C ± 2°C	IEC 60068-2-78
			Humidity	+93% ± 3% RH noncondensing	
			Duration of exposure	4 days	
5	Dry heat storage and transportation	7K3	Temperature	70°C	IEC 60068-2-2
			Relative humidity	a water vapor pressure up to 50 hPa	
			Duration	16 h	
6	Cold storage and transportation	7K3	Temperature	-25°C	IEC 60068-2-1
			Relative humidity	≤15% RH noncondensing	
			Duration	16 h	

Table 3.6 Examples of Environmental Mechanical Tests

No.	*Type of Test*	*Class as in IEC TR 60721-4-7*	*Test Parameters (Based on Classes from IEC TR 60721-4-7)*					*Standard for Testing*
1	Stationary sinusoidal vibration*	7M2	Frequency range		10 Hz ≤ f ≤ 57 Hz	57 Hz < f ≤ 150 Hz		IEC 60068-2-6
			Vibration amplitude		0.075 mm			
			Acceleration amplitude†			1g		
			Duration of vibration: 10 sweep cycles per axis on each of three mutually perpendicular axes					
2	Stationary broadband random vibration* ‡	7M1 and 7M2	Frequency range		10 Hz to 100 Hz	100 Hz to 200 Hz	200 Hz to 2 000 Hz	IEC 60068-2-64
			Acceleration amplitude		1.0 $(m/s^2)^2/Hz$	−3 dB per octave§	0.5 $(m/s^2)^2/Hz$	
			Duration		30 min per perpendicular axis (three total)			
3	Shock	7M1	Peak acceleration		150 m/s^2 (15 g)			IEC 60068-2-27
			Duration		11 ms			
			Pulse shape		Half-sine			
			Number of shocks		Three shocks per direction per axis (18 total)			
4	Shock	7M2	Peak acceleration		300 m/s^2 (30 g)			IEC 60068-2-27
			Duration		6 ms			
			Pulse shape		Half-sine			
			Number of shocks		Three shocks per direction per axis (18 total)			
5	Shock	7M3	Peak acceleration		1,000 m/s^2 (100 g)			IEC 60068-2-27
			Duration		6 ms			
			pulse shape		Half-sine			
			Number of shocks		Three shocks per direction per axis (18 total)			
6	Free fall	7M2	Mass	≤ 1 kg	> 1 kg and ≤ 10 kg	> 10 kg and ≤ 50 kg	> 50 kg	IEC 60068-2-31
			Fall height	0.25 m	0.1 m	0.05 m	0.01 m	
			Number of falls: two from each specified height, in a hard surface					

*To evaluate the behavior of the product, accelerometers are attached to the desired parts and their frequencies and amplitudes are recorded during vibration.

†During the sweeps, the acceleration is the maximum acceleration observed at any point on the product.

‡The random vibration more closely simulates the real field environment, although resonant frequencies and the frequencies of component damage are more easily obtained with the frequency sweep (sinusoidal).

§The octave is defined as the interval of two frequencies having a basic ratio of 2.

healthcare products,tests 2, 4, and 5 are for transportable home healthcare products, and Test 6 is for portable and mobile home healthcare products).

3.2.1 HALT and HASS

Highly accelerated life testing (HALT) is a qualitative and destructive test used as a reliability method to identify the product weaknesses and to verify the design process. HALT is not a qualification test; thus there are no predetermined pass/fail criteria. The goal of HALT is to quickly induce failures and then to determine their root causes. HALT is performed to determine the operational limits for low temperature, high temperature, vibration, and combined temperature and vibration. The stress levels in HALT are typically far beyond those experienced by the product in its normal operating environment. These higher-than-normal stresses accelerate the time to failure and induces defects more rapidly than under real in-use conditions. The product under test is in operation during HALT and is continuously monitored for operational failures. The best time to begin HALT testing during the product development process is when prototypes first become available. Testing typically takes 3 to 5 days.

The HALT procedure is divided into five stages [9, 10]:

- *Stage 1:* Thermal step stresses, in which the temperature is decreased (or increased) in steps of a defined increment followed by a temperature dwell of a defined length. Cold step stresses start at 20°C (68°F) and are decreased in 10°C increments until the lower operating limit is determined or the chamber minimum temperature of –100°C (–148°F) is reached. Hot step stresses start at 20°C (68°F) and are increased in 10°C increments until the higher operating limit is determined or the chamber maximum temperature of 200°C (392°F) is reached. The dwell time at each step is a minimum of 10 minutes following stabilization of the product at the setpoint temperature as determined by the product thermocouple response. Functional testing will occur during this stabilization period.
- *Stage 2:* Rapid thermal transitions stresses, temperature cycles with rapid transition rates (ramps), are applied to the product. The minimum thermal cycle temperature range is within 10°C above of the lower operating limit (LOL) to 10°C below of the upper operating limit (UOL). The dwell time is a minimum of 5 minutes following stabilization of the product at the setpoint temperature as determined by the product thermocouple response. Five cycles are applied. For example, if the LOL was determined to be –55°C and the UOL determined to be 90°C, the thermal transition range would be –45°C to +80°C.
- *Stage 3:* Vibration step stresses, a broadband vibration spectrum applied through the HALT chamber table. The HALT chamber table should apply random vibration energy at setpoint of 5g, as measured RMS over a 10-Hz to 10-kHz bandwidth in 6 degrees of freedom (DOF). Vibration step stresses will start at 5g and increase in 5g steps until either the operating, destruct limits, or chamber maximum vibration level (e.g., 60g) is reached. At 40g levels and above, the vibration step is returned to 10g for 1 minute to detect failures that could be hidden by extreme forces occurring at higher vibration levels. Dwell time at each step is 15 minutes to accumulate fatigue damage. This test is performed at a room temperature of 20°C (68°F) to 25°C (77°F).
- *Stage 4:* Combined environmental (temperature and vibration) stresses, in which all of the stresses used previously in the HALT test are combined and applied simultaneously. The hot and cold temperatures are the same as those used in Stage 2. The dwell time at each hot and cold temperature is the same as used in Stage 2. Vibration level is fixed during each temperature step and

begins at 10g and increases in 10g steps until either the operating or destruct limits or the chamber maximum vibration level (e.g., 60g) is reached. The minimum thermal cycle temperature range is within 10°C of above the lower thermal operating limit and 10°C below the upper thermal operating limit as discovered during thermal step stress testing. A minimum of five combined environment cycles are required unless a destructive failure forces the early termination of the stressing. Each combined environment cycle will consist of one thermal cycle that is conducted while the vibration is held at a single setpoint. The starting vibration level for the five required cycles is determined by dividing the vibration destruct limit or vibration operating limit by five. The vibration level is increased by the same number during each subsequent thermal cycle. Therefore, if the product has a vibration operating limit of 30g, the initial test cycle would be conducted at a vibration level of 6g. The vibration level would be increased by 6g after each complete thermal cycle—Cycle 1: 6 g, Cycle 2: 12 g, Cycle 3: 18 g, Cycle 4: 24 g, Cycle 5: 30 g.

- *Stage 5:* Temperature destruct limits. The cold temperature destruct limit is found by starting at the lower operating limit (found in Stage 1) and decreasing the temperature in 10°C increments until either the low temperature destruct limit or the chamber minimum temperature of –100°C (–148°F) is reached. The hot temperature destruct limit is found by starting at the upper operating limit (found in Stage 1) and increasing the temperature in 10°C increments until either the hot temperature destruct limit or chamber maximum temperature of 200°C (+392°F) is reached. The dwell time established in Stage 1 is used; however, dwell times may be reduced if the product stops operating or if failures occur. If the product fails to operate, the temperature will be reduced or increased toward 20°C (68°F) to see if the product recovers. If the unit is nonoperational after stabilizing at 20°C (68°F), the product should be repaired (if practical) so that the test temperatures can be expanded. If it is not practical to repair the product, Stage 5 will be terminated.

Highly accelerated stress screening (HASS) is performed during manufacturing on products as a screening method used to identify manufacturing defects that would cause a failure in normal environments during use, storage, and transportation. HASS stresses are higher than traditional environmental stress screening (ESS).

Types of stresses used in HASS are similar to those used in HALT and take from one hour to a few hours. HALT must be performed before HASS.

3.2.2 Equipment Used to Perform Environmental Testing

Climatic chambers are used to conduct temperature and humidity tests. Such tests are dry heat and damp heat. The typical controllable temperature/humidity range for most climatic chamber is 5°C (41°F) to 85°C (185°F) (the operating temperature can be more extended depending on the manufacturer) with 10% to 98% RH, limited by a 7°C (44°F) dew point. Since the amount of moisture varies at every temperature (the amount of moisture in the air at 20°C (68°F) and 50% RH is not the same as 10°C (50°F) and 50% RH), the chamber manufacturers use dew point to describe the RH limitation. There are various types of humidity systems used on test chambers such as a water bath, boiler/steam generator, and atomizing system. Deionized (DI) water is recommended for use with humidity systems. Water should be provided within 0.05 to 2MΩ resistivity. Distilled water or reverse osmosis (RO) water exceeding these limits may cause corrosion. If tap water is used, an optional demineralizer filtration system should be used to remove water impurities and minerals that can harm the test chamber. If using a city water line, a water pressure regulator is recommended to lower the water pressure to 25 pounds per square inch gauge (PSIG) for steam generator/boiler systems and 10 PSIG for atomizing humidity systems. High water pressure may cause the top or bottom cap (white) on the water filter to crack. This may result in water leakage that can damage the humidity chamber. If the water supply has a lot of particulate matter, an additional inline prefilter, such as a 5 or 25-micron polypropylene prefilter may be needed.

To perform a rapid change of temperature testing (thermal shock), equipment with a single chamber or multiple chambers may be used. The temperature ranges for a thermal shock chamber are between –160° C to +200°C). When using single-chamber thermal shock equipment, the products or samples remain in one chamber and the chamber air temperature is rapidly (from 15°C per minute) cooled and heated. This usually results in a slower rate of change in the product response temperature as the entire chamber must be cooled down and heated up. However, larger products can be tested in single-compartment chambers. Some equipment uses separate hot and cold chambers with a hydraulic elevator mechanism that transports the products between two or more chambers. This results in a more rapid rate of change in the air temperature. The correct temperature limits must be chosen to not exceed the operating limits or material property limits of the product. It is therefore important that these temperatures be properly measured and monitored during the test through the careful placement of thermocouples on and around the product.

A modern vibration system is composed by three main devices: a vibration exciter, an amplifier, and a control system. The vibration exciter (shaker) (Figure 3.1) is the means of providing the force input to the test object, which is analyzed by applying alternate acceleration with known frequencies, either by a sine or a broadband random signal. The amplifier transfers the power from the line source and transforms it on electrical signals of desired frequency and amplitude. The controller monitors the vibration test and the output exciter acceleration, and supplies an adequate input signal for the amplifier.

In a vibration shaker, the test object must be mounted on the exciter table by its normal means of attachment. In cases where the test object cannot be mounted directly on the exciter table, a special fixture is used for fastening the test object. The fixture must be stiff enough to transmit the generated force or motion uniformly to the test object without introducing any resonances and allowing the product to be vibrated along the specified axes.

The main work operation principle of the shaker may be based on combustion engines, hydraulic engines, pneumatic, or electrodynamic (ED).

The devices used for a vibration test on fixed sine frequency obtain the input signal, which needs to be amplified, from a signal generator. Some auxiliary equipment is also used, such as those for vibration amplitude control, devices for coil

Figure 3.1 Vibration test shaker in vertical configuration.

displacements and acceleration measurements, and devices that quickly move the shaker's table in order to simulate pulse damped vibrations.

For the random vibration the best way to have input signals is by the acceleration power spectral density (PSD) obtained from an actual acceleration. This PSD is defined by frequency bands of the whole considered frequency domain. For such tests, amplifiers with the following specifications must be used: Output power by channel: 350W (AC: 230V); amplifier class: AB; frequency response: 15 Hz to 40 kHz; total harmonic distortion + noise (THD + N) in 8 Ω: less than 0.05% from 20 Hz to 1 kHz and less than 0.1% from 20 Hz to 20 kΩ; input impedance (unbalanced): 20 kΩ; signal-to-noise ratio (without weighting): 90 dB, can be used.

An ED shaker is a vibration system consisting of a table structure that is connected to a ferrous armature surrounded by a coil. This coil is separated by an air gap from a second coil in the body of the shaker system. Controlled current is passed through the coils to create an electromagnetic force between them that moves the table in a single axis. Electrodynamic vibration systems are capable of performing many different tests, such as sine, random, shock, sine-on-random, random-on-random.

A classical shock impulse is created when the shock table changes direction abruptly. This abrupt change in direction causes a rapid velocity change that creates the shock or acceleration impulse.

If shock tests are performed on an ED shaker, the shaker can reverse polarity and perform the shock along both directions of each axis without rotating the fixture and test object.

When performing shock testing on a shock machine, the machine can only apply shock in one axis and one direction. Performing shock tests on a shock machine is much different than using an ED shaker. To create an impulse on a shock machine, the shock table is raised upward to a predetermined height. The table is released, then falls or accelerates downward, impacting a shock programmer. The table bounces off the shock programmer, changing its direction and velocity very rapidly. The shock programmer is typically rubber, felt material, or lead pellets. The magnitude and duration (period) of the velocity change is what determines the peak G level and duration of the shock impulse. The type of shock impulse is a result of the type of shock programmer that is used.

Repetitive shock (RS) vibration originates from a repeated shock impulse excitation created from pneumatic actuators impacting a rigid or semirigid table supported by springs to which test objects are attached.

In order for the environmental test equipment to perform repeatable accurate tests, all instrumentation must be calibrated by a laboratory accredited to the ISO/IEC 17025 standard.

Calibration of a climatic chamber in principle is the determination of the deviation between the indication on the display values of the chamber and the corresponding values measured in the chamber. Using at least one sensor for temperature and/or humidity in close proximity with the tested product in the chamber will usually provide much more reliable data than the indication of the measured values for the whole the climatic chamber. It is crucial to clearly state in the calibration certificate the coverage of the calibration (in one or more points as spatial distribution in the chamber). The calibration of a climatic chamber requires determination of a few parameters (e.g., air temperature/humidity spatial distribution in the specified calibration volume and air temperature/humidity temporal stability over a representative period of time), which will also be considered in the uncertainty budget.

The system required to perform HALT testing must be a single chamber capable of providing both thermal and vibration energy stresses independently or combined within the same chamber at the same time. Temperature and vibration profiles are programmed to run automatically, allowing the temperature rate of change to be controlled during temperature transitions. The chamber can also be run in manual mode with the operator controlling the temperature and vibration set points when operational or destruct limits are determined.

The thermal system used in HALT must be capable of providing a rapid change rates of air temperature, hot and cold, when changing temperature at its maximum possible rate, and must have a minimum thermal range of −100°C (−148°F) to +200°C (+392°F).

The vibration system used should be capable of delivering at 60g (measured RMS from 10 Hz to 10 kHz) into an empty vibration table. It should use repetitive shock vibration to provide 6 degrees of freedom (DOF) quasi-random vibration, with broadband energy from 10 Hz up to 5 kHz or higher.

Product response data (thermal, vibration, and functional performance) is acquired by use of thermocouples and accelerometers during the HALT process. The measurement of acceleration levels at various points on the product and fixture is done by using accelerometers with low mass type, with frequency response capability of 10 Hz to 10 kHz or higher, and a measurement range of ± 500g.

3.3 Environmental Impact from Electrical Products

There is a need to develop electrical products that are in line with environmental management concepts (included in the ISO 14000 family of standards) in order to minimize the environmental problems that can result from the use and disposal of these products. In short, this means that electrical products must be environmentally friendly. The EcoDesign EuP directive (2005/32/CE) states the ecodesign requirements for energy-using products. Recyclability of materials, limitation of environmental hazards, adequate eco-labeling, selection of components with low levels of hazardous substances, energy efficiency, and waste minimization have become important challenges for each manufacturer of electrical products. Life cycle analysis (LCA) is the method used to evaluate the environmental impact of products on the environment. The IEC 62430 Environmentally Conscious Design for Electrical and Electronic Products (based on IEC Guide 114 and ISO TR 14062) describes principles, specifies requirements, and provides guidance for organizations intending to integrate environmental aspects into their design and development in order to minimize the adverse environmental impacts of their products.

3.3.1 RoHS

Restriction of hazardous substances (RoHS) refers to restricted materials that are hazardous to the environment and pollute landfills and are dangerous in terms of occupational exposure during manufacturing and recycling.

The EU Directive 2002/95/EC, also known as RoHS 1, restricts the use of six specific hazardous materials: lead (Pb), cadmium (Cd), mercury (Hg), hexavalent chromium (Hex-Cr), polybrominated biphenyls (PBB), and polybrominated diphenyl ethers (PBDE), found in nine categories of electrical and electronic equipment (EEE).

In 2011 the Directive 2011/65/EU (replacing the previous regulations) was published, which is known as RoHS-Recast or RoHS 2 and RoHS compliance now being required for CE marking of products (the CE mark should be affixed on the product) and to add the RoHS directive, in addition to the rest of the applicable EU Directives, in the Declaration of Conformity issued for the product. The RoHS Regulations use self-declaration as the basis of the compliance. RoHS 2 also added the restriction of hazardous substances on product Categories 8 and 9 (exempted in RoHS 1), and has additional compliance recordkeeping requirements.

The EU Directive 2015/863 published in 2015 (amending the previous regulations RoHS 2) is known as RoHS 3 or RoHS 2 amended, which adds four additional restricted substances (phthalates) to the original list of six restricted substances and the new category 11 to the original list of 10 product categories impacted by RoHS. The RoHS 3 deadline was July 22, 2019. For medical devices and/or monitoring and control tools, there is an additional 2 years to become fully compliance.

EEE is defined as any equipment with a voltage rating not exceeding 1,000V for AC and 1,500V for DC that requires electric currents or electromagnetic fields to work, or equipment used for the generation, transfer, and measurement of electric currents and fields. EEE can be a finished product or a part, component, or assembly used in this [11].

The EU RoHS 2 amended specifies maximum concentration levels for 10 restricted hazardous substances, as listed in Table 3.7 [12].

Actually, being considered in RoHS Pack 15 is the assessment of other substances, such as beryllium, cobalt (dichloride and sulphate), diantimony trioxide, indium phosphide, medium-chain chlorinated paraffins (MCCPs), nickel (sulfate and sulfamate), and tetra-bromo-bis-phenol A (TBBP-A), to be included in the future in an amendment of the RoHS list of restricted substances.

Table 3.7 Maximum Concentration Level for Restricted Hazardous Substances in EEE

Hazardous Substance	*MCV Less than 100 ppm or Less than 0.01% of the Substance by the Weight of Any Homogeneous Part of the Product*	*MVC Less than 1000 ppm or Less than 0.1% of the Substance by the Weight of Any Homogeneous Part of the Product*
Cd	X	
Pb		X
Hg		X
Hexavalent chromium (Cr VI)		X
PBB		X
Polybrominated diphenyl ethers (PBDE) including decabromodiphenyl ether (Deca-BDE)		X
Bis(2-Ethylhexyl) phthalate (DEHP)		X
Butyl benzyl phthalate (BBP)		X
Dibutyl phthalate (DBP)		X
Diisobutyl phthalate (DIBP)		X

MCV = maximum concentration value; ppm = parts per million.

The limits apply to each homogeneous part, so if any one contains more than the allowed concentration, the whole product fails to comply. A homogeneous part is any part that has a uniform composition throughout, or any component or material of the finished product that cannot be removed, detached, or mechanically disjointed into different parts by unscrewing, cutting, crushing, grinding, abrasive processes and similar procedures.

The product categories included in Table 3.8 are EEE under the RoHS 2 amended (RoHS 3) Directive [12].

Compliance deadline for Category 8,9 products for RoHS 3 phthalate restriction is July 22, 2021.

The restriction of DEHP, BBP, and DBP shall not apply to toys that are already subject to the restriction of these substances through Regulation (EC) No 1907/2006.

Annex III of the Directive RoHS 2 amended [12] specifies substances (mercury, lead, cadmium, and hexavalent chromium) on many specific applications that are exempted from the RoHS requirements. Additional exemptions for categories 8 and 9 are contained in a separate list in Annex IV of the RoHS 2 Directive.

Products categories for which at the moment the RoHS requirements do not apply (excepted) are specified in Table 3.9.

RoHS establishes legal obligations on the following, which make or trade EEE from any of restricted products categories:

- Manufacturer;
- Manufacturer's European Authorized Representative (EAR);

Table 3.8 Product Categories to which RoHS Regulations Apply

Category	*Product*	*Examples*
1	Large household appliances	Large cooling appliances, refrigerators, freezers, other large appliances used for refrigeration, conservation and storage of food, washing machines, clothes dryers, dish washing machines, cookers, electric stoves, electric hot plates, microwaves, other large appliances used for cooking and other processing of food, electric heating appliances, electric radiators, other large appliances for heating rooms, beds, seating furniture, electric fans, air conditioner appliances, other fanning, exhaust ventilation and conditioning equipment
2	Small household appliances	Vacuum cleaners, carpet sweepers, other appliances for cleaning, appliances used for sewing, knitting, weaving, and other processing for textiles, irons and other appliances for ironing, mangling, and other care of clothing, toasters, fryers, grinders, coffee machines, and equipment for opening or sealing containers or packages, electric knives, appliances for hair cutting, hair drying, tooth brushing, shaving, massage, and other body care appliances, clocks, watches, and equipment for the purpose of measuring, indicating or registering time, scales
3	Computing (IT) and telecommunications equipment	Mainframes, minicomputers, printers, personal computers (CPU, mouse, screen, and keyboard included), laptop computers (CPU, mouse, screen, and keyboard included), notebook, notepad, printers, copying equipment, electrical and electronic typewriters, pocket and desk calculators, other products and equipment for the collection, storage, processing, presentation or communication of information by electronic means, user terminals and systems, facsimile machine (fax), telephones, cordless telephones, cellular telephones
4	Consumer electronics	TVs, DVD players, stereos, radio set, video cameras, video recorders, hi-fi recorders, audio amplifiers, musical instruments
5	Lighting	Luminaires for fluorescent lamps with the exception of luminaires in households, straight fluorescent lamps, compact fluorescent lamps, high-intensity discharge lamps, including pressure sodium lamps and metal halide lamps, low-pressure sodium lamps, other lighting or equipment for the purpose of spreading or controlling light with the exception of filament bulbs
6	Electrical and electronic tools	Drills, saws, sewing machines, equipment for turning, milling, sanding, grinding, sawing, cutting, shearing, drilling, making holes, punching, folding, bending, or similar processing of wood, metal, and other materials, tools for riveting, nailing, or screwing or removing rivets, nails, screws, or similar uses, tools for welding, soldering, or similar use, equipment for spraying, spreading, dispersing, or other treatment of liquid or gaseous substances by other means, tools for mowing or other gardening activities
7	Toys, leisure. and sports equipment	Electric trains or car racing sets, hand-held video game consoles, video games, computers for biking, diving, running, rowing, etc., sports equipment with electric or electronic components, coin slot machines
8	Medical devices, including IVDs	Equipment for radiotherapy, cardiology, dialysis, nuclear medicine equipment, pulmonary ventilators, laboratory equipment for in vitro diagnosis, analyzers, freezers, fertilization tests, other appliances for detecting, preventing, monitoring, treating, alleviating illness, injury, or disability

Table 3.8 (continued)

Category	*Product*	*Examples*
9	Monitoring and control equipment, including industrial applications	Smoke detector, heating regulators, thermostats, measuring, weighing or adjusting appliances for household or as laboratory equipment, other monitoring and control instruments used in industrial installations
10	Automatic dispensers	All appliances that automatically deliver all types of products, vending machines, ATM machines
11	Other EEE not covered by any of the categories above	Two-wheeled electric vehicles; electronic nicotine delivery systems (ENDS) such as e-cigarettes, cannabis vaporizers, and vape pens; electrical cables that are less than 250V working voltage

RoHS also applies to all wire, cables, and associated connectors and spare parts for repairing, reusing, updating, or upgrading a product both internal and external.

Table 3.9 Product Categories for which RoHS Do Not Apply (Excepted)

Product Not Included in RoHS	*Example*
Military	Equipment used for defense or national security such as arms, munitions, and materials intended for specifically military purposes
Space	Equipment sent into space, such as satellites, telescopes, spacecraft
Transportation	Vehicles used for persons or goods transport
Fixed-location PV panels	For public, commercial, industrial, or residential use
Research and development equipment	Specifically designed for professional use
Non-road mobile machinery (NRMM)	For professional use, such as agricultural, railway, waterway, construction machinery
Large-scale fixed installations (LSFI)	Electrical distribution, heating, ventilation, and air conditioning (HVAC), robotic equipment and lines
Large-scale stationary industrial tools (LSSIT)	CNC, milling, metal-forming, testing machines; cranes
Active Implantable Medical Devices (AIMD)	Pacemakers, implanted defibrillators, implantable LVAD
Sub-equipment for out-of-scope equipment	Equipment specifically designed to be part of another type of equipment that is exempted
Batteries	Individual cells
Compact fluorescent light bulbs/lamps	

- Importer in EU;
- Distributor (including retailers) in the EU.

Manufacturers are responsible for compiling technical documentation, known as the technical file, to demonstrate compliance with the regulations. This should include information on the design, manufacture, and operation of the EEE, which together make it possible to assess whether the product meets RoHS requirements.

The technical file must contain the following:

- Description of the product, how it operates, and design structure information (conceptual design, results of design calculations and examinations);
- Risk assessment of materials, parts; and subassemblies;
- Conformity information on materials, parts, and subassemblies (test reports, declaration from the manufacturers of the raw materials specifying conformity with RoHS Directive requirements, conformance/compliance certificates from all suppliers, etc.);
- Manufacturing documentation and records (manufacturing drawings and schemes of components, subassemblies and circuits, with descriptions and explanations);
- Conformity procedures, specifications, and standards used.

All types of businesses, except distributors, will have to keep relevant compliance documentation for a minimum of 10 years.

The IEC 63000: 2016 Technical Documentation for the Assessment of Electrical and Electronic Products with Respect to the Restriction of Hazardous Substances is the relevant standard that is applicable to requirements for documentation for conformity with RoHS. The IEC 63000, based on the EN 50581:2012 standard, was developed to implement the EU's Directive 2011/65/EU and specifies the technical documentation that the manufacturer compiles in order to declare compliance with the applicable restrictions in chemical substances that have been identified as dangerous for the environment. The implementation of the standard ensures the assessment of electrical and electronic equipment, with respect to the restriction of hazardous substances, and therefore increases trust in EEE manufacturers.

The manufacturers need to have and to implement a QMS Procedure referring to RoHS Directive requirements; additionally, they can draft a Declaration Letter or a Certificate of Compliance consisting of a statement that claims conformity with the RoHS Directive. From each vendor the manufacturer of an EEE shall request a Material Declaration and on each order sent to suppliers, the following needs to be specified: "Please provide objective evidence that the ordered part (material) comply with the RoHS EU Directive requirements." For drafting the Material Declaration the following documents can be used: IEC 62474 Material Declaration for Electrical and Electronic Equipment, which specifies the procedure, content, and form relating to material declarations for products of companies operating in and supplying the electrotechnical industry, IPC-1751A WAM 1 Generic Requirements for Declaration Process Management, which provides the basic supplier/requestor details for declarations necessary between members of a supply chain relationship, and IPC-1752A WAM 1, 2 & 3 Materials Declaration Management Standard, which is the materials declaration standard for companies in the supply chain to share information on materials in products [13]. The restricted substances may therefore be used in the production process as long as they do not violate other regulations and the finished EEE does not contain the substance above the maximum concentration values.

To evaluate of the restricted RoHS substances in a material, parts or whole product are used few testing techniques. The IEC 62321 series of standards consists of test methods and sample preparation (chemical and mechanical) for determination of certain substances in electrotechnical products. One of most used testing methods is the nondestructive X-ray fluorescence spectroscopy (XRF) described in ASTM F2617-15 Standard Test Method for Identification and Quantification of Chromium, Bromine, Cadmium, Mercury, and Lead in Polymeric Material Using Energy Dispersive X-ray Spectrometry, which detects only elemental substances and must be used in conjunction with other analytical methods to achieve a full compliance determination. Other testing methods used include Fourier transform infrared spectroscopy (FTIR), combustion-ion chromatography (C-IC), wet chemical analyses, ultraviolet and visible absorption spectroscopy (UV-VIS), cold vapor atomic absorption spectroscopy (CV-AAS), cold vapor atomic fluorescence spectrometry (CV-AFS), inductively coupled plasma optical emission spectrometry (ICP-OES), inductively coupled plasma mass spectrometry (ICP-MS), and scanning electron microscopy (SEM). For presence of phthalates evaluation, the solvent extraction is analyzed using gas chromatography with mass spectrometry (GC/MS) or with flame ionization detection (GC/FID) [12]. Useful information for RoHS assessment can also be found in IEC/TR 62476 Guidance for Evaluation of Products with Respect to Substance Use Restrictions in Eectrical and Electronic Products, which provides a framework for the use of standards and practices to evaluate electrical and electronic products with respect to restricted substances.

An official specific marking request for designating compliance with the EU RoHS Directive (except the CE Mark) was not introduced as mandatory. Due to this, manufacturers have the liberty to add on a voluntary basis, in the components, PCBs, assemblies, parts, or product labeling a pictogram that indicates conformity with RoHS requirements. This pictogram (see Figure 3.2) can be used as an option to replace the wording “RoHS Compliant” on labels or wherever practical (package, accompanying documents, flyers, etc.). It is recommended the pictogram be a minimum of 22 mm by 25 mm with a minimum diameter of the circle of = 18 mm. The background can be white or black and the pictogram and letters should be of a contrasting color. The color red should be avoided as red suggests the presence of a hazard. Figure 3.2 includes examples from the many RoHS pictograms actually used in the market.

In additon to RoHS, restrictions exist for the use of other hazardous materials and substances in an electrical product.

Asbestos and arsenic (As) must not be present in parts, components, materials, or products. Arsenic must also not be present in inks. Additional requirements referring to the use of chemicals are included in the EU REACH Regulation (see the following section).

Figure 3.2 Examples of RoHS pictograms used in labeling of an EEE.

In the global market many countries have implemented their own version of RoHS requirements as well (Figures 3.3, 3.4); some of these are included in Table 3.10.

Other countries, such as as Brazil, Mexico, Chile, Columbia, Ecuador, Costa Rica, and Thailand, are in advanced stages of adoption legislations referring to the restricted use of hazardous substances.

3.3.2 REACH

The EU general regulation (EC) No. 1907/2006 on the Registration, Evaluation, Authorization and Restriction of Chemicals (REACH) addresses the production and use of chemical substances and their potential impact on human health and the environment. REACH covers all industries products, materials, and substances manufactured in or imported into the EU.

The REACH Regulation, as the name suggests, consists of four actions:

- Registration;
- Evaluation,
- Authorisation;
- Restriction of chemicals.

Figure 3.3 Chinese marking for (a) RoHS-compliant, and (b) noncompliant parts.

(a)

(b)

Figure 3. 4 Japanese marking for (a) RoHS-compliant and (b) noncompliant products.

Table 3.10 RoHS in the Global Market

Country	*Local Regulation*	*Similarity with*	*Marking Requested*	*Status of Noncompliant Product*	*Special Aspects*
China	Administrative Measures for the Restriction of the Use of Hazardous Substances in Electrical and Electronic Products	EU RoHS 2 with six restricted substances and 12 products categories and no exemptions; concentration limits as in [15].	Yes (Figure 3.3)*	Accepted*	Substance concentrations restricted for each homogenous material in the product, at the component or at the subassembly level
Eurasian Economic Union	CU TR 037/2016 or EAC RoHS	EU RoHS 2	EAC Mark	Not addressed	Member states: Armenia, Belarus, Kazakhstan, Kyrgyzstan, and Russia
India	Common legislation for RoHS and WEEE	EU RoHS 2	No	Not addressed	Information to be included in the accompanying documents
Japan	J-MOSS combining Japanese Recycling Law and JIS C 0950 standard	EU RoHS 2 only for seven product categories	Yes (Figure 3.4)†	Accepted†	
Korea	Act for Resource Recycling of Electrical and Electronic Equipment and Vehicles	EU RoHS 2	No	Not addressed	
Singapore	Environmental Protection and Management Act (SG RoHS or Singapore RoHS)	EU RoHS 2 for mobile phones, laptops, flat-panel TVs, refrigerators, air conditioners, and washing machines	No	All other products, including spare parts for the in-scope products listed below, are exempted	A RoHS DoC must be submitted to the National Environmental Agency (NEA), and a Technical File as per IEC 63000 or EN 50581
Taiwan	CNS 15633	EU RoHS 2	No	Accepted‡	
Turkey	Atık Elektrikli ve Elektronik Eyaların (AEEE) Kontrolü Yönetmelii covers both RoHS and WEEE	EU RoHS 2	No	Not addressed	Documentation maintained only for 5 years§
UAE#	UAE Cabinet Decision No. 10 of 2017	EU RoHS 3	ECAS or ESMA quality mark on label and obtain a certificate to put products on the UAE market	Not addressed	Needs national UAE RoHS Declaration of Conformity and associated documentation
Ukraine	Technical Regulation Decree No. 139	EU RoHS 3	Ukrainian Mark	Not addressed	
United States	California** SB 50	EU RoHS 1 only for lead, mercury, cadmium, and hexavalent chromium	Warning requested in product and packaging	Not addressed	Mainly concerns video display products

While RoHS restricts substances present in electrical/electronic equipment (wiring, components, printed circuit boards, displays, subassemblies, etc.), REACH controls all chemicals that might be used to manufacture the product,

Table 3.10 (continued)

Country	*Local Regulation*	*Similarity With*	*Marking Requested*	*Status of Noncompliant Product*	*Special Aspects*
Vietnam	Circular 30/2011/ TT-BCT	EU RoHS 1 with no exceptions	No	Not addressed	Information to be included in the accompanying documents, manufacturer website, marking and labeling of packaging.

*The products compliant for all materials shall be marked with the symbol from Figure 3.3(a). Products and parts that contain restricted substances exceeding limits can be sold in China and marked as in Figure 3.3(b), displaying the Environment Friendly Use Period (EFPU) (in this example, 10 years), which means the period of time in years before any of the RoHS substances are likely to cause possible harm to health or the environment. Along with the noncompliance mark, a hazardous substance table must be supplied with the product in accompanying documents, which lists each part that is out of compliance.

† A compliant product shall be marked with the green G mark shown in Figure 3.4(a). Products that contain restricted substances exceeding limits can be sold in Japan and shall be marked with the orange R mark shown in Figure 3.4(b) and needs to disclose this noncompliance in the packaging, accompanying documents, and on the product website in Japanese.

‡ Need a "table of presence condition of restricted substances" indicating which parts are noncompliant and must add the table to the product, packaging, stickers, and accompanying documents of these products.

§ Manufacturers must submit a Conformity Declaration Form to the Turkish Ministry of Environment and Forestry (TMEF) every year.

Similar regulation applies to the Gulf Cooperation Council for the Arab States (GCC) including Bahrain, Kuwait, Oman, Qatar, Saudi Arabia, and Yemen.

** Other U.S. states that have enacted RoHS-like and e-waste regulations include Colorado, Illinois, Indiana, Minnesota, New Jersey, New Mexico, New York, Rhode Island, and Wisconsin.

including enclosures, brackets, coatings, paints, solvents, and chemicals used during manufacture.

Annex XVII of REACH Regulation consists of a list of restricted chemical substances (currently 74 substances as of August 2020) that shall not be used on the products. Note that all the RoHS restricted substances are also on the REACH restricted list. Exemptions from the substance restrictions in RoHS 2 may not be granted if they result in a weakening of the environmental and human health protection afforded by REACH (the lowest maximum concentration apply).

REACH makes manufacturers and importers of chemical substances (as such and in mixtures) responsible for assessing and managing the risks posed by handling these chemicals, registering those chemicals in a central database of the European Chemicals Agency (ECHA) in Helsinki, and providing the appropriate safety information (included detailed information regarding the composition of all chemicals) to ECHA and their end users. Without registration, substances cannot be manufactured or imported into the EU.

ECHA, monitors REACH, manages the databases necessary to operate the system, coordinates the identification and evaluation of suspicious chemicals referred to as substances of very high concern (SVHC), and is building up a public database in which consumers and professionals can find hazard information.

ECHA identifies the SVHC that are initially included in the candidate list (currently 209 substances as of June 2020) [16] and after evaluation (if the subject substances pose an unacceptable risk to health or the environment) some are included in Annex XIV (substances subject to authorization) of the REACH Regulation

[17]. Once included in that Annex, they cannot be placed on the market or used after a date to be set (referred to as the sunset date) unless the company is granted an authorization.

The SVHC candidate list of substances and the list of substances subject to authorization (Annex XIV), which are not yet restricted, shall not be confused with the list of restricted chemicals from Annex XVII of REACH.

Substances on the SVHC list have been identified as being carcinogenic, mutagenic, reprotoxic, bioaccumulative, toxic, endocrine disruptors (mimics or inhibits the effects of hormones), or if there is scientific evidence of potentially serious effects to human health or the environment.

The products shall not contain SVHC above the 0.1% by weight threshold specified by REACH. When chemical substances are already in use in a product and was added after to the SVHC Candidate List, the manufacturer of the product shall initiate work to identify, validate, and implement alternatives while maintaining product performance.

Inclusion on the SVHC candidate list does not mean that a chemical's use is forbidden. The substance concerned can be used with the appropriate safety and application conditions. REACH calls for the progressive substitution of the most dangerous chemicals SVHC when suitable safer alternatives have been identified.

The manufacturers need to have and to implement a QMS Procedure referring to REACH Regulation requirements. Additionally, they can draft a Declaration Letter or a Certificate of Compliance consisting of a statement that claims conformity with REACH Regulation.

3.3.3 WEEE

Electrical and electronic equipment manufacturers are responsible for the separate collection, treatment, depollution, transport, end-processing, safe disposal, and reuse requirements of their products at the end of working life. This requirement is included in the Waste Electrical and Electronic Equipment (WEEE) EU Directive (2002/96/EC) reformed by WEEE 2 (Directive 2012/19/EU) [18], which mandates the treatment, recovery, reuse, and recycling (referred to as the recycling passport) of electric and electronic equipment. RoHS regulates the hazardous substances used in the manufacture of EEE, while WEEE regulates the disposal of this equipment. The product categories covered by the WEEE Directive are presented in Table 3.11.

WEEE is electrical or electronic equipment that is waste (any substance or object that the holder discards or intends or is required to discard), including all components, subassemblies, and consumables that are part of the product at the time of discarding (end of working life).

In the EU, the six product categories of electrical and electronic products specified in Table 3.11 shall comply with the WEEE Directive and be marked with a crossed-out wheelie bin symbol (Figure 3.5) in the product label and accompanying documents [18]. This marking shown on the product and its literature indicates that it should not be disposed of with other waste at the end of its working life. To prevent possible harm to the environment or human health from uncontrolled waste disposal, the holder needs to separate this from other types of waste and recycle it responsibly to promote the sustainable reuse of material resources.

Table 3.11 Product Categories Covered by the WEEE Directive

Category	*Product*	*Example*
1	Temperature exchange equipment (TEE)	Refrigerators, freezers, equipment that automatically delivers cold products, air conditioning equipment, dehumidifying equipment, heat pumps, radiators containing oil
2	Monitors and equipment with large screens	Screens, monitors, and equipment containing screens having a surface greater than 100 cm^2, televisions, LCD photo frames, laptops, notebooks
3	Lamp bulbs	Straight fluorescent lamps, compact fluorescent lamps, fluorescent lamps, high-intensity discharge lamps, including pressure sodium lamps and metal halide lamps, low-pressure sodium lamps, LED
4	Large equipment	Washing machines, clothes dryers, dish washing machines, cookers, electric stoves, electric hot plates, luminaires, equipment reproducing sound or images, musical equipment (excluding pipe organs installed in churches), appliances for knitting and weaving, large computer-mainframes, large printing machines, copying equipment, large coin slot machines, large medical devices, large monitoring and control instruments, large appliances that automatically deliver products and money, photovoltaic panels
5	Small equipment	Vacuum cleaners, carpet sweepers, appliances for sewing, luminaires, microwaves, ventilation equipment, irons, toasters, electric knives, electric kettles, clocks, and watches, electric shavers, scales, appliances for hair and body care, calculators, radio sets, video cameras, video recorders, hi-fi equipment, musical instruments, equipment reproducing sound or images, electrical and electronic toys, sports equipment, computers for biking, diving, running, rowing, etc., smoke detectors, heating regulators, thermostats, small electrical and electronic tools, small medical devices, small monitoring and control instruments, small appliances that automatically deliver products, small equipment with integrated photovoltaic panels
6	Small IT/computer/communications equipment	ITE with external dimension no more than 50 cm, such as mobile phones, GPS, pocket calculators, routers, personal computers, printers, fax machines, telephones

According to WEEE regulation, manufacturers need to provide environmental information to each EU national regulatory body (the point of contact for every member state is available at https://europa.eu/) regarding their products (e.g., materials and composition, location of hazardous materials and their removal route, special handling concernsm and dismantling instructions). The EEE manufacturers must create dismantling guides and recommendations for easy dismantling, depollution, and recovery of WEEE. Such documentation includes (1) tools and equipment needed for dismantling, (2) procedures for disassembly, (3) instructions for removing batteries, (4) end-processing of metals, and (5) recommendations for recovery and reuse [19]. The national legislation for WEEE is quite different in the EU countries and shows many differences in the transposition process. Manufacturers should contact either the retailer or their local representative for details of where and how they can take the products at end-of-working-life from the users for environmentally safe recycling. These conditions shall be included in the terms and conditions of an agreement between manufacturer and the European representative, retailer, or importer/distributor.

Figure 3.5 The WEEE symbol.

The Regulation (EU) 2017/699 [20] establishes in Annex I a methodology for calculation of the weight of EEE placed on the market of each EU Member State and in Annex II a methodology for the calculation of the quantity of WEEE generated by weight in each Member State.

The "weight of EEE" means the gross weight of any EEE within the scope of the WEEE Directive, including all electrical and electronic accessories, but excluding packaging, batteries/accumulators, instructions, manuals, and nonelectric/electronic accessories and consumables, in contrast with "WEEE generated," which means the total weight of WEEE resulting from EEE at end of working life (estimation based on lifespan of EEE), prior to any activity such as collection, preparation for reuse, treatment, recovery, including recycling, or export. Minimum collection rate shall be 65% calculated on the basis of the total weight of WEEE collected in a given year in the Member State, expressed as a percentage of the average weight of EEE placed on the market in the three preceding years in that Member State.

EEE manufacturers should have the capacity to operate a take-back solution in a country and to provide take-back logistics. In the EU these are known as compliance and take back schemes (CTBS). Take-back operators and recycling vendors shall be audited by the manufacturers' EEE on a regular basis. In addition, manufacturers shall finance take-back systems, treatment, and recycling operations. They must also demonstrate capacity to finance such operations. The EN 50625 series of standards Collection, Logistics & Treatment Requirements for WEEE and EN 50614 Requirements for the Preparing for Re-use of WEEE specifies the logistics aspects related at WEEE [19].

WEEE excludes the applicability to some categories of products. This is similar to RoHS exemptions from Table 3.9 without photovoltaic panels, which are not excluded from applicability of WEEE, and add to the exemption of medical devices and in vitro diagnostic medical devices, where such devices are expected to be infective prior to end of life.

Even if the packaging does not fall in the scope of the WEEE Directive, the manufacturers need to pay attention to recycling the packaging covered by Packaging Waste Directive 94/62/EC amended by 2005/20/EC Directive. The application of this Directive is done by compliance to the following EU standards:

- EN 13427 Packaging–Requirements for the Use of European Standards in the Field of Packaging and Packaging Waste;
- EN 13429 Packaging–Reuse;
- EN 13430 Packaging–Requirements for Packaging Recoverable by Material Recycling;
- EN 13432 Packaging–Requirements for Packaging Recoverable through Composting and Biodegradation–Test Scheme and Evaluation Criteria for the Final Acceptance of Packaging.

References

[1] IEC 60601-1:2012 +A1:2005+A2:2020, *Medical Electrical Equipment Part 1: General Requirements for Basic Safety and Essential Performance*, Geneva, 2020.

[2] IEC 60950-22, *Information Technology Equipment–Safety–Part 22: Equipment to Be Installed Outdoors*, Geneva, 2016.

[3] IEC 62368-1, *Audio/Video, Information and Communication Technology Equipment–Part 1: Safety Requirements*, Geneva, 2018.

[4] IEC 60721-3-3 + Amnd.1+ Amnd.2, *Classification of Environmental Conditions–Third Part: Classification of Groups of Environmental Parameters and Their Severities, Stationary Use at Weather-Protected Locations*, Geneva, 1994.

[5] IEC 60721-3-4 + Amnd.1, *Classification of Environmental Conditions–Third Part: Classification of Groups of Environmental Parameters and Their Severities, Stationary Use at Non-Weather-Protected Locations*, Geneva, 1995.

[6] ETSI EN 300 019-1-0, *Environmental Engineering (EE); Environmental Conditions and Environmental Tests for Telecommunications Equipment; Part 1-0: Classification of Environmental Conditions; Introduction*, Geneva, 2003.

[7] IEC TR 60664-2-1, *Insulation Coordination for Equipment within Low-Voltage Systems–Part 2-1: Application Guide–Explanation of the Application of the IEC 60664 Series, Dimensioning Examples and Dielectric Testing*, Geneva, 2011.

[8] IEC 60068-1, *Environmental Testing–Part 1: General and Guidance*, Geneva, 2013.

[9] Fries, R. C., *Reliable Design of Medical Devices*, Boca, Raton, FL: CRC Press, 2006.

[10] *Qualmark HALT Testing Guidelines*, Denver, CO, 2010.

[11] National Measurement and Regulation Office, *Guidance to RoHS Directive 2011/65/EU, Teddington*, UK, 2015.

[12] Directive 2011/65/EU Restriction of the Use of Certain Hazardous Substances in Electrical and Electronic Equipment, The European Parliament and of The Council, Brussels, 2011.

[13] IPC Material Declaration, https://www.ipc.org/3.0_industry/3.4_ehs/materials-declaration/ipc-1752a-general-information.pdf.

[14] IEC 62321-1, Determination of Certain Substances in Electrotechnical Products–Part 1: Introduction and Overview, Geneva, 2013.

[15] GB/T 26572 *Requirements for Concentration Limits for Certain Restricted Substances in Electronic and Electrical Products*, Beijing, 2011.

[16] https://echa.europa.eu/candidate-list-table, ECHA, Helsinki.

[17] Regulation (EC) 1907/2006 Concerning the Registration, Evaluation, Authorisation and Restriction of Chemicals (REACH), Annex XIV, The European Parliament and of The Council, Brussels, 2006.

[18] EU Directive 2012/19/EU on Waste Electrical and Electronic Equipment (WEEE), The European Parliament and of The Council, Brussels, 2012.

[19] RoHS Guide, https://www.rohsguide.com/rohs-weee.htm.

[20] Regulation (EU) 2017/699, *Common Methodology for the Calculation of the Weight of Electrical and Electronic Equipment (EEE)*, The European Parliament and of The Council, Brussels, 2017.

Selected Bibliography

ETSI ETR 035: *Equipment Engineering (EE); Environmental Engineering Guidance and Terminology, Geneva*, 1997.

EN 50614 Requirements for the Preparing for Re-use of WEEE.

EN 50625 series Collection, Logistics & Treatment Requirements for WEEE.

Euramet, *Calibration of Temperature and/or Humidity Controlled Enclosures*, Braunschweig, 2015.

https://ec.europa.eu/environment/waste/rohs_eee/legis_en.htm.

https://echa.europa.eu/.

http://everyspec.com/library.php.

https://www.desolutions.com.

https://www.nist.gov/standardsgov/compliance-faqs-rohs.

https://www.rohsguide.com/.

CHAPTER 4

Materials Considerations

In the construction of electrical products, a wide range of materials are used that have an influence on safety due to the properties and phenomenon that can act on them during the life cycle of the product. In time, during all the phases that a product will go through during its life cycle, all of these materials or parts may play a major role in adding to or subtracting from the safety elements that were initially designed for the equipment.

We found it interesting that materials may determine safety characteristics even if those parts are not classified and/or considered in all situations as critical components from the beginning of the life of that product. We rarely see materials considered as critical components within the safety and evaluation reports.

The rusting of a screw, aging of an adhesive, modification of the structure of the materials in time due to functionality correlated with the surrounding environment, or using toxic or hazardous materials are elements that can generate major issues for the equipment during use, and as well in the last phase of its life during recycling.

The elements and material properties described below are relevant to the influence that materials used within electrical equipment may have on the overall electrical safety of equipment.

4.1 Corrosion

Corrosion represents a chemical and/or an electrochemical process of converting the metal to an oxide, hydroxide, or sulfide. The main factors that contribute to corrosion are temperature, humidity, pollution of the air (the presence and concentration of sulfur dioxide in the environment, generally halides), and environment salinity. None of these factors can be singled out as the main contributor to corrosion; they work together, in pairs, in groups, or even alone.

Corrosion is the reaction of a metal with the existing environment in which the electrical equipment is designed to be used, and normally, corrosion acts mainly in a destructive manner [1].

The safety and reliability of an electrical product can be dramatically impacted by corrosion, and at the same time may bring unexpected supplementary significant costs to the end user and continuously raise the presence of new and unexpected safety hazards.

The following is a brief review of a few types of corrosion that may affect an electrical product:

1. Uniform (general) corrosion occurs where the environment has similar access to all locations of the exposed metal surface, and the resulting corrosion is evenly distributed over this surface at a consistent rate. Atmospheric corrosion is the most common example of corrosion, including the tarnishing of silver and rusting of iron; it is the most common form of corrosion and is visible. It starts as a cosmetic degeneration, and over time, uniform (general) corrosion may result in equipment failures that impact the performance as well the safety of the equipment by reducing creepage distances, air clearances, and even damaging the mechanical strength.
2. Galvanic corrosion, sometimes referred to as bimetallic or dissimilar metal corrosion, occurs when two metals are in electrical contact in the presence of an electrolyte (conductive solution). This system creates an electrochemical cell or reaction: the electrons are allowed to flow through the electrolyte from the anode (metal being corroded) to the cathode (metal being protected). The loss of electrons from the anode material is a process called oxidation. When two metals are in intimate contact, the metal that is corroded is determined by the relationship between the two metal's relative corrosion potential: the more cathodic or noble metal will be protected and the more anodic metal will corrode or it will be sacrificed. Typically, the corrosion is more aggressive near the junction of the two metals due to lower electrical resistance and a short electrolyte path. The galvanic corrosion reaction is dependent on the conductivity of the electrolyte; higher conductivities will produce more aggressive attacks.
3. Localized corrosion occurs when a small part of a component experiences corrosion or comes in contact with specific corrosion and may cause stresses. The small local area corrodes at a much faster rate than the rest of the component, and the corrosion works alongside other processes such as mechanical stress and fatigue, and the end result is much worse than the result of stress or fatigue alone. Environmental factors such as pH, temperature, and availability of oxygen have a significant impact on the aggressiveness of this type of corrosion.
4. Caustic agent corrosion: occurs when impure gas, liquids, or solids wear a material down. Although most impure gases do not damage the metal in dry form, when exposed to moisture they dissolve and form harmful corrosive droplets. Hydrogen sulfide is an example of one such caustic agent. Although chemicals can corrode metals directly, most of the corrosion affects metals that hold or are submerged in water, or that are subjected to moisture-forming films due to atmospheric exposure.

Corrosion is a natural deteriorating process that affects almost all metals at different rates (in time and qualitative), which can have potential multiple effects on the safety characteristics of equipment and can never be stopped. However, the presence and effects of corrosion may be slowed by selecting the proper materials, and where necessary, the appropriate coatings, by using an appropriate design and technology. Additionally, a robust inspection and maintenance program will help evaluate when corrosion may affect critical functions that can be addressed prior to safe operations being compromised.

The safety standards (e.g., IEC 62368-1) have stipulated clearly practical aspects that need to be followed to avoid the negative effects of corrosion [2]:

- Corrosion-resistant materials shall be used.
- Metallic enclosure and fittings should be protected against corrosion on inside and outside surfaces, unless the metal is inherently resistant to indoor atmosphere corrosion.
- Iron and steel parts can be provided with corrosion protection such as enameling, galvanizing, plating, or other equivalent means.
- Conductive parts in contact at the main protective earthing terminal, protective bonding terminals, and connections shall be selected in accordance with the supplied tables within the standards, in such a way that the potential difference between any two different metals is 0.6V or less (some standards consist of such relevant tables).
- The screws of a terminal block, bolts used for assembling the enclosure, and similar elements, shall be resistant to corrosion for the total service life of the equipment.
- It must be understood how the materials interact with the rest of the involved parts of the equipment, in all possible use conditions, including service and repairs, and how the beginnings of a destructive corrosion process can be identified during regular maintenance.
- A design shall be used for an enclosure (e.g., a clamped enclosure with minimal bolts), which allows easy access for preventative maintenance and service in case of failure of the hosted equipment.
- Bolts and screws, proper torque of the involved parts within the equipment, electrical connections, or fasteners shall be designed and used with an awareness of the multiple ways in which corrosion can interfere with these elements.
- Improperly torqued lugs, nuts, and screws combined with elements of corrosion will lead to early failures of an electrical equipment due to a high risk of breaking off upon routine opening of the door of an enclosure.
- Stainless steel bolts driven into aluminum castings can cause bolt seizure if not adequately lubricated and will rust. The rust may assist in breaking that screw during a service or maintenance operation, or even corrode it if left in place because it could not be removed without damaging it, or it will damage the location where the screw or bolt was used, generating an immediate hazard.
- A combination of aluminum with brass, copper, nickel-plated brass, or copper and gold-plated copper in 1% salt spray, generates severe galvanic corrosion.
- Anodic, cathodic, organic, precipitation, and vapor-phase inhibitors are substances that slow or prevent corrosion when added to an environment in which a metal usually corrodes and are designated as corrosion inhibitors.

- Coatings protect the metal by imposing a physical barrier between the metal substrate and the environment. Common types of coatings are organic (paints, resins, lacquers, and varnishes), inorganic (porcelain enamel, glass linings, and conversion coating-anodization), and metallic.
- Geometrical form is basic in the prevention of corrosion by minimizing or avoiding situations such as stagnation (nondraining structures, dead ends, badly located components, poor assembly) or sustained fluid flow (erosion and cavitation in components contacted by fluids) that worsen corrosion [3].
- Insulation can create an area for potential corrosion attack due to poor installation.

Standards, specifications, recommended practices, and test methods are essential sources of information on corrosion, starting with corrosion fundamentals, the forms of corrosion, and how corrosion is detected and measured, and ending with the multiple ways to protection against corrosion damage. Selected titles of standards are included in Table 4.1.

Additionally, the ASTM's corrosion standards (e.g., G50-10, G91-11, G101-04, G116-99, G69-20,B117-19) provide the appropriate procedures for carrying out corrosion tests on specified metallic materials and alloys. These tests are conducted to examine and evaluate the behavior, susceptibility, and extent of resistance of certain materials to stress corrosion cracking, intergranular corrosion, pitting and crevice corrosion, exfoliation corrosion, and atmospheric and galvanic corrosion.

Instructions and recommendations regarding the maintenance and prevention of corrosion are strongly recommended. We can find many examples of instructions that go something like this: "To be resistant against the effects of ambient influences such as temperature, dampness, dust, gasses, steam, and mechanical stress,

Table 4.1 List of Standards Referring to Corrosion

Standard	*Description*
ISO 11303	Corrosion of Metals and Alloys. Guidelines for Selection ff Protection Methods Against Atmospheric Corrosion
ISO 11844 series	Corrosion of Metals and Alloys–Classification of Low Corrosivity of Indoor Atmospheres
EN 12501 series	Protection of Metallic Materials Against Corrosion
IEC 60512-11-6	Connectors for Electronic Equipment. Tests and Measurements. Climatic Tests. Test 11f. Corrosion, Salt Mist
IEC 60512-11-7	Connectors for Electronic Equipment. Tests and Measurements. Flowing Mixed Gas Corrosion Test
IEC 61701	Salt Mist Corrosion Testing of PV Modules
IEC 60068-2-60	Environmental Testing. Test Methods. Test Ke. Flowing Mixed Gas Corrosion Test
ISO 7539 series	Corrosion of Metals and Alloys–Stress Corrosion Testing
ISO 10271	Dental Metallic Materials. Corrosion Test Methods
UL 50	Enclosures for Electrical Equipment, Non-Environmental Considerations
UL 514 A	Metallic Outlet Boxes*
UL 514 B	Standard for Conduit, Tubing, and Cable Fittings*

*Used for determination of acceptability of corrosion protection.

and to avoid the risk of fire or corrosion, batteries shall be kept clean and dry." Unfortunately, general instructions are often disregarded by the end user.

Instead, we propose providing the end user with information about the extremes of all environmental conditions likely to be encountered and that will affect the safety of the equipment during its life cycle, including storage, transportation, testing, installation, maintenance, and service, during operation of that equipment.

4.2 Adhesives

Adhesives and the use of adhesives are not only an important part of the manufacture of an electrical product, but the chemicals used have a strong impact on the reliability and longevity of the equipment. Adhesives together with contacts and use of rubber parts represent one of the biggest problems, unsolved until now, for manufacturing: aging and loss of the intended function over time. Many adhesives lose or significantly reduce their additive properties over time, generating defects, sometimes with serious consequences. For this reason, close attention must be paid to their selection and use.

Adhesives are utilized to form bonds or joints between adherend materials that can be metal, wood, or plastic.

The classification of adhesives is based on the following [4]:

- Chemical type (epoxy, silicone, acrylic, polymeric-based, resin, etc.);
- Origin (natural, synthetic);
- Physical form (films, tapes, pastes, one- or multipart components);
- Curing method (heat, moisture, radiation, etc.);
- Functional type (structural, pressure-sensitive, hot-melt).

The major uses of adhesives in electrical equipment in general includes bonding of surface-mount components (SMCs), potting or encapsulating components, and conformal coating of circuit boards and assemblies.

Surface-mount adhesives serve two functions: they act as a processing aid (e.g., holding a part temporarily until it can be permanently attached by soldering), and they can also provide stress relief to solder connections during service to prevent premature failure of the electrical connections. After soldering, the now-redundant adhesive must not affect the circuitry in any way. In the manufacturing of electrical equipment, adhesives are often used as assembly parts or even as safeguards to secure parts within the electrical equipment and also to secure the equipment in the location where it will be used, on the walls or the ceiling.

Adhesive mounts provide a quick, economical, and dependable method of supporting, routing, and protecting wires or cables. Some are used with cable ties and others are used independently. Adhesive-backed mounts adhere to a variety of surfaces. This alternative to mechanical fasteners offers the advantage of a lower installed cost with safe, easy-to-use, quality products. The pressure-sensitive adhesive (foam tape) mounts are intended to secure wire bundles or other light objects to smooth surfaces. These mounts are not designated to support excessive loads and should not be used when the maximum expected load exceeds the rated capacity of the mount. There are two types of double-coated polymeric foam adhesives tape: rubber-based and acryl-based, which is better for outdoor and higher-temperature applications. Epoxy adhesive mounts are used on rough indoor or outdoor surfaces that require mixing a two-part epoxy adhesive before application.

When adhesives (e.g., an adhesive tape) are used to secure equipment in a location, there is the possibility that the equipment will fall from the intended location, generating mechanical hazards, choking hazards, and the hazard of opening up, which would allow unauthorized access to the battery compartment (e.g., the battery may come out during the impact or a small coin battery may become accessible to small children). These sources of potential harm are the result of adhesives that did not meet their expected performance.

Tests included in the standard UL 746C Polymeric Materials for use in Electrical Equipment Evaluations confirms the adhesive bond does not excessively degrade when exposed to various environmental conditions, including temperature, humidity, and cold, and will provide adequate performance when used within the limitations of the test program parameters (substrates that can be aluminum, stainless steel, cold-rolled steel, or thermoplastics; indoor or outdoor; temperature range), but the adequate adhesive bond strength must be determined for the particular application of the end product. Some adhesives are listed by UL in conformity with the ANSI/UL 879, Electric Sign Components standard.

The safety standards provide detailed requirements regarding adhesives when serve as safeguards, including situations when parts that may defeat a safeguard fail.

The standards help to identify properties that include viscosity, adhesion, shear strength, and shear modulus, and help to identify adhesive bond or joint mechanical properties that include strength, creep, fracture, and fatigue. They also cover the effects of environmental factors on adhesive bonds and joints, determining various applications of adhesives. Examples of standards referring to adhesives are presented in Table 4.2.

Why does an adhesive fail? Assuming the correct installation and/or use, the most important aspect regarding adhesives is their process of aging. Information data provided by adhesive manufacturers include many attributes of adhesives, but it is difficult to find details related to the aging of adhesives and the expected lifetime of the bonding.

In order to meet and even exceed the expectations of the design team and the manufacturer, adhesives must have adequate bonding properties throughout the entire life of the equipment.

When an adhesive is used, the equipment will be fully tested as per the standard requirements; but this step will not be sufficient.

In order to assure that the adhesive will not fail over time, some attentive construction review must be performed: an examination of the construction in which the adhesive is used shall be performed, and the available technical specification of the adhesive shall be reviewed.

If the data about the adhesive properties are not available, compliance of the adhesive with the applicable requirements of that standard is performed by subjecting it to a conditioning test that will be applied to the complete equipment and parts that use the adhesive. This supplementary conditioning will simulate more intense stress under more severe thermal conditions and relative humidity conditions

Table 4.2 List of Standards Referring to Adhesives

Standard	*Description*
ASTM D1304	Standard Test Methods for Adhesives Relative to Their Use as Electrical Insulation
ASTM D1828	Standard Practice for Atmospheric Exposure of Adhesive-Bonded Joints and Structures
ASTM D3310	Standard Test Method for Determining Corrosivity of Adhesive Materials
ASTM D3808	Standard Test Method for Qualitative Determination of Adhesion of Adhesives to Substrates by Spot Adhesion
ASTM D4800	Standard Guide for Classifying and Specifying Adhesives
ASTM D6105	Standard Practice for Application of Electrical Discharge Surface Treatment (Activation) of Plastics for Adhesive Bonding
ASTM D6463	Standard Test Method for Time to Failure of Pressure Sensitive Articles Under Sustained Shear Loading
ISO 9142	Adhesives–Guide to the Selection of Standard Laboratory Ageing Conditions for Testing Bonded Joints
ISO 10365	Adhesives–Designation of Main Failure Patterns
ISO 16525	Adhesives–Test methods for Isotropic Electrically Conductive Adhesives
ISO 29862	Self adhesive tapes–Determination of Peel Adhesion Properties
UL 746C	Polymeric Materials for Use in Electrical Equipment Evaluations

than the rated environmental conditions for which the equipment was thought to comply.

This accelerated aging process may be relevant for how the adhesive will behave in normal conditions for the duration of the life of that equipment. After the accelerated aging test, as a condition of acceptability stated in the standards, a part secured by adhesive shall not fall off or partly dislodge. Clearances and creepage distances shall not be affected.

There are applications of adhesives in the very sensitive field of the manufacturing of medical electrical equipment, which must meet even more restrictive conditions: adhesives, sealants, and coatings are formulated especially for the assembly of medical devices. These specialty compounds are designed to adhere to dissimilar substrates to resist prolonged exposure to sterilization. For such applications, there are special medical grade adhesives, sealants, coatings, and potting/encapsulation compounds; these need to be nontoxic and show compatibility with blood and body fluids. Additionally, they must be biologically inert. There are special standard tests used to determine their suitability: U.S. Pharmacopeia (USP) Class VI and/or ISO 10993 Biological Evaluation of Medical Devices–Part 1: Evaluation and Testing within a Risk Management Process. These documents apply to the evaluation of materials (and thus the adhesives/sealants are part of this investigation along with the medical devices that are expected to have that adhesive and as well are expected to have direct contact with the patient's body during normal use of the device).

As specified in ISO 10993, the following needs to be considered:

- Categorization based on the nature and duration of the contact with the body;
- The evaluation of biocompatibility;

- The use of material in contact with the human body shall be evaluated based on risk analysis.

The adhesives, sealants, and coatings when used within medical and/or laboratory electrical equipment, along with their adhesives and sealing properties, shall meet supplementary requirements from the point of view of the disinfection and sterilization (to resist against attack from disinfectants, detergents, isopropyl alcohol, and bleaches), and they have to resist multiple cycles of such disinfection and sterilization cycles during the life cycle of the equipment. The processes of disinfection and sterilization may involve a multitude of methods as follows: in an autoclave, ethylene dioxide, radiation, or by using chemical sterilant detergents or disinfectants. Adhesive coatings offered a great value to medical packaging materials, which is demonstrated by improved performance in the closure, sterilization, security, and openings. Coat-weight consistency, adhesive anchorage, seal range porosity, and bond strength are a few of the important attributes of adhesive-coated systems.

Adhesives and sealants must have as well good resistance in terms of mechanical strength, including vibration/impact, and the stress caused by humidity/temperature must not affect their structural integrity or that of the bonded structure realized with them.

Another area where adhesives are used to integrate into electrical equipment are the methods of encapsulating electronic components: potting, casting, and molding providing additional capabilities of insulation, along with reducing the spaces [5].

- Potting is a method of filling small spaces or surfaces with a material that will protect components from physical and environmental damage. Typical resins used for potting are epoxies, polyurethanes, silicones, and acrylics, the latter usually being UV-curing formulations.
- Casting is a method that uses the same type of adhesive resins as potting, although the container (outer casing) is usually removed after the resin has cured, unlike the potting process where the container becomes an integral part of the component.
- Molding is a method that involves the injection of premelted thermoplastic resins into a mold containing the electronic components or circuitry.
- The above methods of building components where the spacings (air clearance and creepage distances) do not exist, and where the only elements that are considered are the type of used material and distance through insulation are acceptable to be used from the point of view of the safety standard.

Other areas in which adhesives have found a large-scale application are SMT technology and conformal coating.

Surface-mount technology (SMT), which includes surface-mounted devices (SMDs), surface-mount components (SMCs), surface-mount assemblies (SMA), and surface-mount package (SMP) [6], has helped to drastically change the manufacturing electronics industry due to its numerous advantages (e.g., reduced the weight, dimensions, and noise on PCBs), and also had a positive impact on the manufacturing of the equipment compared to through-hole technology: it reduced the price of the PCBs, allowed for less holes, significantly improved handling of the materials, and offered a higher level of control of the process of manufacturing.

Adhesives have an important role in the area of conformal coating. The conformal coating process uses several types of resin (e.g., epoxy, polyurethanes, acrylics, and silicones). This process provides a layer that adheres to and covers all the board and the components on it. The purpose of the conformal coating is to protect the board from factors such as

- Environmental factors, such as moisture, temperature, and implicit corrosion;
- Mechanical and electrical interference, such as scratches, minor damages, and involuntary short circuits.

When coatings are used over terminations to increase the creepage distances and air clearances, safety standards impose conditions for the mechanical arrangement and rigidity of the terminations. These condition must ensure that, during normal handling, assembly into equipment, and subsequent use, those terminations will not be subject to deformation that would crack the coating or reduce the physical separation distances between conductive parts below the required values.

When applying adhesives within electrical and electronic equipment it is very important how they are chosen, because it will determine not only the functionality of the equipment, but also contribute to the reliability performance of the equipment. Depending on the application in which the adhesives are used, the appropriate characteristics for these will be chosen as needed, such as thermally conducted adhesives, electrically conductive adhesive, or adhesives able to provide a high level of insulation (e.g., in many applications, including integrated circuits and surface-mount devices, electrically conductive adhesives are required).

Currently, adhesives in electronics manufacturing are indispensable; protection of the environment, health and safety aspects, the involved costs, and speed of manufacture and to market are the main factors that are driving the development and expansion of the use of adhesives in this industry.

4.3 Insulating Materials

Hazards may raise during installation, use and/or service of the electrical equipment. In order to mitigate the effects of any potential hazards, the designers and manufacturers shall use, along with the correct design and properly rated components, only those materials capable to provide the required insulation where it is required.

When safety is involved, those materials will be included by the design team within the list of critical components (LCC) (see Volume 1, Chapter 8) along with the components that may affect the safety of the equipment.

We want to emphasize that when we are discussing insulating materials in terms of product safety, we are discussing the hazards that may be involved related to electrical equipment and the materials that can mitigate the effect of the appearance of such hazards.

With regard to the safety of equipment, in our opinion, there is no classification for which material can be used to determine which hazard is more important than another, and as well, there is no order in which the hazards must be analyzed and mitigated and when, if possible, to be eliminated.

We now discuss the principles that should guide the judgment of design teams of electrical equipment along with product safety specialists: how to detect a hazard, how to prevent the occurrence of it, and how to mitigate the possible negative effects of it.

A hazard occurs in the zone of the transfer of energy: from the source to the human body or any party that may be affected by the generated hazard.

Once the transfer zone is identified, described, and known, along with all the involved details, solutions to and answers for the necessary protections will surface and will be used.

In terms of insulating materials, we consider, using engineering principles—accepted truths that apply throughout this discipline—how to prevent and/or mitigate the following hazards:

A. Electrical energy may lead to electric shock hazards;
B. Mechanical energy may lead to mechanical energy hazards;
C. Radiant energy may lead to some of the most dangerous hazards, radiation hazards;
D. Thermal energy may lead to fire and/or burn hazards;
E. Sound and vibration energy may lead to acoustic hazard and physical damage hazards.

What exactly is insulation? It may be understood as an act of covering something to stop heat, sound, or electricity from escaping or entering, or the fact that something is covered in this way, and as well, may refer to a material that is used to stop heat, sound, or electricity from escaping or entering.

When talking about insulating materials, there is no clear delimitation between the hazards against which those materials are used. The involved materials must ensure that their properties will fully accomplish the final purpose of use within the equipment; it shall be achieved without exposing to a different or supplementary hazard than the one against whom it was intended to provide protection. A clear example is the use of flame-retardant materials to increase protection against possible fire, but neglecting the potential environmental damage that some toxic substances may generate in turn.

In terms of electrical insulation, the materials used prevent electric shock hazards and also prevent the breakdown of the insulation.

When choosing an insulating material, the following aspects should be taken into account:

- Performance characteristics of the material (e.g., the temperature class of the equipment);
- Environmental conditions for which the end-use equipment will be used, such as humidity, temperature, and corrosive environment;

- Environmental conditions that will be developed during the normal and single-fault operation of the equipment (mainly the developed temperature during the functionality of the equipment);
- Aging characteristics of the insulating material correlated with the service life of the equipment;
- Any other conditions that may be involved in the process of regulatory approvals, such as toxicity and banned materials;
- Electrical characteristics of the material;
- Physical properties of the material.

The properties should be chosen also considering the functionality and reliability of the equipment.

Insulating materials is a very broad subject, so we will focus on the insulation referred to and required by the protection against hazards generated by the forms of energy mentioned above. Insulation should be regarded as a protection against forms and levels of energy. We will present the types of insulation and will correlate each of the standard requirements with the specific type of energy.

We summarize the four types of insulation as these are presented in almost all of the safety standards.

The main four types of insulation are functional, basic, double, and reinforced insulation.

The definitions are as follows:

- *Functional insulation:* Insulation between conductive parts necessary for the proper functioning of the equipment;
- *Basic insulation:* Insulation that protects against electric shock;
- *Supplementary insulation:* A supplementary layer of insulation to the basic insulation that offers protection if the basic insulation fails;
- *Double insulation:* This represents both layers, basic and supplementary insulation, together;
- *Reinforced insulation:* This represents a single layer of insulation that is capable of providing the same protection against electric shocks as the two types of insulation (basic + supplementary) that form double insulation. In order to realize these layers of insulation, insulating materials shall be used for this purpose.

Electrical energy shall be kept under control as applicable with one of the above types of insulation, by providing the necessary separation. In order to do this, designers and manufacturers will try to select the best insulating materials.

Air is one of the materials used to provide insulation, but attention must be paid to the environmental conditions in which the equipment will work. Polluted air will create lower insulation and air that is contaminated will not be the best choice to provide any of the above-required layers of protection. We do not discuss hazardous locations here; that subject is detailed in Chapter 6. The important conclusion is that air cannot be considered a reliable and consistent insulating material.

What type of materials may be used to protect against the negative effects of electrical energy? Special attention should be paid to the some of factors that will influence the behavior of the material during the life of the equipment, such as temperature class of the material, composition, and aging behavior. Table 4.3 presents the thermal classes of insulation of some materials [7].

The maximum hot-spot operating temperature is reached by adding the rated ambient temperature of the equipment, a temperature rise, and a 10°C hot-spot allowance. Electrical equipment is usually designed with an average temperature below the rated hot-spot temperature to allow for an acceptable life. Insulation does not suddenly fail if the hot-spot temperature is reached, but useful operating life declines rapidly; a rule of thumb is a halving of life for every 10°C temperature increase. Not all the insulation is necessarily located at the point of maximum temperature, and insulation with a lower thermal classification may be used in other parts of the equipment.

Electrical insulating materials are used for a multitude of components and parts that in turn will be used within the end-use electrical equipment as insulators: coated products, felt, varnished glass cloths, insulating compounds, and so on.

We now briefly review some of these categories of insulating materials [8, 9]:

Coated cloth products: Coated fabric products consist of woven and nonwoven cloth with a coating or resin applied to the surface or saturated into the material,

Table 4.3 Thermal Classes of Insulation of Some Materials

IEC 60085 Thermal Class	*NEMA Class*	*NEMA/UL Letter Class*	*Maximum Hot-Spot Temperature Allowed*	*Typical Materials*
90			90°C	Unimpregnated paper, silk, cotton, vulcanized natural rubber, thermoplastics that soften above 90°C
105	105	A	105°C	Organic materials such as cotton, silk, paper, and some synthetic fibers
120			120°C	Polyurethane, epoxy resins, polyethylene terephthalate, and other materials that have shown usable lifetimes at this temperature
130	130	B	130°C	Inorganic materials such as mica, glass fibers, and asbestos, with high-temperature binders, or others with a usable lifetime at this temperature
155	155	F	155°C	Class 130 materials with binders stable at the higher temperature, or other materials with a usable lifetime at this temperature
180	180	H	180°C	Silicone elastomers, and Class 130 inorganic materials with high-temperature binders, or other materials with a usable lifetime at this temperature
200		N	200°C	As for Class B, including Teflon
220	220	R	220°C	As for IEC class 200
		S	240°C	Polyimide enamel or polyimide films
250			250°C	As for IEC class 200; further IEC classes are designated numerically at 25°C increments

and are used to provide heat insulation, water or chemical protective coverings, and numerous other applications in electronics, filtration, and heating, ventilation, and air conditioning (HVAC) applications. This category of materials includes felt and varnish glass cloths.

Felt: Inexpensive, durable materials made from a variety of pressed fibers that offer resistance to fraying and versatility. Felt is available in either sheets or pads. These sheets and pads are available with or without epoxy resin adhesive backings. Fibers can be made of either natural, synthetic, or mixed materials such as untreated polyester, Nomex (fiber and sheet forms used as a fabric where resistance from heat and flame is required), needle, and natural fibers. Felt is commonly used in a variety of industries as a vibration dampener; this feature makes felt ideal for use in electrical equipment that requires cushioning and padding between moving parts.

Varnished glass cloths: Thin pieces of insulation material coated in varnish. These materials are used for insulation where thermal insulation is necessary; varnished glass cloths have a high-temperature resistance and durability. These cloths come in a variety of material types; sheet felt is often an essential tool in gasketing and sound absorption, and for use in bearing seals and ink roles, where the insulator property is necessary.

Insulating compounds: Materials designed to seal and lubricate. These compounds have outstanding resistance to corrosion, moisture, and heat. Insulating compounds are used in numerous applications such as sealants and lubricants in electrical or assembly mechanisms. This can include operations that involve heavy machinery or industrial equipment. Insulating compounds are available in a wide variety of colors and many forms. An insulating epoxy hardener is also available for casting, coating, or potting of sensitive electronic components.

Insulation paper and film: Paper materials used for electrical insulation in applications such as wedges, coils, and/or motors. Insulation papers and films offer thermal stability and conductivity, low moisture absorption, and high varnish absorption.

Rigid laminate: These materials offer chemical, corrosion, and high-temperature resistance, as well as a significant tracking resistance. Solid rod and bar products are fiberglass-reinforced thermoset polyester shapes that exhibit properties desirable for a wide range of structural and electrical applications, pressboards used mainly for a wide variety of transformers, and sheet laminate materials that are ideal for electrical applications and equipment, switches and appliance insulation, and electric motor insulation or gaskets within the electrical equipment.

Sleeving and tubing: Sleeving and shrinkable tubing are used when additional insulation and/or protection is necessary for wiring and cables. Sleeving and tubing also protect against harsh environments, corrosion, and/or heat. There are several types of sleeving: heat-treated sleeving, silicone resin-coated sleeving, silicone rubber-coated sleeving, and uncoated sleeving. The heat-shrink tubing is available as a heat shrink tubing or for some special applications even in molded shapes. Heat shrink tubing is plastic tubing that is designed to shrink and wrap around wires to protect and insulate them. This tubing shrinks radially when the heat is applied to it, creating a tight, conformed wrapping. Different features can be added such as flame resistance, a higher level of insulation, and identification based on the color. Within the same category of insulators, there are heat-treated sleeves. These are braided fiberglass sleeves that have been heat-treated to remove all organic contaminants and then saturated with various materials for saturation: acrylic and vinyl. The saturation increases ease of use and helps to prevent fraying over time. These features allow heat-treated sleeves to perform in a wide variety of high-temperature applications. Heat-treated sleeves are designed to withstand heat and mechanical stress while maintaining insulating capabilities and still allowing easy handling. The flexibility of these sleeves allows them to be used in diverse locations and in an environment with moving parts keeping the wiring at a distance from those parts. Heat-treated sleeves are typically used for applications such as coil and winding in motors or generators, lighting fixtures, recreational vehicles, industrial equipment such as tow motors and commercial cookware, as well as in the automotive industry, and much more.

Materials for standoff/insulators: These provide ancillary support to many types of electrical systems; the materials used shall combine high mechanical strength and shatter resistance with high arc resistance and dielectric properties at elevated temperatures and humidity and shall provide a good mechanical resistance during the service and maintenance of the equipment. Standoff/insulators may come in different design forms with a variety of different insert types, including steel, brass, or aluminum. Standoff/insulators are ideal in industries that include electrical appliances, the HVAC industry, control panels, and ITE and audio-video equipment.

Materials for surge ropes: Cords of fiber braiding that are designed to be used in vacuum pressure impregnation (VPI) systems. The VPI system is a process that seals gaps in the materials of generators and high-voltage motors with either varnishes or resins through the use of either compression or vacuum methods. Among other uses, surge ropes enhance this process by providing support for the windings in motor applications. Surge ropes are available in a variety of different resin treatments including epoxy resins and Teflon. Electrical insulating surge ropes are recommended for use as replacements for both steel and laminated fiberglass surge rings, spacers on DC armatures, as winding head supports on random wound machines, and as support systems for the nose sections of AC form wound coils.

Varnish/resin: A special category of insulating varnishes and impregnating resins able to ensure that electric motors, generators, transformers, and sensors will function properly and reliably. From environmental protection (moisture and dust) to added strength, varnish, and resin support proper functionality and prevents system failure by providing insulation from many points of view, electrical, mechanical, and environmental. In this category are included:

Resins which are an important material in ensuring the protection of the wire windings and coils. Electrical resin offers protection from mechanical damage, vibration dampening, moisture prevention, and heat prevention.

Solvents, thinners, and inhibitors are insulating materials that are protecting against corrosion. It is known that the primary cause of electrical and electronic failure is the loss of continuity caused by corrosion.

Varnishes, which play an important role in electrical insulation. Liquid varnishes are ideal for applications where wire size must remain the same while insulation levels are enhanced.

Wedges: Stator wedges, rotor wedges, armature wedges, formed slot wedges and liners, magnetic wedges, and rotor pole collars all should be made from high dielectric withstanding voltage materials such as resin-reinforced fiberglass polyester or semiconductive glass reinforced laminate.

Ceramics and polymers: These are utilized as insulators. Many of the ceramics, including glass, porcelain, steatite, and mica, have dielectric constants within the range of 6 to 10. These materials also exhibit a high degree of dimensional stability and mechanical strength. Typical applications include powerline and electrical insulation, switch bases, and light receptacles. Materials based on mica, which is a naturally occurring mineral, play a central part in the design and manufacture of high-voltage rotating machines. At present the industry uses predominantly mica paper, which when suitably supported by a woven glass or film backing and impregnated with epoxy or a similar resin, is used to insulate the copper bars that make up the stator winding of the machine.

Elastomers and thermoplastics: There is a very wide range of polymeric and rubber-like insulation materials. Some elastomers such as silicone have found application in sleeving, traction systems, and increasingly as overhead line insulators, but the main use of their application continues to be related to cables. The leading materials are polyvinyl chloride (PVC), medium density polyethylene (MDPE), cross-linkable polyethylene (XLPE), and ethylene propylene rubber (EPR). All PVC polymers are degraded or reduced in quality by heat and light. Special stabilizers added during manufacture help to retard this degradation at high temperatures.

All of these insulating materials reviewed above should share common properties and should be selected based upon meeting or exceeding application guidelines:

- High insulation resistance to avoid leakage current;
- High dielectric strength and low dielectric constant (relative permittivity) to avoid electrical breakdown;
- Good mechanical properties (strength, tenacity, and elasticity);
- Immune to attacks by acids and alkalis over a range of temperatures;
- Nonhydroscopic, because the dielectric strength of any material drastically decreases with the moisture content;
- Low coefficient of thermal expansion and good temperature endurance;
- Fungus resistance;
- Ozone resistance;
- Flammability.

As well, to protect the human body and the environment, these materials shall meet the regulatory requirements imposed by each market where the equipment is placed. Some of the applicable standards for insulation material are indicated in Table 4.4.

4.4 Hazardous Materials Information

Modern manufacturing, storage, and use of electrical products have introduced or increased human exposure to many hazardous materials that negatively affect health status. These adverse effects are caused by chemical, biological, radiation, and physical hazards and are greatly influenced by exposure, which can be direct or indirect contacts with toxic sources.

Table 4.4 Applicable Standards for Insulation Materials

Standard	*Description*
ASTM D 3295	Standard Specification for PTFE Tubing, Miniature Beading and Spiral Cut Tubing
ASTM D 4388	Standard Specification for Non-metallic Semi-Conducting and Electrically Insulating Rubber Tapes
BS 7655 (series)	Specification for Insulating and Sheathing Materials for Cables. PVC Insulating Compounds. Hard Grade Types. Hard Grade Types
EN 13467	Thermal Insulating Products for Building Equipment and Industrial Installations. Determination of Dimensions, Squareness and Linearity of Preformed Pipe Insulation
EN 13471	Thermal Insulating Products for Building Equipment and Industrial Installations. Determination of the Coefficient of Thermal Expansion
EN 60684 (series)	Flexible Insulating Sleeving. Methods of Test
IEC 60085	Thermal Evaluation and Classification of Electrical Insulation
IEC 60216	Determination of Thermal Endurance Properties of Electrical Insulating Materials
IEC 60243	Methods of Test for Electric Strength of Solid Insulating Materials
IEC 60371	Specifications for Insulating Materials Based on Mica
IEC 60464	Specifications for Insulating Varnishes Containing Solvent
IEC 60626	Combined Flexible Materials for Electrical Insulation
IEC 60672	Specifications for Ceramic and Glass Insulating Materials
IEC 60684	Specifications for Flexible Insulating Sleeving
IEC 60893	Specifications for Industrial Rigid Laminated Sheets Based on Thermosetting Resins for Electrical Purposes
IEC 61068	Polyester Fiber Woven Tapes
MIL-I-631	Insulation, Electrical, Synthetic-Resin Composition
MIL-I-3158	Insulation Tape, Electrical Glass-Fiber (Resin-Filled): and Cord, Fibrous-Glass
MIL-I-3190	Insulation Sleeving, Electrical, Flexible, Coated, General Specification for
MIL-I-17205	Insulation Cloth and Tape, Electrical, Glass Fiber, Varnished
MIL-I-19166	Insulation Tape, Electrical, High-Temperature, Glass Fiber, Pressure-Sensitive
MIL-I-22076	Insulation Tubing, Electrical, Nonrigid, Vinyl, Very Low Temperature Grade
MIL-I-22129	Insulation Tubing, Electrical, Polytetrafluoroethylene Resin, Nonrigid
MIL-I-23264	Insulators, Ceramic, Electrical and Electronic, General Specification for
MIL-I-24092	Insulating Varnishes and Solventless Resins for Application by the Dip Process
MIL-I-24391	Insulation Tape, Electrical, Plastic, Pressure-Sensitive
MIL-I-24768/-1, -2	Insulation, Plastics, Laminated, Thermosetting, Glass Cloth, Epoxy-Resin (GEE, GEB)
ANSI/NEMA FI 3	Calendered Aramid Papers Used for Electrical Insulation
NEMA MG 1	Motors and Generators
NEMA RE 2	Electrical Insulating Varnish
SAE AMS 3638	Tubing, Irradiated Polyolefin Plastic, Electrical Insulation Pigmented, Semi-Rigid, Heat-Shrinkable, 2 to 1 Shrink Ratio
SAE AMS 3653	Tubing, Electrical Insulation Standard Wall, Extruded Polytetrafluoroethylene (PTFE)
SAE AMS 3654	Tubing, Electrical Insulation Light Wall, Extruded Polytetrafluoroethylene (PTFE)
SAE AS 81765	Insulating Components, Molded, Electrical, Heat Shrinkable, General Specification

To warn people involved in these industrial activities about the presence of potentially dangerous situations, the manufacturers or suppliers of raw materials or components are obliged to provide clear and documented information about these dangers. This information shall be included in a safety data sheet (SDS) (formerly MSDS-Material Safety Data Sheet). The Hazard Communication Standard (HCS) (29 CFR 1910.1200(g)), revised in 2012, requires that the chemical manufacturer, distributor, or importer provide SDSs for each hazardous chemical to downstream users to communicate information on these hazards [10]. The issuing of the SDS became mandatory in 2015. Safety data sheets are an integral part of Regulation (EC) No 1907/2006 (REACH) (see Chapter 3, Section 3.3.2). In the category of material (substances) for which an SDS needs to be provided, the chemicals, batteries, gas containers intended for propane, butane or liquefied petroleum gas, metals in massive form, alloys, mixtures containing polymers, and mixtures containing elastomers, are included in general.

In the EU, Regulation (EC) No 1272/2008 on classification, labeling, and packaging of substances and mixtures, amending Regulation (EC) No 1907/2006 provide the requirements applicable to dangerous substances covered in the Directive 67/548/EEC [11].

A SDS is a document that contains information on the potential hazards (health, fire, reactivity, and environmental) and how to work safely, especially with a chemical product. It is an essential starting point for the development of a complete health and safety program. It also contains information on the use, storage, handling, and emergency procedures, all related to the hazards of the material. The SDS contains much more information about the material than the label. SDSs are prepared by the supplier or manufacturer of the material. It is intended to tell what the hazards of the product are, how to use the product safely, what to expect if the recommendations are not followed, what to do if accidents occur, how to recognize symptoms of overexposure, and what to do if such incidents occur.

Traditionally the intended readers of SDSs were occupational hygienists and safety professionals. Now the audience also includes employers, workers, supervisors, nurses, doctors, emergency responders, and workers. To ensure that SDS users can quickly find the information that they need, the information should be in an easy-to-read format and written in a clear, precise, and understandable manner.

Typically, the categories of information that must be present on an SDS include [10]:

1. *Material (substance) identification*: Material identifier (name), manufacturer and suppliers' names, addresses, e-mail, and emergency phone numbers; recommended use of the chemical (e.g., a brief description of what it actually does, such as being flame retardant) and any restrictions on use.
2. *Hazardous identification:* The hazard classification; hazard statement; precautionary statement; pictogram (see Table 4.5).
3. *Composition/information on ingredients:* Ingredient(s) contained in the product indicated on the SDS, including impurities and stabilizing additives. This section includes information on substances, mixtures, and all chemicals where a trade secret is claimed. Chemical name. Common name and synonyms; Chemical Abstracts Service (CAS) number and other unique identifiers.

4. *First aid measures and emergency procedures:* Specific first aid measures related to acute effects of exposure to the product; necessary first-aid instructions by relevant routes of exposure inhalation, skin and eye contact, and ingestion; description of the most important symptoms or effects, and any symptoms that are acute or delayed. First aid steps in the correct sequence; information to assist in planning for emergencies.
5. *Fire-fighting measures:* The temperature and conditions that can cause the chemical to catch fire or explode; means of extinction including the type of fire extinguisher required and information about extinguishing equipment that is not appropriate for a particular situation; personal protective equipment required for fire fighting.
6. *Accidental release measures:* Use of personal precautions—such as removal of ignition sources or providing sufficient ventilation—and protective equipment to prevent the contamination of skin, eyes, and clothing; emergency procedures, including instructions for evacuations, consulting experts when needed, and appropriate protective clothing; methods and materials used for containment, e.g., covering the drains and capping procedures; cleanup procedures, e.g., appropriate techniques for neutralization, decontamination, cleaning or vacuuming; adsorbent materials; and/or equipment required for containment/clean up.
7. *Handling and storage:* Safe handling practices and conditions for safe storage of chemicals; precautions for safe handling, including recommendations for handling incompatible chemicals, minimizing the release of the chemical into the environment, and providing advice on general hygiene practices, e.g., eating, drinking, and smoking in work areas is prohibited; recommendations on the conditions for safe storage, including any incompatibilities. Provide advice on specific storage requirements, e.g., ventilation requirements.
8. *Exposure control/personal protection:* The exposure limits, engineering controls, and personal protective measures that can be used to minimize worker exposure; OSHA permissible exposure limits (PELs), American Conference of Governmental Industrial Hygienists (ACGIH) threshold limit values (TLVs), and any other exposure limit used or recommended by the chemical manufacturer, importer, or employer preparing the safety data sheet, where available; appropriate engineering controls (e.g., use local exhaust ventilation, or use only in an enclosed system); Recommendations for personal protective measures to prevent illness or injury from exposure to chemicals, such as personal protective equipment (PPE) (e.g., appropriate types of eye, face, skin or respiratory protection needed based on hazards and potential exposure); any special requirements for PPE, protective clothing or respirators (e.g., type of glove material, such as PVC or nitrile rubber gloves; and breakthrough time of the glove material.
9. *Physical and chemical properties:* Appearance of the material: physical state, e.g., liquid; color; odor; specific gravity; vapor density; evaporation rate; melting point; freezing point; boiling range; flash point; vapor pressure; the higher possible concentration in the air; odor threshold–the lowest airborne concentration of a chemical that can be perceived by smell;

pH reflecting the corrosive or irritant nature of the substance; flammability–solid, gas; UEL–upper explosion limit or UFL–upper flammable limit; LEL–lower explosion limit or LFL–lower flammable limit[1]; relative density; solubility; partition coefficient; auto-ignition temperature; decomposition temperature; viscosity.

10. *Stability and reactivity data:* The chemical stability of the material and its reactions to light, heat, moisture, shock, and incompatible materials that must not be mixed or stored near each other; storage requirements based on the reactivity or instability of the material; the need for disposal before they become extremely reactive; description of the specific test data; description of any stabilizers that may be needed to maintain chemical stability; indication of any safety issues that may arise should the product change in physical appearance; list of all conditions that should be avoided, e.g., static discharge, shock, vibrations, or environmental conditions that may lead to hazardous conditions; list of any known or anticipated hazardous decomposition products that could be produced because of use, storage, or heating.
11. *Toxicological information:* Information on the likely routes of exposure: inhalation, ingestion, skin and eye contact; the SDS should indicate if the information is unknown; health effects short-term–acute and long-term–chronic, of exposure to the material; how the product is likely to enter the body and the organs which can be affected; the exposure to maximum average concentration limits: TWA–time-weighted average–safe exposure for an 8-hour workday or 48-hour workweek, STEL–short-term exposure limit–safe exposure for a period of up to 15 minutes[2]; C–ceiling–not be exceeded at any time, even for an instant[3]; the numerical measures of toxicity LD50; description of the symptoms including the symptoms associated with exposure to the chemical from the lowest to the most severe exposure; indication of whether the chemical is listed in relevant documents specifying induction of diseases.
12. *Ecological information:* Provides information to evaluate the environmental impact of the chemical(s) if it were released to the environment (non-mandatory).
13. *Disposal consideration:* Provides guidance on proper disposal practices, recycling or reclamation of the chemical(s) or its container, and safe handling practices (non-mandatory).
14. *Transport information:* Provides guidance on classification information for shipping and transporting of hazardous chemical(s) by road, air, rail, or sea (non-mandatory).
15. *Regulatory information:* Identifies the safety, health, and environmental regulations specific for the product that is not indicated anywhere else on the SDS (non-mandatory).

1. For more explanation about UEL, UFL, LEL, and LFL, refer to Chapter 6, Section 6.1.
2. The STEL is higher than the TWA. It may not be sustained more than four times a day.
3. The C is higher than the STEL.

16. *Other information:* Who is responsible for preparation and date of SDS preparation; other useful information also may be included here.

Information from SDS may be withheld to protect industries' rights to protect confidential business information. This information is referred to as trade secrets. The producer of the material can withhold the name and concentration of any ingredient and the name of relevant toxicological studies. Once a claim is filed to withhold information the material label must state the date the exemption filed and the claim registration number. The SDS must state that "an exemption has been granted on the date of X with registry number X for the following material hazards...." Doctors and nurses can access withheld information; however, this information remains confidential.

At the national levels lists of controlled materials are issued that need to be accompanied by an SDS. Normally, an SDS for a controlled material must not be more than 3 years old. If somebody is using a material that was bought more than 3 years ago, they need to contact the manufacturer or supplier to ask for a newer version of the SDS. If the product is new, and significant information becomes available before the 3 years have elapsed, the supplier is required to update the material label and SDS. Table 4.5 provides descriptions, characteristics, and pictograms of hazardous materials (hazmat) [10, 12].

The collection of LD50 and LC50 values is in the database Registry of Toxic Effects of Chemical Substances (RTECS) that is available by subscription on the internet.

Due to the risks of harm caused by hazardous materials, the use of risk management or system theoretic accident model and processes (STAMP) as methods to prevent accidents should be considered.

SDSs do not in themselves constitute a risk assessment, but are merely the starting reference point for such an assessment, as the SDS only gives information about the material itself—the risk must be assessed from use of the material in the work activity, including amounts, concentrations, and other information.

The steps that need to be taken when a risk assessment is conducted to prevent or control exposure to substances hazardous to health are [13]:

- Identify hazards intrinsic to substances to be used (SDS);
- Assess the risks to health arising from the use of hazardous substances in the work activity;
- Decide what precautions and control measures are necessary to reduce the risk;
- Implement the control measures;
- Ensure control measures are used and maintained;
- Monitoring exposure of users (if necessary);
- Consider whether biological monitoring and/or health surveillance is appropriate or required;
- Ensure the users have sufficient information, instruction, and training to perform the work safely and competently;

Table 4.5 List of Hazardous Materials

Class of Hazardous Materials (HazMat)	*Description*	*Characteristics*	*Pictogram*
Class 1	Explosive	Division 1.1: Substances and articles that have a mass explosion hazard Division 1.2: Substances and articles that have a projection/blast hazard but not a mass explosion hazard Division 1.3: Substances and articles that have a fire hazard and either a minor blast hazard or a minor projection hazard or both, but not a mass explosion hazard Division 1.4: Substances and articles that are classified as explosives and present a major fire hazard Division 1.5: Blasting agents Division 1.6: Extremely insensitive explosives	 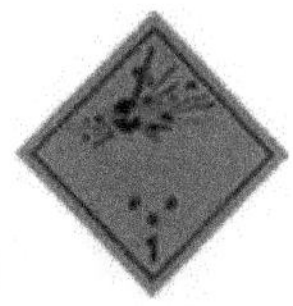
Class 2	Gases	Nonflammable Gas In packaging exerts absolute pressure of 280 kPa (41 psi) or greater at 20°C (68°F) Asphyxiant gases: gases that dilute or replace the oxygen normally in the atmosphere Oxidizing gases: gases that may, generally by providing oxygen, cause or contribute to the combustion of other material more than air does Flammable gas at 20°C (68°F) or less At 101.3 kPa (14.7 psi) of pressure: Is ignitable when in a mixture of 13% or less by volume with air Has a flammable range with the air of at least 12% points regardless of the lower flammable limit.	 
Class 3	Flammable or combustible liquids	Flammable Liquids Liquids that have a flashpoint of less than 60°C and are capable of sustaining combustion.	

- Ensure adequate procedures are in place to deal with any emergency situation that may foreseeably arise.

Some generic SDSs are available on databases accessible via the internet; the information should be treated with caution as the generic material (substance) may not be identical to the substance you have, and this is particularly important where a hazardous preparation (a mixture of substances) is concerned. In such cases, you should always obtain the dedicated product SDS from the supplier.

Table 4.5 (continued)

Class of Hazardous Materials (HazMat)	*Description*	*Characteristics*	*Pictogram*
Class 4	Flammable solids	Flammable Solids, Self-Reactive Substances, and Solid Desensitized Explosives Solids which, under conditions encountered in transport, are readily combustible or may cause or contribute to fire through friction; self-reactive substances which are liable to undergo a strongly exothermic reaction; solid desensitized explosives which may explode if not diluted sufficiently. Substances Liable to Spontaneous Combustion Substances that are liable to spontaneous heating under normal conditions encountered in transport, or to heating up in contact with air, and being then liable to catch fire. Substances that in Contact with Water Emit Flammable Gases Substances which, by interaction with water, are liable to become spontaneously flammable or to give off flammable gases in dangerous quantities.	  
Class 5	Oxidizer or organic peroxide	Oxidizing Substances Substances which, while in themselves not necessarily combustible, may generally by yielding oxygen, cause or contribute to the combustion of other material. Oxidizing gases, Category 1 Oxidizing liquids, Categories 1, 2, 3 Oxidizing solids, Categories 1, 2, 3 Organic Peroxides Organic substances that contain the bivalent –O–O– structure and may be considered derivatives of hydrogen peroxide, where one or both of the hydrogen atoms have been replaced by organic radicals.	 
Class 6	Poison; toxic	Substances that are presumed to be toxic or corrosive to humans as to pose a hazard to health, with an LD50* value ≤ 300 mg/kg (oral) or ≤ 1,000 mg/kg (dermal) or an LC50† value ≤ 4000 ml/m3 (inhalation of dusts or mists).‡ Poison Gas at 20°C (68°F) or less Pressure of 101.3 kPa (14.7 psi) Poses a health hazard to humans in transportation Presumed to be toxic to humans Acute toxicity (oral, dermal, inhalation), Categories 1, 2, 3	
		Biohazardous Substances	

Table 4.5 (continued)

Class of Hazardous Materials (HazMat)	*Description*	*Characteristics*	*Pictogram*
Class 7	Radioactive	Substances or a combination of substances that emit ionizing radiation (uranium, plutonium, etc.)	
Class 8	Corrosive	Substances that cause full thickness destruction of intact skin tissue in the exposure time of fewer than 4 hours	
		Substances that exhibit a corrosion rate of more than 6.25 mm per year on either steel or aluminum surfaces at 55°C	
		Corrosive to metals, Category 1	
		Skin corrosion, Categories 1A, 1B, 1C	
		Serious eye damage, Category 1	
Class 9	Miscellaneous	Hazardous substances that do not fall into the other categories (asbestos, air-bag inflators, self-inflating life rafts, dry ice, etc.)	

* LD = lethal dose. LD50 is the amount of a material, given all at once, which causes the death of 50% (one-half) of a group of test animals. The LD50 is one way to measure the short-term poisoning potential (acute toxicity) of a material.
† LC = lethal concentration. LC50 values usually refer to the concentration of a chemical in the air that kills 50% of test animals during the observation period.

References

[1] Taylor, J. W., K. W. McFarlan, and S. Ellis, *Corrosion Reduction in Power Distribution, Eaton's Crouse-Hinds Division*, IEEE PCIC-2017.

[2] IEC 62368-1 Audio/Video, Information and Communication Technology Equipment–Part 1: Safety Requirements, IEC, Geneva, 2018.

[3] ASM Handbook, Vol. 13A, *Corrosion: Fundamentals, Testing, and Protection, Materials Park*, OH: ASM, 2003.

[4] Tavakoli, M., "The Adhesive Bonding of Medical Devices," *Medical Device & Diagnostic Industry*, June 2001.

[5] https://www.lboro.ac.uk/microsites/mechman/research/ipm-ktn/pdf/Technology_review/adhesives-in-electronics.pdf.

[6] https://forum.digikey.com/t/surface-mount-abbreviations-smd-smt-sma-smc-and-smp/410.

[7] IEC 60085 Electrical Insulation–Thermal Evaluation and Designation, Geneva, 2004.

[8] Laughton, M. A., and D. F. Warne (eds.), *Electrical Engineer's Reference Book*, 16th Edition, Newnes, 2003.

[9] Fink, D. G., and W. H. Beaty (eds.), *Standard Handbook for Electrical Engineers*, McGraw-Hill, 1998.

[10] OSHA, Hazard Communication Standard 29 CFR 1910.1200(g) (HCS), Globally Harmonized System (GHS), 2012.

[11] EU Directive 67/548/EEC, EU Directive Relating to the Classification, Packaging and Labelling of Dangerous Substances, 1967; replaced by Regulation (EC) No 1272/2008.

[12] Manchester Metropolitan University, *Control of Substances Hazardous to Health (COSHH), Guidance Notes on Risk Assessment*, 2012.

[13] SDS–OSHA, 2012.

Selected Bibliography

Campo, E. A., *Selection of Polymeric Materials: How to Select Design Properties from Different Standards*, Norwich, NY: William Andrew, 2007.

https://www.ccohs.ca/oshanswers/hsprograms/hazardous_energy.html.

http://www.engineeringtoolbox.com/nema-insulation-classes-d734.html.

https://www.masterbond.com/properties/sterilization-resistant-adhesives.

https://www.niehs.nih.gov/health/topics/agents/flame_retardants/index.cfm.

https://www.thomasnet.com/articles/chemicals/corrosion-resistant-coatings/.

ISO 871, Plastics–Determination of Ignition Temperature Using a Hot-Air Furnace, Geneva, 2006.

Lyon, R. E., "Plastics and Rubber," *Handbook of Materials for Fire Protection*, C. A. Harper (ed.), New York: McGraw-Hill, 2004.

Pilchik, R., "Adhesive and Non-Adhesive Systems for Health-Care Packaging," *Medical Device Technology*, Vol. 11, No. 3, April 2000, pp. 8–10.

CHAPTER 5

Safety of Electronic Product Radiation Sources

This chapter deals with electronic product radiation (nonionizing or ionizing radiation sources and sonic waves) and compliance and safety aspects related to products using this type of radiation.

Defining the electronic product radiation [1] we can refer to: (A) any nonionizing or ionizing electromagnetic or particulate radiation, or (B) any sonic, infrasonic, or ultrasonic wave that is emitted from an electronic product as the result of the operation of an electronic circuit in such product.

The nonionizing and ionizing radiations are included in the generic name of electromagnetic radiation (one of the many ways that energy travels through space, in which the strength of the electric and the magnetic field vary) and can be described in terms of a stream of massless particles (photons) contains a certain amount of energy and traveling in a wavelike pattern at the speed of light (on the earth, it is approximately 3×10^8 m/s). Ionizing radiation is a form of energy that acts by removing electrons from atoms and molecules of materials that include air, water, and living tissue. Ionizing radiation can travel unseen and pass through these materials. Nonionizing radiation exists all around us from many sources (radiofrequency, microwave, infrared, ultraviolet) and can heat the substances. These radiations are located to the right of ionizing radiation on the electromagnetic spectrum (see Figure 5.1). The dividing line between ionizing and nonionizing radiation occurs in the ultraviolet part of the electromagnetic spectrum. The sonic waves as a variation in air pressure are considered to travel at the speed of sound (in dry air at 20°C, which it is approximately 343 m/s).

The different types of electromagnetic radiations are defined by the amount of energy found in the photons. Radio waves have photons with low energies, microwave photons have a little more energy than radio waves, and infrared photons have more than visible ultraviolet X-rays, and the most energetic of all, gamma-rays. The generation of electromagnetic radiation can be classified into two categories: (1) systems or processes that produce radiation covering a broad continuous spectrum of frequencies and (2) those that emit (and absorb) radiation of discrete frequencies, which are characteristic of particular systems. The Sun with its continuous spectrum is an example of the first, while a radio transmitter tuned to one frequency exemplifies the second category.

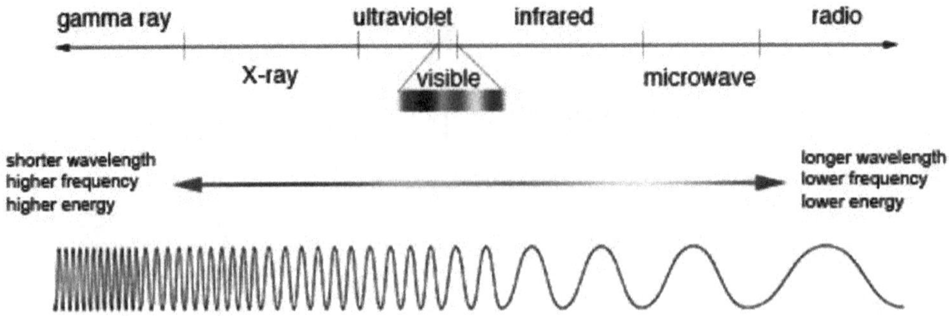

Figure 5.1 Electromagnetic spectrum [2].

All these radiations expressed in terms of energy, wavelength, and frequency are summarized in the electromagnetic spectrum (Figure 5.1) and the sound waves spectrum as presented in Table 5.1.

5.1 Nonionizing Radiation Sources

Nonionizing radiation (NIR) includes all radiations and fields of the electromagnetic spectrum that do not normally have sufficient energy to produce ionization in the matter; characterized by energy per photon less than about 12 eV, wavelengths greater than 100 nm, and frequencies lower than 3 PHz (3×10^{15}) Hz. It consists of extremely low-frequency radiation (ELF), radio wave (RW), microwave (MW), IR, visible light (VL), and UV light.

Various nonionizing radiation survey meters are available for measuring electromagnetic fields. The frequency ranges covered by instruments are: 10 Hz to 300 kHz, 0.5 MHz to 6 GHz, 6 GHz to 40 GHz, and the 2.45 GHz microwave oven frequency. These instruments are capable of measuring the electric field strength (E-field) (e.g., electro field strength meter), magnetic field strength (H-field) (e.g., Tesla meter or Gauss meter), or both depending on the instrument. An electromagnetic field meter (e.g., RF strength meter) can detect either static (DC) permanent (rare-earth) magnetic or dynamic (AC) electromagnetic fields (EMFs). Generally, the EMFs meters are equipped with a spherical triaxial (X, Y, Z directions) isotropic sensor. Depending on the instrument, electromagnetic field strengths from electrical power lines, transformers and wiring for overhead lighting, solar panels, transformers, RF induction heaters, RF heat sealers, radio and television transmitters, microwave ovens, and other sources can be measured.

5.1.1 EMF Radiation

Electromagnetic radiation (EMR) consists of waves of electric and magnetic fields in the frequency range from 3 Hz to 300 GHz, but not all configurations of electric and magnetic fields are described as radiation. Certainly, static fields, like the Earth's magnetic field, are not called radiation.

Electric fields are a vector quantity created by differences in voltage: the higher the voltage, the stronger the resultant field will be. If a voltage V is supplied across a specified distance r, then the electric field formula is $E = V/r$. Magnetic fields are

Table 5.1 Electromagnetic and Sound Waves Spectrums

A. Electromagnetic Spectrum

Type of Electromagnetic Radiation		*Wavelength*	*Frequency*	*Energy per Photon*	*Application*
Extremely low frequency	ELF	10,000–100,000 km	3 - 30 Hz	12.4–124 feV	Communication with submarine
AC power grid		5,000 km or 6,000 km	50 or 60 Hz	206 peV or 248 peV	High-voltage power transmission lines; induction furnaces, electric appliances, electrical wiring, domestic power lines
Super low frequency	SLF	1,000–10,000 km	30 - 300 Hz	124 feV–12.4 peV	Communication with submarine
Ultra-low frequency	ULF	100–1,000 km	0.3–3 kHz	1.24–12.4 peV	Communications in mines; commercial airplanes (400 Hz)
Very low frequency	VLF	10–100 km	3–30 kHz	12.4–124 peV	Communication with submarine; electric railway
Low frequency	LF	1–10 km	30–300 kHz	1.24 neV–124 peV	Long-distance communications; induction heater; RFID tags (125–134 kHz)
Medium frequency	MF	100m–1 km	0.3–3 MHz	1.24–12.4 neV	Medium wave-AM radio; maritime ship-to-shore communication; transoceanic air traffic control
		187–555m	540–1,600 kHz 0.5–4 MHz	2.2–6.6 neV	Broadcast AM- radio electrosurgery
High frequency	HF	10–100m	3–30 MHz	12.4 neV–0.124 μeV	Shortwave AM radio; aviation communication, weather stations, amateur radio; RFID tags (13.56 MHz)
Very high frequency	VHF	1–10m	30–300 MHz	0.124–1.24 μeV	
		3.4–5.5m	54–88 MHz		Broadcast VHF1-TV
		2.8–3.4m	88–108 MHz	3.64 -4.43 μeV	Broadcast FM-radio
		1.35–1.7m	176-222 MHz		Broadcast VHF2-TV
Ultra-high frequency	UHF	100 mm–1m	300 MHz–3 GHz	<1.24 μeV	RFID tags (433 MHz, 860–960 MHz).
			470 MHz–1 GHz		Broadcast UHF-TV EU: 470–862 MHz Australia: 526–820 MHz Japan: 470–770 MHz
Super high frequency	SHF	10–100 mm	3–30 GHz	12.4 μeV	Wi-Fi.
Extremely high frequency	EHF	1–10 mm	30–300 GHz	124 μeV	mmWave, radio-astronomy, military
Microwave	MW	1 mm–1m	300 MHz–300 GHz	1.24 μeV–1.24 meV	Cooking, radar, cellphones

created when electric current flows: the greater the current, the stronger the magnetic field. An electric field will exist even when there is no current flowing. If current does flow, the strength of the magnetic field will vary with power consumption, but the electric field strength will be constant.

Besides natural sources (sun, electric charges in the atmosphere, earth's magnetic field, etc.), the electromagnetic spectrum also includes fields generated by human-made sources (power lines are associated low-frequency electromagnetic

Table 5.1 (continued)

Type of Electromagnetic Radiation		Wavelength	Frequency	Energy per Photon	Application
Infrared	FIR far*	15 μm–1 mm 50 μm–1 mm †	300 GHz (300 x 10^9) Hz to 430 THz (430 x 10^{12}) Hz	1.24 meV–1.7 eV	Solid state physics, space-time spectroscopy
	LWIR* long (Thermal Infra-red)	8–15 μm			Thermal imaging; night vision; IR laser
	MIR mid *	3–8 μm 3-50 μm ‡			Radiation thermometers
	SWIR§ short (reflected IR)	1.4–3 μm			Long distance communication; Fiber optic communication (especially 1,530–1,565 μm)
	NIR near ¶ (reflected IR)	700 nm–1.4 μm 700 nm–3 μm #			Optic fiber communication
Visible		400 (violet)–700 (red) nm	400-750 THz	1.7–3.2 eV	
Ultraviolet	UV-A	320–400 nm	750 THz (750 x 10^{12}) to 3 PHz (3 x 10^{15}) Hz	3.2 eV–12.4 eV	Sterilization, esthetic treatment
	UV-B	280–320 nm			Bacterial purification
	UV-C	100–280 nm			Bacterial purification
X-ray		10 pm–100 nm	3 PHz (3 x 10^{15}) Hz to 30 EHz (3 x 10^{19}) Hz	12 eV–124 keV	Diagnostic imaging
Gamma ray		< 10 pm	> 30 EHz (3 x 10^{19}) Hz	>124 KeV	Sterilization; medical therapy

B. Sound Waves Spectrum

Type of Sound Wave	Symbol	Wavelength	Frequency	Application
Infrasound	IS	>17m	< 20 Hz	Industrial, atmospheric studies, seismology, detection of volcanic eruptions
Audible sound		17m–17 mm	20 Hz–20 kHz	Verbal communication, music, environment
Ultrasound	US	17 mm–1.7 μm	20 kHz–200 MHz	Medical, industrial, laboratory

† Far-infrared according to ISO 20473.
‡ Mid-infrared according to ISO 20473.
§ Band IR-B according to International Commission on Illumination.
¶ Band IR-A according to International Commission on Illumination.
# Near-infrared = NIR+SWIR according to ISO 20473.
* Band IR-C according to International Commission on Illumination.

fields; higher-frequency radio waves are used to transmit information—whether via TV antennas, radio stations, mobile phone base stations, LED lighting units, lasers, or X-ray equipment).

One of the main characteristics that defines an EMF is its frequency or its corresponding wavelength (the higher the frequency, the shorter the wavelength).

5.1.1.1 EMF Exposure

Environmental exposure to man-made electromagnetic fields has been steadily increasing as growing electricity demand, ever-advancing technologies, and changes in social behavior have created more and more artificial sources. Everyone is exposed to a complex mix of weak electric and magnetic fields, both at home and at work, from the generation and transmission of electricity, domestic appliances, and industrial equipment, to telecommunications and broadcasting (see also Section 5.1.4).

Low-frequency electric fields influence the human body just as they influence any other material made up of charged particles. When electric fields act on conductive materials, they influence the distribution of electric charges at their surface. They cause current to flow through the body to the ground. Low-frequency magnetic fields induce circulating currents within the human body. The strength of these currents depends on the intensity of the outside magnetic field. If sufficiently large, these currents could cause stimulation of nerves and muscles or affect other biological processes.

A wide range of environmental influences causes biological effects, but a biological effect does not mean a health hazard.

Research has been done on possible biological effects (damage to tissues due to heat) of exposure. According to the U.S. National Institute for Occupational Safety and Health (NIOSH) research on protecting workers from possible health hazards is due to:

- Radio frequency (RF);
- ELF;
- Static magnetic fields.

The technical document NIOSH Publication No. 98-154 (1998) is a reference manual for industrial hygienists, safety personnel, and researchers who measure occupational exposures to RF, ELF, and static EMF.

RF radiation comes from sources such as radios, cellular phones, the processing and cooking of foods, heat sealers, vinyl welders, high-frequency welders, induction heaters, flow solder machines, communications transmitters, radar transmitters, ion implant equipment, microwave drying equipment, sputtering equipment, and glue curing.

Generating, transmitting, distributing, and using electricity expose people to ELF radiation. Power lines, household wiring, and any electrical devices can generate ELF radiation. How much a person is exposed to electromagnetic radiation depends on the strength of the electromagnetic field, distance from the source of the field, and the exposure length of time. The highest exposure occurs when the person is very close to a source putting out a strong field and stays there for a long period. ELF radiation has even lower energy than other types of nonionizing radiation like radiofrequency radiation, visible light, and infrared.

Examples of devices that produce static magnetic fields are MRI equipment, DC power lines, DC welding, electrolysis devices, magnets, and electric furnaces.

We are exposed to low levels of nonionizing radiation every day. Exposure to intense, direct amounts of nonionizing radiation may result in damage to tissue due to heat.

The International Commission on Nonionizing Radiation Protection (ICNIRP) develops health criteria documents on NIR. Each document includes an overview of the physical characteristics, measurement and instrumentation, sources, applications of NIR, a thorough review of the literature on biological effects, and an evaluation of the health risks of exposure to NIR. These health criteria have provided the scientific database for the subsequent development of exposure limits and codes of practice relating to NIR [3].

Two classes of guidance are provided by the ICNIRP: Basic restrictions, which are restrictions on exposure to time-varying electric, magnetic, and electromagnetic fields that are based directly on established health effects, and reference levels, which are levels provided for practical exposure assessment purposes to determine whether the basic restrictions are likely to be exceeded. These levels are changed in time, on the basis of the statistics, improvement of measurement methods, and identification of a new correlation between the intensity of the radiations with the human body and environment.

Different scientific bases were used in the development of basic exposure restrictions for various frequency ranges [3]:

Between 3 Hz and 10 MHz, basic restrictions are provided on current density to prevent effects on nervous system functions.

Between 100 kHz and 10 GHz, basic restrictions on SAR are provided to prevent whole-body heat stress and excessive localized tissue heating. In the 100-kHz to 10-MHz range, restrictions are provided on both current density and SAR.

Between 10 and 300 GHz, basic restrictions are provided on power density to prevent excessive heating in tissue at or near the body surface.

The ICNIRP notes that the industries causing exposure to electric and magnetic fields are responsible for ensuring compliance with all aspects of the guidance.

Measures for protection must be implemented when exposure in the workplace results in the basic restrictions being exceeded include, but are not limited to:

- Engineering (good safety design, use of interlocks or similar health protection mechanisms) and administrative (limitations on access and the use of audible and visible warnings) controls, to reduce device emissions of fields to acceptable levels;
- Personal protection programs (protective clothing, etc.);
- Medical surveillance;
- Establishment and implementation of rules that will prevent interference with medical electronic equipment and devices (e.g., implantable cardiac pacemakers, etc.), detonation of electro-explosive devices (detonators), fires and explosions resulting from ignition of flammable materials by sparks caused by induced fields, contact currents, or spark discharges.

A common measure of exposure to RF radiation of the human body is the SAR, as the rate of energy absorption in tissue, measured in watts per kilogram (W/Kg) of tissue. The U. S. and European regulations recommend SAR limits as following: FCC limit is 1.6 W/kg averaged over 1 gram of tissue and for the European Union the limit is 2.0 W/kg averaged over 10g of tissue. Practically, the SAR is the dosimetry measure that has been widely adopted at frequencies above 100 kHz. Table 5.2 consists of basic restrictions for electromagnetic field exposure from 100 kHz to 300 GHz, for specific averaging time intervals as specified in ICNIRP 2020 Guidance [3].

In the European Union, the Recommendation 1999/519/EC "Limitation of exposure of the general public to electromagnetic fields (0 Hz to 300 GHz)" and the Directive 2013/35/EU "The minimum health and safety requirements regarding the exposure of workers to the risks arising from physical agents (electromagnetic fields)," represent the documents that specify the limits for exposure of the public and workers to nonionizing electromagnetic radiation [4]. According to these documents the restricted level of the power density S in the frequency range 10–300 GHz is 10 W/m^2, the magnetic flux density at 0 Hz (DC current) is restricted at 40 mT, and the RMS current density basic restriction is presented in Table 5.3.

Due to the importance of the subject many standards developers and regulation bodies have issued standards and guidelines referring to EMF exposure to protect human health by establishing limits for overexposure to EMF levels present in our environment (see Table 5.4).

These standards include nonhazardous levels of EM fields, methods of measurement, type of test instruments, and means of protection.

Although a large safety factor is applied to the level known to cause a health consequence, finding a statistical association between some agent and a specific disease does not mean that the agent caused the disease. Similarly, the absence of health effects could mean that there really are none, however, it could also signify that an existing effect is undetectable with present methods. Results of diverse studies must be considered together before drawing conclusions about possible health

Table 5.2 SAR for Electromagnetic Field Exposure from 100 kHz to 300 GHz

		SAR [W/Kg]		
		Over 30 Minutes	*Over 6 Minutes*	*Over 6 Minutes*
Exposure	*Frequency Range*	*Whole-Body Average*	*Local Head/Torso*	*Local Limb*
Occupational	100 kHz to 6 GHz	0.4	10	20
Occupational	> 6 GHz to 300 GHz	0.4	N/A	N/A
General public	100 kHz to 6 GHz	0.08	2	4
General public	> 6 GHz to 300 GHz	0.08	N/A	N/A

Table 5.3 Basic Restriction of the RMS Current Density

Frequency (f)	>0–1 Hz	1–4 Hz	4 Hz–1 KHz	1 KHz–10 MHz
RMS current density [mA/m²]	8	8 / f	2	f/500

risks of a suspected environmental hazard. Consistent evidence from these very different types of studies increases the degree of certainty about a true effect.

Despite the risks involved by the use of electromagnetic radiation, many human-made sources were developed due to the useful application of EMR in areas as telecommunication, health technologies, and physical investigation. We need to refer to EM radiation in terms of both the positive and negative sides of the effects. Without progress in the positive applications of EM radiations, it would not possible to get the outstanding results seen in the above-mentioned areas of science and technology.

5.1.2 MRI

MRI is a widely used noninvasive technique for taking detailed pictures of the anatomy and physiological processes inside the human body.

Using pulsed RF on the range of 10 to 300 MHz, MRI technology is an example of medical diagnostic exams that do not involve exposure to ionizing radiation. Protons (hydrogen nuclei) in a water molecule of a specific tissue change the orientation of their own rotational polarization vector when placed in a magnetic field. This phenomenon is the basis for nuclear magnetic resonance (NMR), or as is also called, MRI.

MRI technology can be described as follows:

- In a tissue located in a magnetic field and excited by a pulsed radio frequency signal, a nuclear precession occurs of the tissue's own rotational polarization vector, until magnetic resonance (meaning resonant absorption of electromagnetic energy by an ensemble of atomic particles situated in a magnetic field [5]) occurs.
- The NMR occurs when the pulsed radio frequency signal becomes equal with the Larmor precession frequency (specific to each nucleus) determined by the gyromagnetic ratio (a nuclei-specific constant) and the strength of the applied magnetic field.
- To arrive at magnetic resonance, one needs to use a stronger magnetic field. To achieve this goal, in addition to the static magnetic field (generated by the basic magnet), there is a superimposed a gradient (time-varying field) created by gradient coils located in the magnet (actual medical applications have field strengths between 0.2–4 Tesla).
- After obtaining the magnetic resonance, the RF pulses are stopped and the relaxation rotation of the polarization vector of the nuclei (return of the proton spin to equilibrium) starts, which induces an electromagnetic force in a detector coil, generating the signal processed into an image.

Table 5.4 Standards and Guidelines Referring to EMF Exposure

Standard	*Name*
EN 50360	Product standard to demonstrate the compliance of wireless communication devices, with the basic restrictions and exposure limit values related to human exposure to electromagnetic fields in the frequency range from 300 MHz to 6 GHz: devices used next to the ear.
EN 50364	Product standard for human exposure to electromagnetic fields from devices operating in the frequency range 0 Hz to 300 GHz, used in electronic article surveillance (EAS), in radio frequency identification systems (RFID), and similar applications.
EN 50385	Product standard to demonstrate the compliance of base station equipment with radiofrequency electromagnetic field exposure limits (110 MHz–100 GHz), when placed on the market.
EN 50566	Product standard to demonstrate the compliance of wireless communication devices with the basic restrictions and exposure limit values related to human exposure to electromagnetic fields in the frequency range from 30 MHz to 6 GHz: hand-held and body mounted devices in close proximity to the human body.
EN 50663	Generic standard for assessment of low power electronic and electrical equipment related to human exposure restrictions for electromagnetic fields (10 MHz–300 GHz).
FCC-OET Bulletin 65	Evaluating compliance with FCC guidelines for human exposure to radiofrequency electromagnetic fields.
ICNIRP (2020)	Guidelines for limiting exposure to EMFs (100 KHz to 300 GHz).
IEEE C95.1	IEEE standard for safety levels with respect to human exposure to electric, magnetic, and electromagneticfFields, 0 Hz to 300 GHz.
IEEE C95.3	Recommended practice for measurements and computations of radio frequency electromagnetic fields with respect to human exposure to such fields, 100 kHz-300 GHz.
IEEE C95.7	Recommended practice for radio frequency safety programs, 3 kHz to 300 GHz.
IEC 62471	Photobiological safety of lamps and lamp systems.
IEC 61786-1	Measurement of DC magnetic, AC magnetic and AC electric fields from 1 Hz to 100 kHz with regard to exposure of human beings–Part 1: Requirements for measuring instruments.
IEC 61566	Measurement of exposure to radio-frequency electromagnetic fields. Field strength in the frequency range 100 kHz to 1 GHz.
IEC 61786-2	Measurement of DC magnetic, AC magnetic and AC electric fields from 1 Hz to 100 kHz with regard to exposure of human beings–Part 2: Basic standard for measurements.
IEC/IEEE 62209-1528	Measurement procedure for the assessment of specific absorption rate of human exposure to radio frequency fields from hand-held and body-worn wireless communication devices–Part 1528: Human models, instrumentation, and procedures (frequency range of 4 MHz to 10 GHz).
IEC/EN 62209-1	Human exposure to radio frequency fields from hand-held and body-mounted wireless communication devices. Human models, instrumentation, and procedures. Procedure to determine the SAR for hand-held devices used in close proximity to the ear (frequency range of 300 MHz to 3 GHz).
IEC 62209-3	Measurement procedure for the assessment of specific absorption rate of human exposure to radio frequency fields from hand-held and body-mounted wireless communication devices–Part 3: Vector measurement-based systems (frequency range of 600 MHz to 6 GHz).
IEC/EN 62232	Determination of RF field strength, power density and SAR in the vicinity of radiocommunication base stations for the purpose of evaluating human exposure.

Table 5.4 (continued)

Standard	*Name*
IEC 62233	Measurement methods for electromagnetic fields of household appliances and similar apparatus with regard to human exposure.
IEC/EN 62479	Assessment of the compliance of low-power electronic and electrical equipment with the basic restrictions related to human exposure to electromagnetic fields (10 MHz to 300 GHz).
IEC 62311	Assessment of electronic and electrical equipment related to human exposure restrictions for electromagnetic fields (0 Hz–300 GHz).
IEC 62577	Evaluation of human exposure to electromagnetic fields from a stand-alone broadcast transmitter (30 MHz–40 GHz).
IEC TR 62669	Case studies supporting IEC 62232 - Determination of RF field strength, power density and SAR in the vicinity of radiocommunication base stations for the purpose of evaluating human exposure.

MRI is used in medicine as a method for determining the structures of biological molecules on the area of the body being investigated (brain, spine, chest, abdomen, and other organs). Another application of MRI is in laboratory spectroscopy investigations. The main advantage of MRI when compared with other imaging tools (e.g., X-ray computed tomography (CT) scan) is that it does not require exposure of the human body to ionizing radiation, and such, is very safe [6].

The IEC 60601-2-33 is the main safety standard applicable to the medical diagnostic MRI systems covering the safety of patients examined with these systems and personnel involved with its operation [5].

The MRI equipment normally operates in an RF shielded room that is a controlled access area with special safety measures due to the high intensity of the magnetic fields (static and gradient) and pulsed RF radiation. The basic magnets used can be of permanent, resistive, or superconducting (used for magnetic field higher than 0.5 Tesla) type. A schematic structure of an MRI system is presented in Figure 5.2. Other standards related to the use of MRI are presented in Table 5.5.

During the usage of MRI equipment, the user needs to consider many safety related aspects. Below are recommendations to be considered for providing the safe use of MRI devices:

- Due to the attraction power of the strong magnetic field it is very dangerous to introduce objects containing iron or other magnetically active materials parts sensitive to high-intensity magnetic fields (e.g., credit cards), or persons with medical implants (electrical, metallic orthopedic, etc.) into an MRI room. Warning and prohibition signs as specified in the IEC 60601-2-33 amd need to be used.
- For equipment with superconducting magnets, in order to prevent accidents (transition of the electrical conductivity of a coil that is carrying a current from a superconducting state to normal conductivity, resulting in rapid boil-off of fluid cryogen and decay of the magnetic field) one needs adequate provisions for supplies of cryogenic liquid (helium and nitrogen). Leaking helium or nitrogen gas will displace the oxygen. An ambient air oxygen

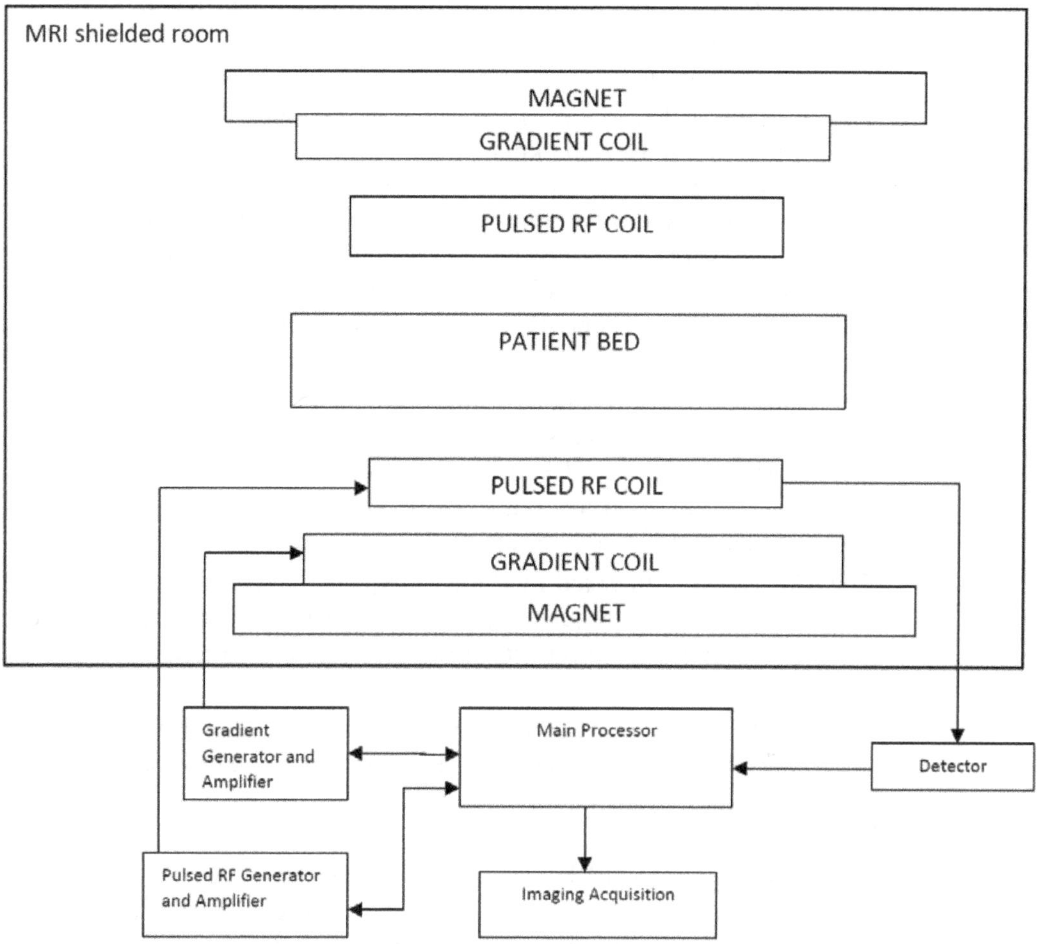

Figure 5.2 Structure of an MRI system.

Table 5.5 Standards for MRI Equipment

Standard	*Title of Standard*
ASTM F2052	Standard test method for measurement of magnetically induced displacement force on medical devices in the magnetic resonance environment.
ASTM F2119	Standard test method for evaluation of MR image artifacts from passive implants.
ASTM F2182	Standard test method for measurement of radio frequency induced heating on or near passive implants during magnetic resonance imaging.
ASTM F2213	Standard test method for measurement of magnetically induced torque on medical devices in the magnetic resonance environment.
ASTM F2503	Standard practice for marking medical devices and other items for safety in the magnetic resonance environment.
IEC 60601-2-33	Medical electrical equipment–Part 2-33: Particular requirements for the basic safety and essential performance of magnetic resonance equipment for medical diagnosis.
ISO/TS 10974	Assessment of the safety of magnetic resonance imaging for patients with an active implantable medical device.

concentration of less than 17% to 18% is not sufficient for human respiration [5].

- The high rates of change of currents passing through the gradient coils in a static magnetic field produce vibrations in the audible frequency range. These are often manifested as loud knocking high-pressure sounds that are hazardous for hearing. This is the reason hearing protection for an unweighted peak sound pressure level of a maximum 140 dB referenced to 20 μPa is in any accessible area, and must be used during an MRI procedure. For exposure of persons to noise, a suitable warning sign is specified in the ISO 7731 "Ergonomics–Danger signals for public and work areas–Auditory danger signals."
- Another safety concern is the heating of the patient [5]. Although heating by the induced electric currents is presently low, it is additive to that caused by the RF exposure in the MRI equipment and the concern relates to the combined effect of the two heating sources. The radio frequency induced heat load can be directly related to the SAR.
- The SAR is also influenced by the RF frequency (increasing approximately as the square of the frequency), the type and number of radio frequency pulses, the duration and repetition rate of pulses, and the type of coil used for transmission.
- The auxiliary equipment (physiological monitoring and RF transmit coils) that has not been specifically tested and approved for use in the environment of the MRI equipment may result in burns or other injuries to the patient.
- According to the ASTM F2503 standard, it is considered MR safe that an item poses no known hazards in all specified MR environments. Such items include nonconducting, nonmagnetic items (e.g., plastic and glass). An item may be determined to be MR safe by providing a scientifically based rationale rather than test data [7].
- The conditions that define a specified MR environment include field strength, spatial gradient, dB/dt (time rate of change of the magnetic field), RF fields, and specific absorption rate (SAR).
- MRI equipment must not be used when the temperature is greater than 24°C or the humidity is greater than 60%.
- In some countries, legislations may exist that cover occupational limits for exposure to static magnetic fields and time varying magnetic fields. The U.S. FDA considers static fields below 4T to be a nonsignificant risk.
- Permanent magnets cannot be deenergized in case of emergency.
- MRI equipment comprised of a superconducting magnet or resistive magnet must be provided with an emergency shut down of the unit.
- A facility must be provided to allow the operator to immediately stop the scan by interrupting the power to the gradient system and that to the RF transmit coil.

5.1.3 RF Radiation

RF radiation is a type of nonionizing radiation consisting of radio waves (RW) and microwaves (MW), covering the electromagnetic spectrum in the frequency ranges RW: 3 kHz–300 GHz, and MW: 300 MHz–300 GHz.

Man-made RF radiation sources include radio waves (broadcasting radio and television signals, Wi-Fi networks, Bluetooth® devices, transmitting signals from cordless telephones, mobile phones, and cellular base stations, satellite phones, and two-way radios) and microwave radiation (microwave ovens, microwave diathermia, and radar) [8].

If RF radiation is absorbed by the body in large amounts, it can produce heat. This can lead to burns and body tissue damage.

5.1.3.1 RW

The RW radiation covers the ranges of radio broadcasting, television broadcast, digital video, and so on.

The CE regulations cover a broad range of products (cell phone handsets and cell base stations, licensed two-way land mobile radios and their base stations, microwave backhaul transmission systems, emergency beacons, radar, radio astronomy, and many others) and set the requirements for the essential health and safety as well for product operability. There are over 20 different new approach directives and many other non-CE directives and EU regulations. manufacturers need to be compliant with all the applicable requirements.

The EU Directive 2014/53/EU (RED) relating to availability on the market of radio equipment define radio equipment as an electrical or electronic product that intentionally emits and/or receives radio waves for the purpose of radio communication and/or radiodetermination (the determination of the position, velocity and/or other characteristics of an object, or the obtaining of information relating to those parameters by means of the propagation properties of radio waves), or an electrical or electronic product which must be completed with an accessory, such as an antenna, so as to intentionally emit and/or receive radio waves for the purpose of radio communication and/or radiodetermination [9].

The RED directive does not apply to radio equipment used exclusively for military, state security, radio amateurs, marine (covered by the Directive 96/98/EC), civil aviation, and research.

Radio equipment shall be so constructed to ensure [9]:

- The protection of health and safety of persons;
- An adequate level of electromagnetic compatibility;
- the efficient use of radio spectrum in order to avoid harmful interference;
- Efficient interworks via networks with other radio equipment;
- Protection of personal data and privacy of the user and of the subscriber;
- Certain features supporting protection from fraud and access to emergency services;
- Certain features in order to facilitate its use by users with a disability.

The radio equipment is a licensed service application, meaning that in addition to the equipment compliance assessment the end user needs to receive a license (exclusive rights) to transmit on a particular frequency band(s), at a specified maximum power, and for a defined geographic location. VHF and UHF business and professional radios are examples of licensed device applications [10].

However, many transmitters are unlicensed (license exempt) and can transmit on certain licensed bands but at power levels that will not interfere with licensed users. Such transmitters include many familiar consumer products such as Wi-Fi, Bluetooth, Zigbee, remote keyless entry, RFID, and garage door controls. These devices typically transmit at less than 200 mW and operate in designated frequency bands with specified powers, bandwidths, and duty cycles. They are sold without the user having to obtain a license.

For most wireless products, the RED conformity assessment process can be summarized in five steps [10]:

1. Step 1 begins by determining the applicable directives, technical standards, and conformity assessment procedures.
2. Step 2 covers the tests and engineering assessments of the design, construction, and safety operations.
3. Step 3 ensures that the reports and findings from the evaluations are prepared, collected, and maintained to substantiate the product's compliance.
4. Step 4 refers to labeling the product and issuing a Declaration of Conformity (DoC) prepared to notify the end user and regulatory agencies that the product is ready for placement in the market.
5. Step 5 requires manufacturers to maintain compliance throughout the life cycle of the product. This includes evaluating continuing compliance in the situations of component replacements, product improvements, and redesigns.

The European Communications Office (ECO) is an organization providing information on the harmonized radio spectrum. The ECO maintains a website portal www.efis.cept.org/ that includes the frequency information system (EFIS) where it details frequency allocation, band tables and restrictions, other utilities that identify operations across the radio spectrum, type of transmission, and country. The ECO website portal also maintains the current revision of the primary guidance document identified as ERC Recommendation 70-03 (summary of the compliance requirements and country-specific harmonized frequencies for common low power short-range transmitters) www.docdb.cept.org/ [10].

The digital video services are delivered to users through two infrastructures:

1. Wireless by digitized video via Motion Picture Experts Group (MPEG) protocol and digital modulation for RF transport;
2. Wired by digitized video via MPEG protocol and internet transport.

To operate these infrastructures, depending on the region are used the delivery technologies described in Table 5.6.

The conformity assessment procedure for the majority of radio equipment is based on an internal production control and testing process according to applicable

Table 5.6 Digital Video Standards per Regions

	Mobile	*Terrestrial*	*Satellite*	*Cable*
Europe, Asia	DVB-H, DAB, DVB-SH	DVB-T/T2	DVB-S/S2	DVB-C
China	T-MMB (CMMB and DAB)	DTMB	DVB-S/ABS-S	DVB-C
South Korea	T/S-DMB	ATSC		
Japan	ISDBT 1seg	ISDB-T	ISDB-S	J.83 Annex C
United States	MediaFLO	ATSC	Mixed	J.83 Annex B

standards (example presented in Table 5.7) that refer to effective use of spectrum, EMC, and product safety.

5.1.3.2 RF Electrosurgery

Electrosurgical (ES) devices (also known as surgical diathermy) use radiofrequency between 250 kHz to 4 MHz and are indicated for general tissue cutting (470 kHz) and/or coagulation (250 kHz). The application of RF electrosurgery in medicine are in many areas (general surgery, dermatology, gynecology, plastic surgery, ophthalmology, otolaryngology, orthopedics, urology, neurosurgery, etc.).

The surgical diathermy is based on the heating effect induced by the RF current. Basically, an ES device consists of an RF power generator (maximum 400W across 500Ω load at 2 kVp-p in the cutting mode and approx. 200W bursts of 10–15s with repetition frequency of 15 kHz in the coagulation mode) connected with two electrodes (one active-cutting electrode and the second as a neutral reference-return electrode) that work in monopolar or bipolar mode. The monopolar mode is used for cutting, coagulation, and hemostasis, while the bipolar mode is used for microsurgery (100W across 500Ω, with 1, 5 kVp-p open circuit). In monopolar mode, the active electrode is in the form of a needle and due to its very small surface area, it allows a high density of current on the tissue point that needs to be cut. The second electrode is a large passive metallic part that establish a large contact area with the patient, such the heat produced is negligible. In bipolar mode, both the electrodes (active and dispersive) are within the ES tip. The current flows only through the tissue between the tips of the two electrodes and returns to the generator without passing the patient [11].

The hazards associated with the use of ES equipment can be burns, interference with the heart muscle, interference with other medical devices used concomitantly with ES (e.g., implantable devices, endoscopic equipment), and explosions due to sparks. The main standard for ES equipment is the IEC 60601-2-2 "Medical electrical equipment–Part 2-2: Particular requirements for the basic safety and essential performance of high-frequency surgical equipment and high-frequency surgical accessories" and the ANSI/AAMI HF18 "Electrosurgical devices."

The following safety concerns need to be considered [12]:

- The applied parts (electrodes) are to be isolated from the earth at both high and low frequencies, but the neutral electrode is referenced to earth at high

Table 5.7 Standards for Equipment Using Radio Waves

Standard	*Title of Standard*
ANSI/AAMI HF18	Electrosurgical devices.
IEC 60215	Safety requirements for radio transmitting equipment. General requirements and terminology.
IEC 60244	Methods of measurement for radio transmitters.
IEC 60315	Methods of measurement on radio receivers for various classes of emission.
IEC 60601-2-2	Medical electrical equipment–Part 2-2: Particular requirements for the basic safety and essential performance of high frequency surgical equipment and high frequency surgical accessories.
IEC 62002	Mobile and portable DVB-T/H* radio access.
IEC 60489	Methods of measurement for radio equipment used in the mobile services.
IEC 60510	Methods of measurement for radio equipment used in satellite earth stations.
IEC 60835	Methods of measurement for equipment used in digital microwave radio transmission systems.
IEC 60945	Maritime navigation and radiocommunication equipment and systems. General requirements and methods of testing and required test results.
IEC TS 61149	Guide for safe handling and operation of mobile radio equipment.
IEC 62106	Specification of the radio data system (RDS) for VHF/FM sound broadcasting in the frequency range from 87,5 MHz to 108,0 MHz.
IEC 62273	Methods of measurement for radio transmitters.
ISO 14443	Identification cards, contactless integrated circuit cards, and proximity cards.
ISO/IEC 15693 series	Cards and security devices for personal identification and contactless vicinity objects. Part 1: Physical characteristics.
	Part 2: Air interface and initialization.
	Part 3: Anticollision and transmission protocol.
ISO 18000-3	Information technology. Radio frequency identification for item management–Part 3: Parameters for air interface communications at 13,56 MHz..
ISO 17367	Supply chain applications of RFID and product tagging.
ISO/IEC 18046 (parts 1 to 3)	Information technology and RFID performance test methods.
ISO/IEC 18092	Information technology, telecommunications, and information exchange between systems
ETSI† EN 300 433-2	Electromagnetic compatibility and radio spectrum matters (ERM); Citizens' Band (CB) radio equipment; Part 2: Harmonized EN covering the essential requirements of article 3.2 of the R&TTE Directive.
ETSI EN 300 113-2	Electromagnetic compatibility and ERM; Land mobile service; Radio equipment intended for the transmission of data (and/or speech) using constant or nonconstant envelope modulation and having an antenna connector; Part 2: Harmonized EN covering the essential requirements.
ETSI EN 301 178-2	Electromagnetic compatibility and ERM; Portable VHF radiotelephone equipment for the maritime mobile service operating in the VHF bands (for non-GMDSS applications only); Part 2: Harmonized EN covering essential requirements of article 3.2.
ETSI EN 300 328	Wideband transmission systems; Data transmission equipment operating in the 2.4 GHz band; Harmonized standard for access to radio spectrum.
ETSI EN 300 220	Electromagnetic compatibility and ERM short radius Devices (SRD) in the frequency band 25 MHz to 1,000 MHz with power levels up to 500 mW.

Table 5.7 (continued)

Standard	*Title of Standard*
ETSI EN 300 330	Electromagnetic compatibility and ERM SRD in the frequency range 9 KHz to 25 MHz, inductive loop systems in the frequency range of 9 kHz to 30 MHz (e.g., Electronic Article Surveillance (EAS), RFID, and NFC).
ETSI EN 300 440	SRD and radio equipment operating in the frequency range from 1 GHz to 40 GHz. Harmonized standard for access to radio spectrum.
ETSI EN 300 489-1	EMC standard for radio equipment and services; Part 1: Common technical requirements.
ETSI EN 300 489-3	EMC standard for radio equipment and services; Part 3: Specific conditions for SRD operating on frequencies between 9 kHz and 246 GHz.
ETSI EN 300 489-17	EMC standard for radio equipment and services; Part 17: Specific conditions for broadband data transmission systems.
ETSI EN 301 893	5 GHz RLAN; harmonised standard covering the essential requirements of article 3.2 of Directive 2014/53/EU.
ETSI EN 301 357	Cordless audio devices in the range 25 MHz to 2,000 MHz; harmonised standard covering the essential requirements of article 3.2 of Directive 2014/53/EU.

* DVB-T/H = digital video broadcasting–terrestrial/handheld mobile.

† ETSI = European Telecommunications Standards Institute.

frequencies by components (e.g., capacitor) satisfying the requirements of a type of BF applied part;

- ES equipment shall incorporate means an output control to enable the output power to be reduced to not more than 5% of the rated output power or 10W, whichever is smaller;
- The rated output power shall not exceed 400W for any operating mode averaged over any period of 1s;
- When the equipment is switched off and on again or the mains supply is interrupted and reestablished, the output power for a given setting of the output control shall not increase by more than 20% and the mode of operation shall not be changed except to a stand-by mode;
- The effective series capacitance with the active electrode should not exceed 5 nF;
- Where cutting and coagulation output may be separately selected, the lights indicator should be yellow for cutting and blue for coagulation.

5.1.3.3 Wi-Fi Networks

Wi-Fi (an acronym for the generic term wireless fidelity) is a wireless networking technology using physical data link layer (PHY), which allows the networking of computers and digital devices without the need for wire connection. It describes a technology for radio wireless local area networking (LAN) of devices based on the IEEE 802.11TM standard (IEEE Standard for Information Technology, telecommunications and information exchange between systems, local and metropolitan area networks, specific requirements, Part 11: wireless LAN medium access control

(MAC) and PHY specifications). Multiple versions (with different parameters and hardware structures) of the IEEE 802.11 standards are used. Wi-Fi networks transmit information over the air using radio waves with wavelengths longer than infrared light, heaving typically the frequency of either 2.4 GHz or 5 GHz. These two Wi-Fi frequency bands are then subdivided into multiple channels with each channel possibly being shared by many different networks. In a Wi-Fi network, a device known as a wireless router first receives the data from the internet via broadband internet connection and then converts it into radio waves. The wireless router then emits the radio waves to the surrounding area and the wireless devices (computer, smartphone, etc.) within the router range captures them and decodes them.

In addition to the working frequency and covered range, another characteristic of a Wi-Fi network is the speed defined as how much data the network can transmit. The speed is calculated in million bits per second (Mbps) (see Table 5.8).

Because Wi-Fi depends on radio waves, Wi-Fi networks can be disrupted by the interference caused by other Wi-Fi networks or various electronic appliances, including microwave ovens, cordless telephones, refrigerators, televisions, transistor radios, and Bluetooth devices. Table 5.8 lists the Wi-Fi network generations for each 802.11 standard type based on its characteristics [13].

Wi-Fi security protocols prevent unauthorized access or damage to digital devices using wireless networks. The most basic wireless security is Wired Equivalent Privacy (WEP), which was replaced in the couple of last years by Wi-Fi Protected Access (WPA), Wi-Fi Protected Access II (WPA2), and Wi-Fi Protected Access III (WPA3), which uses even stronger encryption and mitigates security issues posed by weak passwords [13].

5.1.3.4 Mobile Phone

A mobile phone (cellular phone, smartphone, cell phone, or hand phone) is a portable telephone that can make and receive calls over a radio frequency link while the user is moving within a telephone service area. Mobile phones send signals to (and receive them from) nearby cellular base stations (antennas and auxiliary radio equipment) using RF waves. This is a form of energy in the electromagnetic spectrum that falls between UHF radio waves and microwaves. At this time there is no strong evidence that exposure to RF waves from mobile phones and cellular base

Table 5.8 Wi-Fi Generations

Wi-Fi Generation	*802.11 Standard*	*Maximum Speed*	*Frequency*	*Maximum Range Covered*
Wi-Fi 1	802.11b	11 Mbps	2.4 GHz	45m
Wi-Fi 2	802.11a	54 Mbps	5 GHz	30m
Wi-Fi 3	802.11g	108 Mbps	2.4 GHz	52m
Wi-Fi 4	802.11n	600 Mbps	2.4 or 5 GHz	50m
Wi-Fi 5	802.11ac	3.5 Gbps	5 GHz	70m
Wi-Fi 6	802.11ax	9.6 Gbps	2.4 and 5 GHz	70m

stations causes any noticeable health effects. However, this does not mean that the RF waves from cell phones and cellular base stations have been proven to be absolutely safe. Most expert organizations agree that more research is needed to help clarify this, especially for any possible long-term effects. To reduce the long-term effects of RF radiation, some recommendations are suggested, such as

- Network operators must be instructed to apply reduced cumulative power density near to residential areas, schools, hospitals, office buildings, and so on;
- To install multiple radios and TVs, Wi-Fi, and mobile phone base stations with lower transmitted power so that the power efficiency of the base station will be reduced and the health heating effect will be reduced accordingly.

Analog mobile (cellular) phones were the first generation while the first digital cellular phone was marked as the second generation (2G). The next paragraphs include a short description of the mobile phone generations.

Third-generation (3G) generally includes high data speeds, always-on data access, and greater voice capacity. There are several different 3G technology standards. The most used is Universal Mobile Telecommunications Service (UMTS), which is based on wideband code division multiple access (WCDMA). The specific frequency bands originally defined by the UMTS standard are 1,885–2,025 MHz for the mobile-to-base (uplink) and 2,110–2,200 MHz for the base-to-mobile (downlink). In the United States, 1,710–1,755 MHz, 1.900 MHz, and 2,110–2,155 MHz are used. While UMTS2100 is the most widely deployed UMTS band, in some countries UMTS operators use the 850-MHz (900 MHz in Europe) and/or 1,900-MHz bands (independently, meaning uplink and downlink are within the same band). Some carriers, such as in the United States, use band numbers to identify the UMTS frequencies; for example, Band I (2100 MHz), Band IV (1700/2100 MHz), and Band V (850 MHz).

4G is the fourth generation of cellular phone communications standards. It is a successor of the 3G and provides ultra-broadband internet access for mobile devices. The high data transfer rates make 4G networks suitable for use in USB wireless modems for laptops and even home internet access.

5G is the fifth generation of cellular phone communications standards, is a successor to 4G, and is faster with greater capacity and delivers peak download speeds range from 1 Gb/s to 10 Gb/s. 5G is enabled by having a larger number of smaller multiple input and output (MIMO) antennas closer together. 5G technology can be applied to different bands:

1 Sub-6 is a catch-all for any band that sits below 6 GHz. This is the space where all the previous digital mobile networks (starting from 2G onwards) have been operating so far;
2. Also mmWave, which uses bands much higher than were ever used before for this specific purpose, 26–28 GHz.

A list of standards applicable to cellular and cordless phones is presented in Table 5.9.

Table 5.9 Standards for Mobile and Cordless Phones

Standard	*Title of Standard*
ETSI EN 300 607 series	Digital cellular telecommunications system (Phase 2+); mobile station (MS) conformance specification. Part 1: Conformance specification. Part 4: SIM* Application toolkit conformance specifications.
ETSI EN 300 609-4	Global System For Mobile Communications (GSM). Part 4: Harmonized EN for GSM repeaters covering the essential requirements of article 3.2 of the R&TTE Directive.
ETSI EN 301 406	Digital Enhanced Cordless Telecommunications (DECT)†. Harmonized standard covering the essential requirements of article 3.2 of the Directive 2014/53/EU (RED).
ETSI EN 301 419-1	Digital cellular telecommunications system (Phase 2). Requirements for accessing the mobile stations to the GSM network 900 and DCS‡ 1,800.
ETSI EN 301 419-5	Digital cellular telecommunications system (Phase 2). Requirements for cordless telephone system mobile stations (CTS-MS).
ETSI EN 301 489-52	EMC standard for radio equipment and services. Part 52: Specific conditions for cellular communication user equipment (UE) radio and ancillary equipment.
ETSI EN 301 502	GSM communications. Base Station (BS) equipment. Harmonized standard covering the essential requirements of article 3.2 of the Directive 2014/53/EU.
ETSI EN 301 908-1	IMT§ cellular networks. Harmonized standard covering the essential requirements of article 3.2 of the Directive 2014/53/EU. Part 1: Introduction and common requirements.
ETSI EN 301 908-2	IMT cellular networks. Harmonized standard for access to radio spectrum. Part 2: CDMA¶ direct spread (UTRA FDD) UE.
ETSI TBR 10	DECT. General terminal attachment requirements. Telephony applications.
ETSI TS 134 121-1	UMTS. UE conformance specification. Radio transmission and reception (FDD). Part 1: Conformance specification.
IEC 60118-13	Electroacoustics Hearing aids. Part 13: Requirements and methods of measurement for electromagnetic immunity to mobile digital wireless devices.
IEC 60489	Methods of measurement for radio equipment used in the mobile services.
IEC TS 61149	Guide for safe handling and operation of mobile radio equipment.
IEC 62684	Interoperability specifications of common EPS for use with data-enabled mobile telephones.
ISO/IEC 15431	Information technology. Telecommunications and information exchange between systems. Private integrated services network. Interexchange signaling protocol. Wireless terminal call handling additional network features.

The ETSI EN 300 607-1 describes the technical characteristics and methods of test for mobile stations (MSs), operating in the 400 MHz, 900 MHz and/or 1 800 MHz frequency band (GSM 400, GSM 900 and/or DCS 1,800) within the digital cellular telecommunications system.

* SIM = Subscriber Identity Module.

† Digital Enhanced Cordless Telecommunications (1880-1930MHz).

‡ DCS = Digital Cellular System.

§ IMT = International Mobile Telecommunications.

¶ CDMA = Code division multiple access.

5.1.3.5 Microwave Radiation

Microwaves refers to electromagnetic radiation of sufficiently short wavelength for which practical use can be made of a waveguide and associated cavity techniques in its transmission and reception. The term is taken to signify radiations or fields having a frequency range of 300 MHz–300 GHz and has multiple applications in household ovens, diathermia therapy, radar equipment, cellular phones, and so on.

MW radiation is absorbed near the skin, while RF radiation may be absorbed throughout the body. At high enough intensities both will damage tissue through heating [8].

Microwave ovens are devices designed to heat, cook, or dry food through the application of electromagnetic energy generated by a magnetron at frequencies assigned by the FCC in the normal industrial, scientific, and medical (ISM) application heating bands between 890 MHz–6 GHz. For this application (except industrial use) the power density radiation is limited from 1 mW/cm^2 at 5 cm from the external surface of the oven to a maximum of 5 mW/cm^2. A variety of abnormal operating conditions are specified that require compliance with the power density limit (e.g., operation with the door fixed in any position that allows the generation of microwaves). Other requirements refer to safety interlocks (which shut off microwave generation when the oven door is open) and means to monitor the correct functioning of those interlocks. Table 5.10 lists performance and safety standards applicable to microwave ovens.

Microwave diathermy (MWD) uses microwave EM field for noninvasive thermotherapy treatment (generates heat deep into body tissue, resulting in increased blood flow and speeds up metabolism). MWD is used in physiotherapy (pain,

Table 5.10 Standards for Microwave Ovens

Standard	*Title of Standard*
21 CFR § 1030.10	U.S. Code of Federal Regulation (CFR), Part 1030: Performance standards for microwave and radio frequency emitting products. Section 1030.10 Microwave ovens.
ASTM F1317*	Standard test method for calibration of microwave ovens.
ASTM F1360	Standard specification for ovens, microwave, and electric.
AS/NZS 2895.1	Performance of household electrical appliances, microwave ovens. Part 1: methods for measuring the performance of microwave ovens for household and similar purposes.
CSA C22.2 No. 150	Microwave ovens.
CSA C388	Energy performance and capacity measurement of household microwave ovens.
IEC 60335-2-25	Household and similar electrical appliances, safety. Particular requirements for microwave ovens, including combination microwave ovens.
IEC 60335-2-90	Household and similar electrical appliances, safety. Particular requirements for commercial microwave ovens.
IEC 60705	Household microwave ovens. Methods for measuring performance.
JIS C 9250†	Microwave ovens.

* ASTM = American Society for Testing and Materials.

† JIS = Japanese Industrial Standards (published through the JSA=Japanese Standards Association).

muscle spasm, chronic inflammation, and fibrosis). About 50% of the microwave beam applied to the skin is reflected, the rest is absorbed by tissues at depths of only 6–8 cm. The depth of penetration of MWD is inverse proportional with the applicable frequency. Typically, they use 915 MHz and 2,450 MHz (new models) with a maximum power of 250W. Excessive dosage can cause skin burns. The microwave is generated by a magnetron (as in MW ovens) submitted to an antenna located in an applicator that is in direct contact with the skin in the treated area. A cooling system (flow of liquid or air) is included in the applicator for limiting the skin temperature. As a safety measure, a fuse is inserted in the anode supply circuit of the magnetron for current limitation. Microwaves can cause cataract formation, so protective eyewear must be worn whenever microwave diathermy is used.

The U. S. FDA 21 CFR Part 890 Subpart F Section 890.5275 is the FDA reference document for MWD.

5.1.3.6 Radar

Radio detection and ranging (as known as radar) is a radiofrequency system with an electromagnetic sensor (transceiver) used for detecting, locating, tracking, and recognizing objects of various types at considerable distances. It operates by emitting pulsed microwave signals with frequencies between 400 MHz to 40 GHz toward objects (commonly referred to as targets), and observing the echoes returned from them (Figure 5.3) [14]. The targets may be an aircraft, ships, spacecraft, automotive vehicles, astronomical bodies, or even birds, insects, and rain. Besides determining the presence, location, and velocity of such objects, radar can sometimes obtain their size and shape as well. Radars are used for navigation, weather forecasting, and military applications, as well as a variety of other functions. A radar system can measure velocity instead of (or along with) range. This is done by measuring the shift in frequency (Doppler effect: refers to the change in wave frequency during the relative motion between a wave source and its observer; works on both light and sound wave sources [15]) of a wave caused by an object in motion.

What distinguishes radar from optical and infrared sensing devices is its ability to detect faraway objects under adverse weather conditions and to determine their range, or distance, with precision.

The peak power in the emitted pulse can be high even though the average power may be low. Many radars rotate or move up and down. This reduces the mean power density to which the public is exposed in the vicinity of radars. Even high power, nonrotating military radars limit exposures to below guideline levels at locations of public access.

All parts (radar sensor, radar and/or target tracking functions, ancillary units and display, processing and presentation display) of a radar system are tested to the requirements of radar standards. The performance test includes measurement of the transmission frequency spectrum, controls and signal processing functions, minimum range, discrimination, and accuracy measurements. Radar systems use a dedicated method to designate the frequency bands by letter (e.g., L-band 1,2 GHz-1,4 GHz; X-band 9,2-9,5 GHz; S-band 2,9-3,1 GHz). Table 5.11 lists the standards applicable to radar equipment and applications.

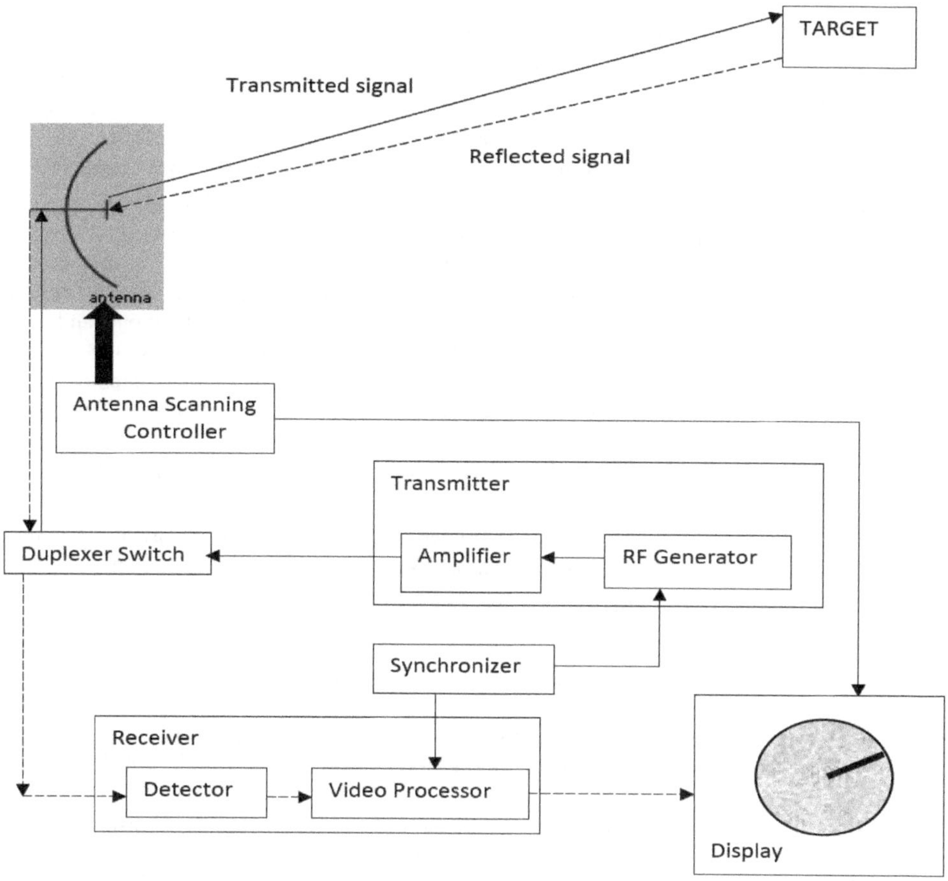

Figure 5.3 Illustration of the radar work principle.

5.1.4 Optical Radiation

Optical radiation includes light emitted from all sources including sunlight. Artificial optical radiation includes light in all its forms including infra-red, laser beams, light emissive diodes, and ultra-violet.

Optical radiation consists of any electromagnetic radiation in the wavelength range between 100 nm and 1 mm. The spectrum of optical radiation is divided into ultraviolet radiation, visible radiation, and infrared radiation (Figure 5.4) [14]:

1. Ultraviolet radiation: optical radiation of wavelength range between 100 nm and 400 nm. The ultraviolet region is divided into UV-A (315-400 nm), UV-B (280-315 nm), and UV-C (100-280 nm);
2. Visible radiation: optical radiation of wavelength range between 400 nm and 780 nm;
3. Infrared radiation: optical radiation of wavelength range between 780 nm and 1 mm. The infrared region is divided into IR-A (780–1,400 nm), IR-B (1,400–3,000 nm), and IR-C (3,000 nm–1 mm).

Table 5.11 Standards for Radar Equipment

Standard	*Title of Standard*
AFM 100-28*	Evaluation of Radar & Radar Type Sensors
ARINC 708†	Airborne Weather Radar
FAA-C-2499	Air Route Surveillance Radar (ARSR-3) Antenna Radome, DOT-FAA‡
FAA-E-2704	Airport Surveillance Radar (ASR-9), DOT-FAA
FAA ORDER 6480.7	Airport Traffic Control and Terminal Radar Approach Control Facility Design Guidelines, DOT-FAA
EN 61097-1	Global maritime distress and safety system (GMDSS)–Part 1: Radar transponder–Marine search and rescue (SART)–Operational and performance requirements, methods of testing and required test results
IEC 62388	Maritime Navigation and Radiocommunications equipment and systems–Shipborne Radar–Performance requirements, Methods of testing and Required test results
IEEE 521	Standard Letter Designations for Radar-Frequency Bands
IEEE 686	Radar Definitions
ISO 19926-1	Meteorology–Weather radar–Part 1: System performance and operation
ISO DIS 23032	Meteorology–Ground-based remote sensing of wind–Radar wind profiler
DI-MISC-81114	Radar Spectrum Management (RSM) Control Plan, NPFC§
R 91	Radar Equipment for the Measurement of the Speed of Vehicles, OIML¶
DO-158	Minimum Performance Standards–Airborne Doppler Radar Navigation Equipment, RTCA#
DO-181	Minimum Operational Performance Standards for Air Traffic Control Radar Beacon System/Mode Select (ATCRBS/Mode S) Airborne Equipment, RTCA
DO-220	Minimum Operational Performance Standards (Mops) for Airborne Weather Radar Systems, RTCA
J3122	Test Target Correlation–Radar Characteristics, SAE**

* AFM= US Air-Force standard.
† ARINC= Standards developed by U.S. Airlines Electronic Engineering Committee (AEEC).
‡ DOT-FAA= US Department of Transportation- Federal Aviation Administration.
§ NPFC= North Pacific Fisheries Commission.
¶ OIML= International Organization of Legal Metrology.
# RTCA= Radio Technical Commission for Aeronautics.
** SAE= Society of Automotive Engineering.

5.1.4.1 Optical Radiation Exposure

Artificial sources of optical radiation are present in a very wide range of workplace environments, work activities, and research applications (biology, medicine, cosmetics, lighting, communication, quantities analysis, nondestructive testing, military, air and water disinfection, hygiene, etc.).

The following are considered light sources: The Sun, incandescent lamps, gas discharge lamps, fluorescent lamps, neon glow lamps, low-pressure mercury lamps, low-pressure sodium lamps, high-pressure xenon lamps, deuterium lamps, flashlamps, LEDs, and lasers.

The discipline that consists of the study of the interaction of optical radiation (natural and artificial) with living organisms is named *photobiology*. This discipline covers all the chemical and biological effects of optical radiation capable to

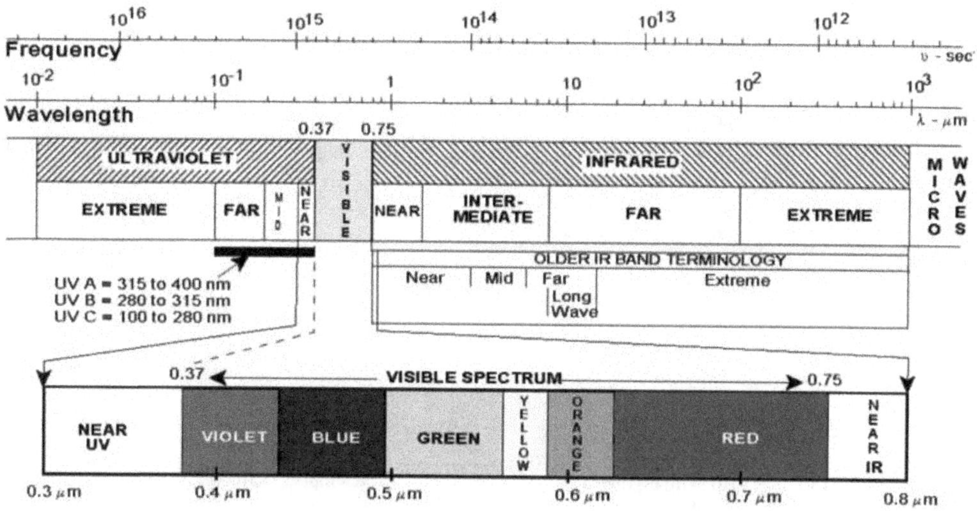

Figure 5.4 Optical radiation spectrum.

induce excited states in molecules due to the absorption of one or more photons. In photobiology topics from the atomic level to that of ecological aspects are covered.

Photoreception (vision) is a phenomenon that results in the formation of an image. This field covers the structure and photochemistry of the visual pigments in the rod and cone photoreceptors of the eyes.

As part of photobiology, photomedicine is concerned with both the harmful and beneficial effects of optical radiation. The potential detrimental consequences (injury hazards) from noncoherent optical radiation are presented in Table 5.12.

Then there is the beneficial area, where light alone (e.g., laser surgery, photo-immunology, infrared physiotherapy, low-level light therapy (LLLT), infrared thermometry and thermography, infrared and ultraviolet spectrophotometry) or sensitizers plus light (e.g., phototherapy), are used to treat certain clinical conditions. One of the main goals of photobiology is to establish the exposure limit (EL) values that represent the level of exposure to noncoherent radiation of the eye or skin that is not expected to result in adverse biological effects. Compliance with these limits will ensure that those who are exposed to such sources are protected against all known adverse health effects.

Within the EU all luminaire and lamp manufacturers need to be aware of the implications of the Artificial Optical Radiation Directive 2006/25/EC [16], which refers to the minimum health and safety requirements regarding the exposure of workers to risks arising from physical agents with a wavelength range between 100 nm and 1 mm (ultraviolet, visible and infrared radiations) to the eyes and skin.

Table 5.12 summarizes the hazards and exposure limit values for noncoherent optical radiation [16, 17].

In addition, the ICNIRP has published a reference document, "ICNIRP: 2013 Guidelines on Limits of Exposure to Incoherent Visible and Infrared Radiation," used as the basis for standards developers (see also Section 5.1.1.1).

The standard IEC 62471 "Photobiological safety of lamps and lamp systems," gives guidance for evaluating the photobiological safety of lamps and lamp systems

Table 5.12 Hazards and Exposure Limit Values for Noncoherent Optical Radiation

Optical Radiation	*Wavelength*	*Hazard*	*Characteristic to Be Considered*	*Permissible Exposure Time [s]*		*Exposure Limit for Noncoherent Optical Radiation*
Actinic ultraviolet	200–315 nm	To eyes: photochemical damage to the cornea, to conjunctiva and to eye lens (cataract); and thermal injury to the retina.	Irradiance E	Exempt RG 0	30,000	E =1 mW/m² for RG0 exposure of 8 hours daily or for effective radiant exposure is 30 J/m² or effective radiant UV power of 2 mW/klm
				RG1	10,000	E= 3 mW/m² for RG1
				RG2	1,000	E= 30 mW/m² for RG2
				RG3	0	
	200–315 nm	To skin: erythema (sunburn); elastosis (wrinkles, aged skin).	Irradiance E	Exempt RG 0	30,000	E =1 mW/m² for RG0 exposure of 8 hours daily or for effective radiant exposure is 30 J/m² or effective radiant UV power of 2 mW/klm
				RG1	10,000	E= 3 mW/m² for RG1
				RG2	1,000	E= 30 mW/m² for RG2
				RG3	0	
UV-A	315–400 nm	To skin: elastosis (wrinkles, aged skin), dark pigment spots, sun allergy.	Irradiance E	Exempt RG 0	1,000	E =10 W/m² for RG0
				RG1	300	E= 33 W/m² for RG1
				RG2	100	E= 100 W/m² for RG2
				RG3	0	
Visible blue light	400–550 nm	To eyes: photochemical damage to retina.	Irradiance E	Exempt RG 0	10,000	E=0.01 W/m² for RG0
				RG1	100	E= 1 W/m² for RG1
				RG2	0.25	E= 400 W/m² for RG2
				RG3	0	This damage mechanism dominates over the thermal damage mechanism for times exceeding 10 seconds.
		To skin: thermal damage.	Irradiance E	Exempt RG 0	10,000	E=0.01 W/m² for RG0
				RG1	100	E= 1 W/m² for RG1
				RG2	0.25	E= 400 W/m² for RG2
				RG3	0	

including luminaires. The exposure limits are specified, as well as the reference measurement technique and classification scheme for the evaluation and control of photobiological hazards from all electrically powered incoherent broadband sources of optical radiation, including LEDs (excluding lasers), in the wavelength range from 200 nm through 3,000 nm.

The IEC 62471 standard also provides specific guidance on the geometrical conditions under which measurements should be made, taking into account

Table 5.12 (continued)

Optical Radiation	*Wavelength*	*Hazard*	*Characteristic to Be Considered*	*Permissible Exposure Time [s]*		*Exposure Limit for Noncoherent Optical Radiation*
Infrared A (NIR)	780-1,400 nm	To eye: thermal damage to the lens (cataract).	Irradiance E	Exempt	1,000	E=100 W/m² for RG0
				RG 0		
				RG1	100	E= 1 kW/m² for RG1
				RG2	10	E= 10 kW/m² for RG2
				RG3	0	
		To skin: thermal damage.	Irradiance E	Exempt	1,000	E=100 W/m² for RG0
				RG 0		E= 1 kW/m² for RG1
				RG1	100	E= 10 kW/m² for RG2
				RG2	10	
				RG3	0	
Infrared B (SWIR)	1,400-3,000 nm	To eye: thermal damage to the lens (cataract); burns to the cornea.	Irradiance E	Exempt	1,000	E=100 W/m² for RG0
				RG 0		
				RG1	100	E= 1 kW/m² for RG1
				RG2	10	E= 10 kW/m² for RG2
				RG3	0	
		To skin: thermal damage.	Irradiance E	Exempt	1,000	E=100 W/m² for RG0
				RG 0		
				RG1	100	E= 1 kW/m² for RG1
				RG2	10	E= 10 kW/m² for RG2
				RG3	0	

biophysical phenomena such as the effect of eye movements on retinal irradiation. A reduced spectral range over which radiance should be considered is 300–1,400 nm, since the retina is essentially protected outside this range due to transmission characteristics of the lens of the eye [17].

In order to determine the risk group of a source, its irradiance (E expressed in W/m² representing the radiant flux density of optical power received by a surface per area unit) or spectral irradiance (F expressed in W/m²nm representing the irradiance per wavelength interval) or radiance (L expressed in W/m²sr representing the irradiance per unit solid angle-expressed in steradian) has to be measured at a specific distance and weighted with action spectra and maximum permissible exposure time (PET) (Table 5.12).

UV and blue-light phototoxicity are the key potential hazards in lamp safety standards. The phototoxicity depends not only on wavelength but also on the exposure duration and exposure geometry of the radiation source. The blue light (part of the visible light spectrum, what the human eye can see, having the shortest wavelength of 380 nm to 500 nm range, and highest energy) hazard is covered in the IEC/TR 62778 "Application of IEC 62471 for the assessment of blue light hazard to light sources and luminaires." In addition to the spectral measurements the possibility to evaluate blue light hazards directly in applications is introduced. About one-third of all visible light is considered high-energy visible, or blue light. For blue light hazard, no further assessment is normally not necessary except in the case of luminaires using narrow beam focusing optics. Inadequate exposure (for a long time or very close to the eyes) to blue light can also contribute to an increased chance in developing myopia, or nearsightedness.

The IEC62471 recommends that detailed measurements are not required for sources (white or broadband sources emitting in the visible region) having a luminance less than 104cd/m^2 (the level considered as one visually comfortable to view). This means that a lighting product is considered safe without the need to provide any warning or advice when it provides optical radiation within the limits (maximum irradiance for a specified permissible exposure time) specified in Table 5.12.

Where the total irradiance of a white-light source exceeds the above levels, and for all other sources, one should proceed with an evaluation of photobiological safety, at the appropriate distance of 200 mm or an illuminance of 500 lux, depending on the intended application of the finished product.

There are four risk groups of lamps and luminaires with associated exposure limits (Table 5.13).

However, the range of 100 nm to 1 mm is often restricted for practical purposes to 200–3,000 nm due to atmospheric absorption below 200 nm, and the negligible effect of low-energy photons in the IR-C (3,000 nm to 1 mm).

The luminaire standard IEC 60598-1 takes into account the possibility of having light sources that emit a level of UV greater than 2 mW/klm. In this case, the lamps are identified by the symbol in Figure 5.5 [18].

This symbol indicates that the lamp cannot be used in open luminaires without any glass protection designed so that the UV emission will correspond with RG 0 of IEC 62471.

Table 5.13 Categorization of Lamps and luminaires According to IEC 62471

Risk Group	*Hazard*	*Comments*	*Examples of Lamp or Lamp System*
RG 3	Hazardous even for momentary exposure.		Professional projection systems; intense pulsed light (IPL); UV from plasma cutting and welding; UV sterilization; IR from welding, etc.
RG 2	Does not pose a hazard due to aversion response of the eye to bright light or thermal discomfort.	Aversion response of the eye is 0.25s.	Desk-top projectors; vehicle headlights; UV insect traps; high-pressure mercury vapor lamps; IR from furnaces, etc.
RG 1	No photobiological hazards under the normal behavioral limitation on exposure.	Includes lamps that emit infrared radiation without a strong visual stimulus and which do not pose a near-infrared retinal hazard within 100 seconds.	Desk lamps; vehicle side, brake, and indicator lights; low-pressure sodium lamps; high-pressure sodium lamps; photocopiers; IR from artificial heaters; LED Lanterns; etc.
RG 0	No photobiological hazards.	Exempt of any evaluation.	Indicator LEDs in electric and electronic equipment; flat- screen LED televisions; LED computer monitor; LED remote control devices; smart phones and tablet screens; etc.

Figure 5.5 Symbol for light source with that emit a level of UV greater than 2 mW/klm.

Projection lamps, photographic lamps, special purpose, and stage lighting lamps shall be marked with the symbols shown in Figure 5.6 on the packaging or accompanying information [18].

For this kind of lamp, it is necessary to refer to the manufacturer's documentation for risk assessment (only for IR).

Table 5.14 summarizes the main standards related to photobiological safety and measurement for non-coherent light sources.

A risk assessment focusing on the particular hazard(s) must identify controls to minimize the risk of harmful exposure to as low as reasonably practicable. Control measures to consider when managing artificial optical radiation risks are

- Use of an alternative safer light source that can achieve the same result;
- Use of filters, screens, remote viewing, curtains, safety interlocks, dedicated rooms, and remote controls;
- Training workers in best practice and giving them appropriate information;
- Organizing work to reduce exposure to workers and restricting access to hazardous areas;
- Use of protective equipment, such as clothing, goggles, or face shields;
- Use of relevant safety signs.

Figure 5.6 Symbols for lamps with photobiological hazard (do not stare at the light source).

Table 5.14 Photobiological Safety and Measurement Standards

Standard	*Title of Standard*
ANSI/IESNA* RP27 (series)	Recommended Practice for Photobiological Safety for Lamps and Lamps Systems:
	- Part 1: General Requirements
	- Part 2: Recommended Practice for Photobiological Safety for Lamps and Lamp Systems-Measurement Techniques
	- Part 3: Recommended Practice for Photobiological Safety for Lamps
CIE† 53	Methods of characterizing the performance of radiometers and photometers
CIE 63	Spectro-radiometric measurement of light sources
CIE 105	Spectro-radiometry of pulsed optical radiation sources
CIE S009/E	Photobiological safety of lamps and lamp systems
GB/T 20145	Photobiological safety of lamps and lamp systems
IEC 60601-2-57	Medical Electrical Equipment–Part 2-57: Particular requirements for the basic safety and essential performance of non-laser light source equipment intended for therapeutic, diagnostic, monitoring and cosmetic/aesthetic use
IEC/EN 62115	Electric Toys - Safety
IEC 62471 (series)	Photobiological safety of lamps and lamp systems
	- Part 2: Guidance on manufacturing requirements relating to non-laser optical radiation safety
	- Part 4: Measuring methods
	- Part 5 Image projectors
IEC TR 62778	Application of IEC 62471 for the assessment of blue light hazard to light sources and luminaires.
JIS C 7550	Photobiological safety of lamps and lamp systems

* IESNA = Illuminating Engineering Society of North America.

† CIE = International Commission on Illumination.

Products that have a risk group higher than 1 need to be accompanied by installation instructions and symbols.

The standard will require symbols to not stare at the light source or installation distances.

5.1.4.2 Infrared Radiation

IR that covers the electromagnetic range between 300 GHz (300×10^9) Hz to 430 THz (430×10^{12}) Hz (in wavelength units from 700 nm to 1 mm) is divided into five subranges near infrared radiation (IR) (700–1,400 nm), short IR (1,400–3,000 nm), mid IR (3–8 μm), long IR (8–15 μm), and far IR (15 μm–1 mm). The IR has a wide utilization in medicine, military, communications, thermal imaging, night vision, and different techniques of laboratory measurements and analyses (thermometry, spectrophotometry, etc.) (see also Table 5.1).

When humans are exposed to IR this radiation is absorbed by skin and eyes as heat and when is excessive may be hazardous, resulting in skin damage or blindness. Since this radiation is invisible and manifests itself through heat sensation and pain, special protective clothing and IR-proof goggles must be worn in the presence of such radiation. Sources of IR radiation include furnaces, heat lamps, IR lasers, and so on.

The intensity of an emitted infrared energy is proportional to its temperature. By understanding how this radiation depends on temperature we can make accurate measurements of temperature over a wide range.

Infrared radiation thermometers. The 0.7 to 14 μm (micron) range is used for infrared temperature measurement (noncontact IR thermometer scanner) that consists of advanced optic systems and detectors. Most radiation thermometers are of the type known as spectral band thermometers. The noise equivalent temperature difference (NETD), which is a measurement of thermal resolution or sensitivity, must be less than or equal to 0.10°C at 30°C (0.18°F at 86°F).

Infrared imaging system. The IR imaging system has an operating spectral range between 2 and 15 μm and must have the gain or contrast set to be able to distinguish the target from the other parts at a distance, which permits the recognition of thermal anomalies. Another characteristic of the IR Imaging system is the field of view (FOV). FOV is the open observable area a person can see via an optical device. In the case of optical devices and sensors, FOV describes the angle through which the devices can pick up the infrared radiation. FOV is normally expressed in angular degrees and a value of approximately 25 degrees is suggested. The imaging system must have a means of recording images seen on the camera's screen. The images may either be in a video format or in individual still frame images. The detector and lens combination of the infrared imaging system must have sufficient resolution. For practical purposes, the camera's detector array must have no less than 120 × 120 pixels. The thermal imaging camera consists of the following components: an optic system, sensor (detector), amplifier, signal processing electronics, and display. Table 5.15 is a list of standards for infrared radiation applications.

5.1.4.3 LED

LED means a p-n junction solid-state device of which the radiated output, either in the infrared region, the visible region, or the ultraviolet region is a function of the physical construction, the material used, and exciting current of the device. LEDs are semiconductor light sources that combine a P-type semiconductor (larger hole concentration) with an N-type semiconductor (larger electron concentration). Applying a sufficient forward voltage will cause the electrons and holes to recombine at the p-n junction, releasing energy in the form of light. Compared with conventional light sources that first convert electrical energy into heat, and then into the light, LEDs convert electrical energy directly into the light, delivering efficient light generation with little-wasted electricity. As a current-driven device in LEDs, the level of light is a function of the current, increasing the light output.

All radiation-emitting (including light-emitting) products are subject to regulation by the FDA. LEDs are different from lasers in that they are less intense, do not have as narrow bandwidth as a laser, and do not emit coherent radiation.

Two types of LEDs are available, a lamp type (with leads) and a chip type (surface mount). Users can select the ideal type based on set requirements. The LED color (emission wavelength) will change depending on the materials used.

The basic materials for LEDs are: aluminum nitride (AlN), aluminum gallium nitride (AlGaN), aluminum gallium indium nitride (AlGaInN), indium gallium

Table 5.15 Standards for Infrared Radiation Applications

Standard	*Description*
ASTM E1655	Standard Practices for Infrared Multivariate Quantitative Analysis; These practices apply to analyses conducted in the NIR spectral region (roughly 780 to 2,500 nm) through the mid-infrared (MIR) spectral region (roughly 4,000 to 400 cm^{-1})
ASTM E1862	Standard for Measuring and Compensating for Reflected Temperature Using Infrared Imaging Radiometers
ASTM E1897	Standard for Measuring and Compensating for Transmittance of an Attenuating Medium Using Infrared Imaging Radiometers
ASTM E1933	Standard Test Methods for Measuring and Compensating for Emissivity Using Infrared Imaging Radiometers
ASTM E1934	Standard Guide for Examining Electrical and Mechanical Equipment with Infrared Thermography
ASTM E2758-10	Standard Guide for Selection and Use of Wideband, Low Temperature Infrared Thermometers
IEC 60335-2-27	Household and similar electrical appliances–Safety–Part 2-27: Particular requirements for appliances for skin exposure to optical radiation (wavelength 100 nm to 1 mm)
IEC 61920	Infrared free air applications (700 nm to 1,600 nm)
IEC 80601-2-59	Medical electrical equipment Part 2-59: Particular requirements for the basic safety and essential performance of screening thermographs for human febrile temperature screening
IEC TS 62492 (series)	Industrial process control devices - Radiation thermometers - Part 1: Technical data for radiation thermometers - Part 2: Determination of the technical data for radiation thermometers
ISO 6781	Thermal Insulation—Qualitative Detection of Thermal Irregularities in Building Envelopes—Infrared Method
ISO 18251-1	Non-destructive testing—Infrared thermography—System and equipment—Part 1: Description of characteristics
ISO 18434-1	Condition monitoring and diagnostics of machines—Thermography—Part 1: General procedures
JSA - JIS K 0117	General rules for infrared spectrophotometric analysis; 2.5 μm to 25 μm
JSA - JIS T 8141	Personal eye protectors for optical radiations; infrared and ultraviolet
SAE J359	Infrared Testing
UL 482	UL Standard for Safety Portable Sun/Heat Lamps (Ultraviolet/Infrared)

nitride (InGaN), gallium arsenide (GaAs), aluminum gallium arsenide (AlGaAs), gallium arsenide phosphide (GaAsP), aluminium gallium indium phosphide (AlGaInP), and gallium phosphide (GaP).

LEDs are used in general lighting service (GLS) and low voltage circuits as indicator devices.

The general lighting service sources are defined as white-light sources used to illuminate spaces. The LEDs used in GLS are made based on two technologies: phosphor-converted (PC) and color-mixed LEDs. White LEDs are usually created through the use of a blue LED and a phosphor. (Combining a blue LED with yellow phosphor, which is a complementary color, will yield white light. This method is easier than other solutions and provides high efficiency, making it the most popular choice on the market.) Color-mixed LEDs combine the three primary colors (red, green, blue) resulting in the white light. Generally, this method is not used for lighting, but for full-color LED devices [19].

The non-GLS category takes into account all types of LEDs, through the spectrum from UV to IR, including white LEDs. Depending on the application, the optical output of such LEDs can vary significantly from very low-level indication to high-power LEDs, for example, used in industrial and signaling applications.

LEDs have become more widely available to consumers in recent years and are expected to replace traditional light sources (e.g., fluorescent and high-intensity discharge lamps) due to their higher energy efficiency.

When selecting LEDs for GLS it is necessary to understand the datasheet specifications so that the optimum LED part can be chosen for the particular application. A few of these specifications are [19]

- LED color;
- Color rendering index (CRI), which is the indicator for how well a light source performs at representing colors;
- Correlated color temperature (CCT) index (to be less than 4,000K);
- Spatial uniformity of color;
- LED light intensity (the lumen output);
- LED luminance (in cd/m^2);
- LED current/voltage/power;
- LED junction temperature and thermal resistance;
- LED beam angle (LED viewing angle defined in degrees);
- Case temperature (to not exceed 55°C);
- LED operational lifetime.

Note that specifications should match the luminance to background ratio based on visual ergonomics criteria.

To determine representative values for LED lamps that are currently manufactured or distributed into commerce within the United States, manufacturers must follow DOE test procedure methods specified at the US10 CFR 430, Appendix BB.

If LEDs are connected directly to the AC mains supply without any current limiting LED driver or LED power supply, they will immediately fail open-circuit and exhibit severe physical damage, including breakage and burn marks. If LEDs are powered via a LED driver or LED power supply, but the current and/or voltage is too low, the LEDs will appear too dim or fail to light at all. If the current and/or voltage is too high the LEDs could either age prematurely or fail catastrophically. Depending on time and ambient temperature, the amount of LED light reduces exponentially and the color of white LEDs tends to become bluer. The hotter the environment, the shorter the LED life. LEDs are electrically polarized and will only operate correctly when their positive terminal (anode) is connected to the supply positive, and their negative terminal (cathode) is connected to the supply negative.

While LEDs are highly efficient, there is still an amount of electrical power that is converted to heat, not just light. LED overheating can be caused by many rather obvious design factors such as inadequate heat sinking and excess LED power. But a well-designed LED luminaire can be equally overheated by installing it in an environment for which it was not designed. To get the very best performance from the LEDs the excess heat needs to be conducted away from the LEDs to avoid exceeding the recommended maximum operating temperatures. LEDs that overheat will degrade more rapidly in terms of

- Light output, also referred to as lumen maintenance;
- Color change with time.

In extreme cases overheating LEDs can result in physical damage to the LEDs themselves, plastic lenses, wires, connectors, printed circuit boards, and drive electronics. Thermal design is intrinsic to every element of design including LED selection, PCB design and layout, and electronics design. A heatsink transfers the generated heat from the LED board to the ambient air using convection and radiation.

Typical sources of human exposure to LED light include general illumination at home or work, tablets and E-readers, and outdoor lighting. The low visual response elicited from low-power white or colored LEDs leads to the conclusion that no photochemical safety concerns exist. Due to the narrow-band emission of LED chips and the limited emission range of LED phosphors, one can restrict consideration to the visible region, no risks are posed in the UV or the IR. For UV-emitting LEDs, irradiance is important for assessing the potential hazards to the cornea, conjunctiva, lens, or skin. The ICNIRP provides guidelines for the assessment of potential risks from ultraviolet radiation. Radiance (or brightness) is important for assessing the potential retinal hazards of any bright optical source that can be imaged on the retina. Because of their limited radiance (compared to lasers, for example), currently available LEDs are not likely to pose a retinal thermal hazard.

The potential adverse effects of exposure to humans to the light from LEDs GLS are:

- Flicker effect;
- Glare (especially at night) from street lighting, due to increased scatter of the shorter wavelength emissions compared to traditional street lighting and excessive blue light emitted;
- Disruption of circadian rhythm due to the high blue content of white LEDs, leading to disturbed sleep that can have short and long-term health consequences;
- Visual impairment due to high CCT index (more than 4,000K).

Note that no guidelines exist in Europe or the United States to limit flicker in LED-based products.

Concerns regarding the above have been voiced by professional regulators:

- The French Agency for Food, Environmental, and Occupational Health and Safety (ANSES) noted that the IEC 62471 needs to be updated to provide more guidance on evaluating LEDs, specifically as [20]:
  - Limit LEDs for general public to < RG1;

- Provide distance at which product < RG0;
- Require that LED systems of > RG1 be installed only by professionals;
- Recommending that manufacturers design lighting systems that provide only indirect light to reduce glare.

- The American Medical Association (AMA) makes the following recommendations [21]:
  - Encourage use of < 3,000K CCT lighting for outdoor installations;
  - Encourage use of proper shielding of LEDs to reduce glare;
  - Encourage use of dimming in off-peak times.

The representative standards applicable to LEDs are presented in Table 5.16.

5.1.4.4 LASER Radiation

Light amplification by stimulated emission of radiation (laser) represents any device that can be made to produce or amplify electromagnetic radiation in the optical radiation wavelength range (180 nm–1 mm) primarily by the process of controlled stimulated emission than occurs when a photon causes a molecule in an excited state to emit a second photon. Typically, laser products are used for the demonstration of physical and optical phenomena, materials processing, data reading and storage, and transmission and display of information. Such systems have found use in industry, business, entertainment, research, education, medicine, and consumer products. Lasers emit optical (UV, visible light, IR) radiations and are primarily an eye and skin hazard.

Laser light has some characteristics that make it different from ordinary light [22]:

1. Laser light is coherent (all waves of light energy emitted are in phase with each other) in the sense that the light constitutes very long wave trains, contrary to ordinary light, where each photon can be regarded as a limited wave packet independent of other photons.
2. Laser light can be made very collimated (all rays are very parallel and travel a long distance with little spread).
3. Laser light is usually very monochromatic (has a single color or wavelength with a very narrow spectral bandwidth) or consists of a small number of very narrow bands.
4. Laser light may be (but is not necessarily) plane-polarized.
5. The light from some types of lasers is given off in extremely short pulses of extremely high power (energy per time unit). However, this does not hold for all lasers. Some lasers emit light continuously and have very feeble power.

The radiation produced by a laser exhibits high temporal and spatial coherence. In order to begin the process of stimulated emission, the lasing medium absorbs the energy from a pump source. The atoms in the lasing medium are excited to a higher energy state. These atoms will eventually return to their ground state. A large number of atoms that are excited to higher states create a population

Table 5.16 Standards Related to LED and LED Devices

Standard	*Title of Standard*
10 CFR 430 Appendix BB	DOE Energy Conservation Program for Consumer Products
81 FR 76877	DOE Energy Conservation Program: Test Procedures for Integrated Light-Emitting Diode Lamps
83 FR 47806	DOE Energy Conservation Program: Test Procedures for Integrated Light-Emitting Diode Lamps
ANSI/NEMA* C78.50	Electric Lamps–Assigned LED Lamp Codes
ANSI/NEMA C78.51	Electric Lamps–LED (Light Emitting Diode) Lamps–Method of Designation
ANSI/NEMA C78.53	Electric Lamps–Performance Specifications for Direct Replacement LED (Light Emitting Diode) Lamps
ANSI/NEMA C78.377	American National Standard for Electric Lamps–Specifications for the Chromaticity of Solid-State Lighting (SSL) Products
ANSI/NEMA C78.62612	Electric Lamps– Self-Ballasted LED Lamps — Performance Specifications
ANSI/UL 8750	LED Equipment for Use in Lighting Products
AS/NZS 61347.2.13	Particular requirements for DC or AC supplied electronic control gear for LED modules
CSA C871	Performance of LED replacement lamps
CAN/CSA-C22.2 No. 250.13	LED equipment for lighting applications
DIN 15780	Entertainment technology–LED for scenic lighting applications
EN 4706	Aerospace series–LED color and brightness ranking
IEC 63013	LED packages–Long-term luminous and radiant flux maintenance projection
IEC 62031	LED modules for general lighting–Safety specifications
IEC 62717	LED modules for general lighting–Performance requirements
IEC 61347-2-13	Lamp control gear - Part 2-13: Particular requirements for DC or AC supplied electronic control gear for LED modules
IEC 62560	Self-Ballasted LED Lamps for GLS by voltage>50V – Safety specifications
IEC 62838	LEDs lamps for general lighting services with supply voltages not exceeding 50V a.c. r.m.s. or 120V ripple free d.c–Safety specifications
IEC 62776	Double-capped LED lamps designed to retrofit linear fluorescent lamps–Safety specifications
IEC 60601-2-50	Medical electrical equipment–Part 2-50: Particular requirements for the basic safety and essential performance of infant phototherapy equipment
IES† LM-80-20	Approved Method: Measuring Luminous Flux and Color Maintenance of LED Packages, Arrays, and Modules
IES TM-21-11	Projecting Long Term Lumen Maintenance of LED Light Sources
IES LM-80-08	Method for Measuring Lumen Maintenance of LED Lamps, Light Engines, and Luminaires
IES TM-28	Prediction of Lumen Maintenance of LED Lamps and Luminaires
IESNA‡ LM-79-08	Electrical and Photometric Testing of Solid-State Lighting Devices
ISO 20894	Aircraft - LED based taxiing light system–General requirements
ISO 13207-1	Road vehicles — LED lamp characteristics for bulb compatible failure detection–Part 1: LED lamps used as direction indicators
SAE J 2650	Performance Requirements for LED Road Illumination Devices

Table 5.16 (continued)

Standard	*Title of Standard*
SAE J 1889	LED Signal and Marking Lighting Devices
SAE USCAR§-33-1	Specification for Testing Automotive LED Modules

* NEMA = National Electrical Manufacturers Association.
† IES = Illuminating Engineering Society
‡ IESNA = Illuminating Engineering Society of North America.
§ USCAR = United States Council for Automotive Research.

inversion. Population inversion describes the number of atoms in the excited state versus the number of atoms in the ground state. In order for the atoms to return to their ground state, they must release energy. This energy is released in the form of photons. The energy that must be released by the atom to return to the ground state will direct the wavelength of the photon emitted. If all the excited atoms released the same amount of energy to return to their ground state, the released photons would all have the same wavelength and would be considered fully monochromatic. Most lasers do not emit a single wavelength but a range of slightly differing wavelengths.

The lasing medium may be a solid, a gas, liquid, or plasma (Figure 5.7) [14], as well as including gas, chemical, dye, fiber-based, solid-state, and semiconductor lasers. The laser radiation can be output in a continuous wave (CW) or a pulsed wave. A continuous wave laser emits light that maintains a steady amplitude and frequency. A pulse wave will vary in amplitude and is also characterized by the system's pulse repetition frequency (PRF).

Gas lasers can be pulsed or in a CW. The gas dynamic laser obtains its inverted population through a rapid temperature rise produced by accelerating the gas through a supersonic nozzle. In chemical lasers, the inversion is produced by a chemical reaction. In the electric discharge laser, the lasing medium is electrically pumped. The gas can also be optically pumped. In an optically pumped gas laser, the lasing medium is contained in a transparent cylinder. The cylinder is in a resonant cavity formed by two highly reflective mirrors. Many gas lasers use carbon dioxide (CO_2) as the lasing medium (actually a mixture of CO_2 and other gases). In other configuration, the lasing medium is a mixture of helium and neon gas in a gas discharge or plasma tube. This is the basis for most high-energy or high-power lasers. The first gas laser was an optically pumped CW helium-neon laser.

The common laser pointer (usually battery-powered) is designed for pointing at objects by illuminating them with a collimated visible laser beam and contains a small GaInP/AlGaInP laser diode operating in the red (630 nm-670 nm), spectral region. Other laser pointers emit green (520 nm and 532 nm) or even blue or yellow light and normally contain a small diode-pumped solid-state laser with a nonlinear crystal for frequency doubling. Green laser pointers are usually based on a miniature Nd:YVO_4 laser with a potassium titanyl phosphate (KTP) crystal for intracavity frequency doubling. It is well established that humans are most visually sensitive to green light. Humans are also far more sensitive to green light at night. As a result, green laser pointers (working at 532 nm) are a much more significant

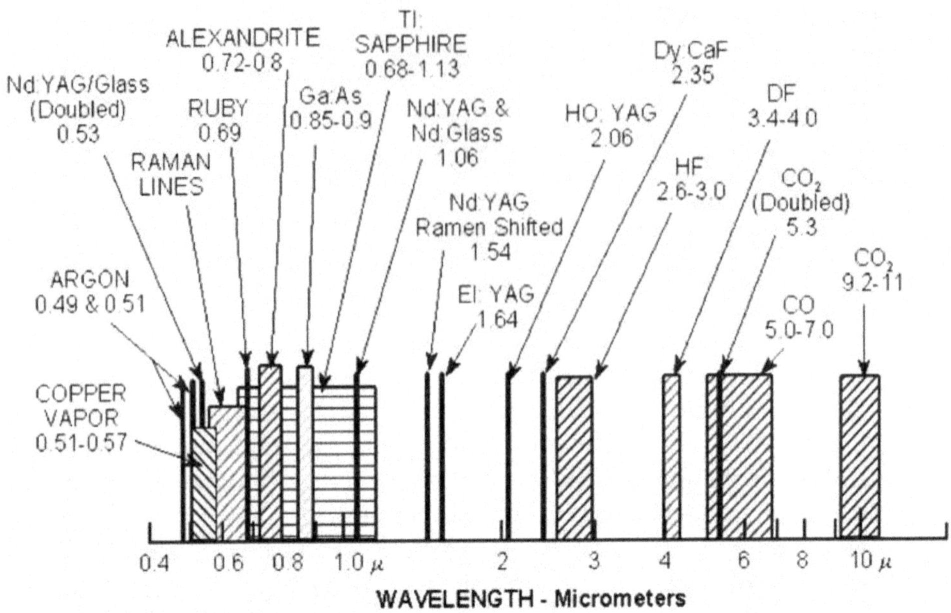

Figure 5.7 The spectral output of several laser types.

safety hazard than red laser pointers. Due to a 50 nm shift in color sensitivity toward the blue wavelengths and away from the red wavelengths at night, blue light also appears to be much, much brighter than red light at a comparable power output. The laser pointers should not be confused with lamps containing LEDs, which emit a much more diffuse beam (with much lower spatial coherence, similar to that of an incandescent lamp) and can also emit light with different colors, or white light.

The dye laser is an example of a laser using a liquid for the lasing medium. The lasing medium is an organic dye dissolved in a solvent, such as an ethyl alcohol. Dye lasers operate from the near UV to the near IR, are optically pumped, and are tunable over a fairly wide wavelength range.

Another type of laser is the semiconductor or injection laser (known as a laser diode). The junctions of most semiconductor diodes will emit some radiation if the devices are forward biased. This radiation is the result of energy released when electrons and holes recombine in the junction. There are two kinds of semiconductor diode emitters: (1) LED, which produces incoherent spontaneous emission when forward biased and has a broad (800 angstrom) spectral output, and (2) the laser diode, which maintains a coherent emission when pulsed beyond a threshold current and has a narrow spectral width (< 10 angstroms). In the laser diode, the end faces of the junction region are polished to form mirror surfaces. They can operate CW at room temperature, but the pulsed operation is more common.

Fiber lasers use fibers that are doped with rare-earth elements as the pumping medium. These rare-earth elements include elements, such as erbium (most common), ytterbium, and neodymium. There are other elements, such as thulium, that are used for doping purposes. Erbium-doped fiber lasers can emit in the 1.5-to-1.6-micron wavelength, which is important due to eye safety concerns in this part of the spectrum. Other wavelength emissions for erbium include 2.7 and 0.55 microns. Fiber-based laser systems are beneficial in many ways. The fiber gain medium is compact compared to many other types of the gain medium and is highly efficient. The fiber gain medium can also be physically manipulated to save space. Fiber-based lasers are able to achieve high output powers. The gain medium of a fiber laser can extend for several kilometers to achieve these higher power outputs. The fact that the light is already propagating in a flexible fiber can also allow for system designs that implement a gain cavity in one location and then deliver the output in another location. Characteristics and applications of various types of laser are summarized in Table 5.17.

Aspects related to safety, performance, and application are included in a wide range of standards. The IEC 60825 series of standards covers a system of classification of lasers and laser products emitting radiation according to their degree of optical radiation hazard to aid hazard evaluation and to aid the determination of user control measures, as well as establish requirements for the manufacturer to supply information so that proper precautions can be adopted through labels and instructions. This will provide an adequate warning to individuals of hazards associated with accessible radiation from laser products (see Figure 5.8 for laser labels) and provide means to reduce the possibility of injury by minimizing unnecessary accessible radiation and to give improved control of the laser radiation hazards through protective features. For the United States there are similar recommendations, but with specific U.S. differences, which are included in the ANSI Z 136 series of standards, in Code of Federal Regulation (CFR) Title 21 Part 10, and in the FDA Publication 86-8260 "Compliance Guide for Laser Products."

The manufacturers of laser products or incorporating a laser for acceptance in the United States' market need to fill the FDA-CDRH Form 3632 dated 09/2020 "Guide for Preparing Product Reports for Lasers and Product Containing Lasers" [23] based on measurements conducted on the product. The filed form accompanied by the requested documentation needs to be sent to FDA-CDRH for review and after acceptance will be included in the CDRH database. When a product

Table 5.17 Characteristics and Application of Various Types of Lasers

Active medium	*Examples*	*Wavelength (nm)*	*Applications*
Gas	He-Ne	543; 632; 1,152	Alignment, barcode scanning, printing, measurement
	CO^2	10,600	Cutting, welding, surgery
	Argon-ion	488; 514	Entertainment, surgery, printing, measurement
Liquid	Dye laser	310-1,200	Entertainment, medical diagnostic, measurement
Solid	Neodymium: YAG	1,064; 532	Cutting, welding, entertainment, surgery
	Ruby	694	Holography, surgery
Semiconductor	Various	600–29,000	Communications, pointers, CD, DVD

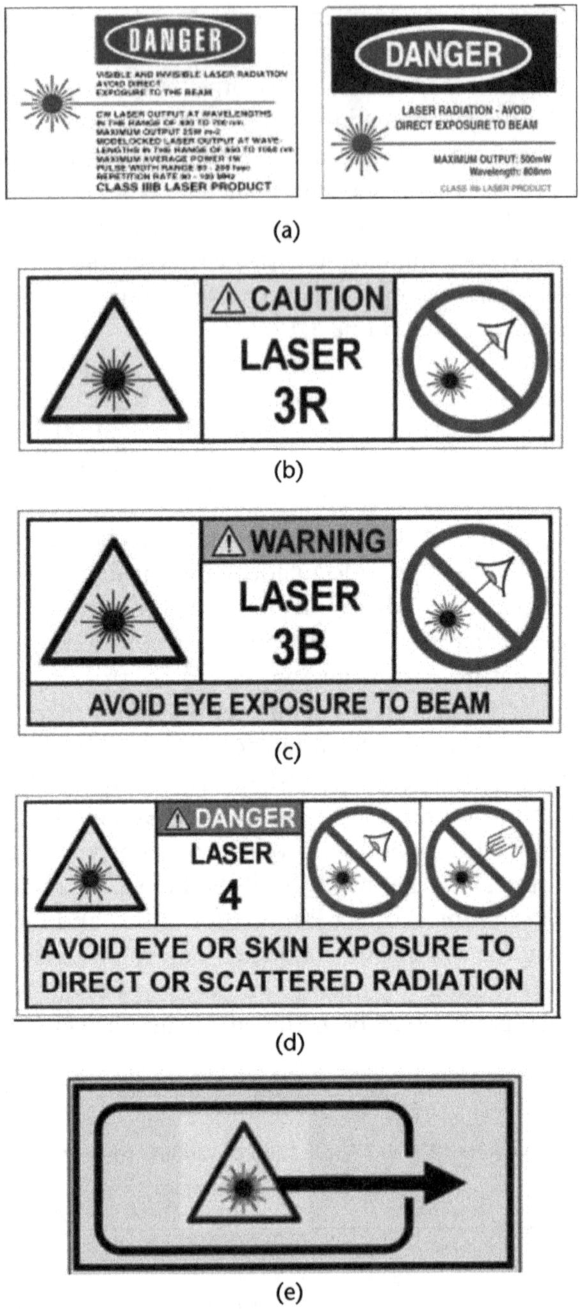

Figure 5.8 Sample of safety labels used in laser equipment: (a) FDA-CDRH , (b) IEC safety label for class 3R, (c) IEC safety label for class 3B, (d) IEC safety label for class 4, and (e) IEC label for Aperture (for classes 3R, 3B, 4).

report is received by CDRH, a unique identification number called an accession number is assigned to the report. Table 5.18 summarizes part of the standards for laser equipment.

Table 5.18 Standards for Laser Equipment

Standard	*Title of Standard*
21 CFR 1040.10	FDA Title 21-Chapter I-Subchapter J-Part 1040-Performance Standards for Lighting-Emitting Products- Sec.1040.10 Laser products.
21 CFR 1040.11	FDA Title 21-Chapter I-Subchapter J-Part 1040-Performance Standards for Lighting-Emitting Products- Sec.1040.11 Specific purpose laser products.
ANSI Z136.1	Safe Use of Lasers, Maximum Permissible Exposure levels.
ANSI Z136.3	Safe Use of Lasers in Health Care.
ANSI Z136.4	Recommended Practice for Laser Safety Measurements for Hazard Evaluation.
ANSI Z136.5	Safe Use of Lasers in Educational Institutions.
ANSI Z136.6	Safe Use of Lasers Outdoors.
ANSI Z136.8	Safe Use of Lasers in Research, Development, or Testing.
ANSI Z136.9	Safe Use of Lasers in Manufacturing Environments.
IEC 60825	Safety of laser products:
series	- Part 1: Equipment classification and requirements.
	- Part 2: Safety of optical fiber communication systems (OFCS).
	- Part 3: Guidance for laser displays and shows.
	- Part 4: Laser guards.
	- Part 5: Manufacturer's checklist for IEC 60825-1.
	- Part 8: Guidelines for the safe use of laser beams on humans.
	- Part 12: Safety of free space optical communication systems used for transmission of information.
	- Part 13: Measurements for classification of laser products.
	- Part 14: A user's guide.
	- Part 17: Safety aspects for use of passive optical components and optical cables in high power optical fiber communication systems.
IEC 60601-2-22	Medical electrical equipment–Part 2-22: Particular requirements for basic safety and essential performance of surgical, cosmetic, therapeutic and diagnostic laser equipment.
IEC 60335-2-113	Household and similar electrical appliances–Safety–Part 2-113: Particular requirements for cosmetic and beauty care appliances incorporating lasers and intense light sources.
ISO 11151 series	Lasers and laser-related equipment–Standard optical components–Part 1: Components for the UV, visible and near-infrared spectral ranges (170 nm to 2,100 nm).
	- Part 2: Components for the Infrared Spectral Range (near-infrared to midinfrared, from wavelengths 2,1 μm to 15,0 μm).
ISO 11252	Lasers and laser-related equipment–Laser device–Minimum requirements for documentation.
ISO 13694	Optics and photonics–Lasers and laser-related equipment–Test methods for laser beam power (energy) density distribution.
ISO 11553-1	Safety of machinery–Laser processing machines–Part 1: Laser safety requirements.
ISO 11146	Lasers and laser-related equipment–Test methods for laser beam widths, divergence angles and beam propagation ratios.
ISO 17526	Optics and optical instruments–Lasers and laser-related equipment–Lifetime of lasers.

Lasers are classified numerically to communicate the eye and skin hazard from the laser emission. There are two hazard classification systems applicable laser performance standards, the FDA's system, and the International Electrotechnical Commission's (IEC) system. Table 5.19 shows how the systems are comparable and how they differ. Each hazard class has a maximum power limit in milliwatts (mW) except class IV (4) which has no limit.

The following products are not included in the above class limits: laser illuminated projectors (LIPs), remote controlled mobile laser products (RCMLPs), laser light distance and ranging (lidar/ladar) mapping and navigation, laser rangefinders, speedometers, laser pointers, and laser illuminators. For the specific purpose laser products, the hazard limit is included in dedicated regulations.

Under the 21 CFR 1040.10 "Laser Products," laser products have certain surveying, leveling, or alignment (SLA) uses. This regulation applies to lasers, products containing lasers, and products intended to contain lasers, specifying classification and user precautions based on radiation accessible during use. Also are specified limits of radiation from viewing optics, interlocks/labels based on radiation accessible during maintenance and service, hazard classes, radiation indicators, and safety aperture label, beam attenuator, emission indicator (some with time delay), remote door interlock, key control, scanning safeguards, required user, maintenance and service manuals, and so on.

Under the 21 CFR 1040.11 "Specific Purpose Laser Products," SLA products are limited to no higher than Class IIIa or the CDRH recognized equivalent, IEC Class 3R. This regulation requires an indication of power levels on medical lasers with +/- 20% accuracy, also referring to demonstration lasers, including display or entertainment (note that variances, with extensive human access limitations, are often granted for laser light shows) [24].

Table 5.19 Hazard Classification for Laser Products

FDA/CDRH	*IEC*	*Classification Description*
I	1	Not recognized as hazardous.
	1M	Do not expose users of telescopic optics.
	1C	Users must follow instructions.
IIa		Hazardous when looking directly for long periods. Supermarket scanners.
II	2	Hazardous–do not stare into the beam; Emit less than 1 mW visible CW radiation. Not considered hazardous for momentary (<0.25 sec) unintentional exposure.
	2M	Hazardous–do not stare into the beam or expose users of telescopic optics.
IIIa		Avoid exposure to the beam. Visible lasers with 1–5 mW power output, invisible lasers, and those having 1–5 times the accessible emission limit (AEL) of class I lasers.
	3R	Avoid exposure to the beam.
IIIb	3B	Direct eye or skin exposure hazard. Includes all other class 3 lasers at all wavelengths which have a power output less than 500 mW.
IV	4	Hazardous for direct or scattered exposure. High power; eye and skin hazard; potential diffuse reflection hazard or fire hazard.

Laser lamps are a type of white-light source that can be more efficient than LED at higher output levels. A type of solid-state lighting (SSL), like LEDs, laser lamps are finding increased use in lamps used in cinema and other image projectors, spotlights, stadium lighting, and car headlights.

Laser lamp technology also presents the laser exposure-related health risks encountered with other laser applications as the risk of exposure to source hot spots or coherent laser light. A logical concern is whether laser lamps also produce the same harmful effects as LEDs used in GLS.

Remote phosphor pumped by a blue laser lamp works by using a blue laser to pump a remote yellow phosphor to produce white light. This is very similar to LED lamp technology, but the laser-pumped lamp can be brighter due to the higher output of blue lasers. LEDs tend to have phosphors that are located very close to the blue light source. The phosphors in these lamps can be located at a considerable distance from the blue light source because beam collimation is possible. Fiber optics can be used to deliver the blue pumped-laser light to the phosphor for conversion to white light. Uses for laser-pumped phosphor illuminators include LIPs, stadium lighting, and automotive headlamps.

Direct multicolor laser lamps work by directly combining three or four laser colors to produce white light. Direct combination results in higher efficiencies and no limited lifetime issues as compared to lamps that use phosphors. Currently, higher-power green lasers are produced by up-conversion using a nonlinear optical crystal, but direct high-power green lasers are now available. Individual laser colors can be delivered through optical fibers into medical instruments, such as endoscopes for viewing internal structures during medical procedures. For general illumination applications, recent US DOE studies showed that four-color (tetrachromic) laser illumination was visually pleasing despite preconceived notions that the quality of illumination would be perceived as unpleasant.

NIR laser illuminators for use in surveillance systems are a solid-state lighting system working in CW mode between 810–940 nm with a power between 100 mW^{-5} (Class IIIb or IV) and have the following characteristics:

- Variable divergence;
- Nominal ocular hazard distance (NOHD): 1m–2.5 km.

NIR lasers illuminators used in surveillance applications are categorized as SLA lasers and may present an ocular hazard by exposing the public to the produced radiation. If people are exposed to NIR laser within the NOHD, they can potentially suffer eye injury. Long-range surveillance (more than 50m) in environments with low ambient light and/or atmospheric interference is a useful capability for many applications, such as traffic monitoring, airport/seaport security, building complex security, general public space monitoring, police surveillance, aid in firefighting, professional sports, border protection, and military/defense. NIR surveillance systems can be fixed in location, (e.g., mounted on a building or bridge), or can be mobile (e.g., mounted on a vehicle, aircraft, or unmanned aerial vehicle (UAV)).

Laser light distance (or detection) and ranging (lidar) are laser products for distance, detection, or ranging measurements. Lidar products generate a range-based dataset in 1, 2, or 3 dimensions. During operation, some of these products may intentionally or unintentionally expose people to hazardous levels of laser radiation.

Lidar platforms include fixed and rotary-wing aircraft, marine vessels, ground-based, and autonomous vehicles. Lidardevices are not considered specific purpose laser products.

Laser hazard class limit was not designed for Lidar products, so the public may be exposed to these emissions. Also, for these products currently premarket safety review of the products do not exist, so they enter into commerce as designed by the manufacturer. Therefore, it is difficult to detect or associate injuries with the product due to the exposure of invisible emissions.

As a safety plan, manufacturers may instruct the Lidar operator to terminate the emissions when are identified unsafe conditions, but this is an administrative control and is not reliable.

5.1.4.5 Ultraviolet Radiation

UV radiation covers nonionizing electromagnetic energy in the wavelength range of 100–400 nm, and is divided into three bands: UV-A (320–400 nm), UV-B (280–320 nm), and UV-C (100-–280 nm). UV radiation comes naturally from the sun, but it can also be created by artificial sources (black lights, welding arcs, UV lasers, fluorescent UV lamp for tanning, UV sterilizers, photo-flash light, etc.) used in industry (spectrophotometry, curing adhesives, coatings, inks, varnishes, decorative glazes, and lacquers, food processing, disinfection of HVAC environment, and healthcare (sterilization, hygiene, etc.). UV radiation has a high photon energy range and is particularly hazardous because there are usually no immediate symptoms of excessive exposure. Too much UV radiation can cause damage to the body's immune system, skin burns, premature aging of the skin (wrinkles), eye damage (cataracts), and skin cancer, but short-term (10–15 minutes) exposure to solar UV radiation can be helpful by producing vitamin D in the human body. Most UV-B (280–320 nm) and all UV-C (100-–280 nm) are absorbed by ozone (O_3) molecules in the upper atmosphere(stratosphere). Consequently, 99% of the solar UV radiation reaching the Earth's surface is UV-A (320–400 nm).

UV-C lamps are light sources that emit optical radiation in the 100- to 280-nm wavelength range. Traditional low-pressure, mercury-based lamps have a peak emission at 253.7 nm. The widespread availability of UV-emitting LEDs is a cause for concern. UV-C lamps are also known as germicidal lamps due to their ability to kill bacteria and other organisms. Exposure to UV-C can cause sunburn of the cornea (photokeratitis), skin sunburn/erythema, and DNA damage leading to long-term effects (e.g., skin cancer) and irritation to the respiratory system (due to the ozone produced). Another potential danger from UV-C lamps is the risk of incorrect installation. There have been reports of incorrect replacement of UV-A lamps with the more dangerous UV-C lamps, usually in insect trap fixtures, with resultant injuries. There have also been warnings about the potential for incorrect replacement of UV-C lamps into UV nail curing appliances that are designed to use UV-A lamps. A dedicated standardization for lamp holders and fixtures for UV-C lamps will avoid such problems in the future. Common uses for UV-C lamps are air and water disinfection, food processing hygiene, laboratory hygiene, medical device sterilization, and consumer use for home sterilization, both for air and on surfaces.

Products using UV-C light need to meet the requirements for safety with proper labels to warn users of the dangers of exposure and photobiological assessment to address personal injury concerns for UV-C. The UV safety spectacles are comfortable and are made from hard, durable non-UV transmissive plastic with close fitting side shields to prevent UV radiation from reaching the eyes.

It is important to remember that product safety certifications of the electrical products address hazards like an electric shock, mechanical, chemical, electromagnetic interferences, fire, and others, but do not address performance and efficacy claims. All products are eligible for performance testing such as photobiologic, photometric, and environmental testing. Performance can be assessed with or without a product safety certification and will not include a certification mark. To measure the effective irradiance from a specific UV lamp, a radiometer-photometer equipped with a sensor in accordance with a sensitivity curve provided by guidance (e.g., American Conference of Governmental Industrial Hygienists (ACGIH) or ICNIRP) or by a standard (e.g., IEC 62471) should be used. The evaluated points should be taken at the closest point to the lamps that the operator's hand or head may reach. The measurement results will be compared with the allowable exposure levels specified in the guidance or the standard.

Requiring that all lamp manufacturers comply with the applicable standards (see Table 5.20), and provide consumers with information about the potential hazards of these lamps through labeling, would ensure that consumers are informed about potential hazards from UV, intense visible, or infrared emissions from lamps.

5.2 Ionizing Radiation Sources

Ionizing radiation is a form of energy that acts by removing electrons from atoms and molecules of materials that include air, water, and living tissue. Ionizing radiation can travel unseen and pass through these materials. Ionization radiation can cause changes in the structure of atoms or molecules by ionization. The ionizing radiations have a frequency ranging from 1 PHz (petahertz) to 10 ZHz (zettahertz), which includes X-rays and gamma rays. This range of the electromagnetic spectrum covers the superior part of ultraviolet radiations (part of UV-C spectrum), X-Ray radiation, alpha, beta, and gamma rays.

Ionizing radiation can be divided into two categories: natural and man-made [25].

Ionizing radiation from natural sources is radiation that comes from natural sources and typically is at low levels. This means that the usual amount of ionizing radiation from natural sources absorbed by human bodies (dose) is small.

In nature, sources of ionizing radiation include radiation from space (cosmic and solar radiation, from the earth (terrestrial radiation), and from building materials.

The cosmic radiation is produced by the stars, including the Sun, and consists of high-energy charged particles, X-rays, and gamma rays that react with the earth's atmosphere to produce secondary radiation which reaches the earth. Higher

Table 5.20 Standards for Ultraviolet Radiation

Standard	*Title of Standard*
ASTM D4329	Standard Practice for Fluorescent UV Lamp Apparatus Exposure of Plastics.
ASTM D5208	Standard Practice for Fluorescent UV Exposure of Photodegradable Plastics.
ASTM E2193	Standard Test Method for Ultraviolet Transmittance of Mono-ethylene Glycol (using Ultraviolet Spectrophotometry).
ASTM E3006	Standard Practice for Ultraviolet Conditioning of Photovoltaic Modules or Mini-Modules Using a Fluorescent U) Lamp Apparatus.
ASTM G154	Standard Practice for Operating Fluorescent UV Lamp Apparatus for Exposure of Non-metallic Materials.
BS 8466	Hats. Protection against solar UVR. Method of test and performance requirements.
CIE 155	Ultraviolet Air Disinfection.
CIE 187	UV-C Photo carcinogenesis Risks from Germicidal Lamps.
EN 13758-1	Textiles. Solar UV protective properties. Method of test for apparel fabrics.
EN 13758-2	Textiles. Solar UV protective properties. Classification and marking of apparel.
IEC 60335-2-27	Household and similar electrical appliances–Safety–Part 2-27: Particular requirements for appliances for skin exposure to optical radiation.
IEC 61228	Fluorescent ultraviolet lamps used for tanning–Measurement and specification method.
ISO 15714	Method of evaluating the UV dose to airborne microorganisms transiting in-duct ultraviolet germicidal irradiation devices.
ISO 15727	UV-C devices — Measurement of output of UV-C lamp.
ISO 15858	UV-C Devices — Safety information–Permissible human exposure.
UL 979	Standard for Water Treatment Appliances.
UL 8802	Outline of Investigation for Germicidal Systems.

altitudes mean greater exposure to cosmic radiation. The average annual dose or exposure from cosmic radiation is 0.33 millisievert (mSv) (33 mrem). This yearly amount of radiation is similar to the amount of radiation from three chest X-rays. This low radiation dose is unlikely to affect human health [25].

Terrestrial radiation refers to radioactive materials that are all around us and can be naturally found on the earth. We are exposed to radioactive materials in the earth when they are breathed in (inhaled), eaten (ingested) (internal radiation), or come through our skin (terrestrial external radiation). Radioactive materials from the earth can enter the human body by being inhaled. The naturally occurring radioactive element uranium can break down and change into an odorless, colorless gas known as radon. Radon can be found in rocks, soil, groundwater, and building materials all around us. When radon is released into the environment, we can breathe it in. On average, a person inhales radioactive materials to cause a radiation dose of 2.28 mSv (228 mrem) per year. The ingested natural sources and terrestrial external radiations cause a small dose of about 0.21–0.29 mSv (21–29 mrem) per year. The low radiation doses ingested or from terrestrial external radiation are unlikely to affect human health, but the inhaled radioactive material in the form of radon could affect health, and radon levels in buildings should be tested and fixed if they are elevated. The amounts of radon you are exposed to may differ depending on where you live, which is why you test radon levels at home to determine if an unacceptable risk of radon exposure exists [25].

Building materials (sandstone, concrete, brick, natural stone, gypsum, and granite) contain naturally occurring radioactive elements like radium, uranium, and thorium. These elements can break down or decay into the radioactive gas radon.

Ionizing radiation from man-made sources is used daily and is found in the use of diagnostic medical examinations, in medical therapy, as a by-product of nuclear power generation, in smoke detectors, in disinfection process of medical instruments, and to perform many other tasks in our everyday lives.

Medical therapy and diagnostic techniques are the main man-made sources of exposure to ionizing radiation including radiation therapy, X-ray, beta, gamma radiation, radiography, fluoroscopy, CT, PET, and nuclear medicine procedures (bone scan, cardiac stress test).

5.2.1 X-Ray Radiation and Beta, Gamma Radiation

X-rays are electromagnetic radiation on the range of 3 PHz (petahertz)–30 EHz (exahertz) and normally have sufficient energy to produce ionization in atoms. The energy required for ionization varies with the material (e.g., 34 eV in air, 25 eV in tissue) but is generally in the range of several eV. The energy range of X-ray radiation is between 12 eV–124 keV with a wavelength of 10 pm–100 nm. Gamma radiation has the energy higher of 124 keV, a frequency higher than 30 EHz, and a wavelength less than 10 pm.

An X-ray beam consists of a distribution of photon energies, and the rate at which energy is delivered by a beam is determined by the number of photons of each energy. The photon energy distribution may be varied by changing the voltage. The number of photons emitted may be varied by changing the current.

X-rays are produced when accelerated electrons collide with the target in the designated X-ray tube, in which the electrons are produced by a cathode (a heated filament of tungsten wire). The electrons bombard the anode (target) under the influence of the high voltage applied between the cathode and the anode. The area of the anode where the electron strike is called the focal spot. When electrons hit the anode, they decelerate or brake and emit bremsstrahlung (meaning braking radiation in German) X-rays. Most of the kinetic energy of the electrons striking the target is converted into heat, less than 1% being transformed into X-rays. The total X-ray energy emitted per second depends on the atomic number Z of the target material and the X-ray tube current.

When electrons change from one atomic orbit to another, characteristic X-rays are produced.

Characteristics and properties of X-rays include:

- They are electrically neutral, not experiencing deviation or deflection when inside an electric, magnetic, or combined field.
- They travel in straight lines at the speed of light.
- They produce biological and chemical effects, which means they can affect an organism by producing ionization and/or cellular changes that may be responsible for disorders or further mutations.

- They are not visible to the human eye or animals, so their detection is possible only by using instruments and photographic methods. This is an important consideration in undertaking protective measures for the human body.
- They produce images on photographic film and fluorescence on certain types of crystals, and both phenomena are used as a means to obtain X-rays films and fluoroscopy images on medical displays, respectively.
- They produce secondary radiation and scattered radiation, which means that a biological object receiving X-ray, in turn, produces new rays with different characteristics. These rays are detrimental to the safety of people who work with X-rays.

The main types of X-ray tubes are the stationary mode tube and the rotating anode tube.

In the stationary tube (used mostly in small capacity X-ray devices), the anode and the cathode are sealed into a glass envelope (tube) in a vacuum. The glass X-ray tube is enclosed in a casing made of aluminum and lined with thin sheets of lead to prevent leakage of radiation. The X-rays leave the tube through a plastic-covered window. The cathode, which contains the filament, is made from nickel or stainless steel.

The tubes with rotating anodes are based on the removal of the target from the electron beam before it reaches a temperature that is too high under the electron bombardment and the rapid replacement of it by another cooler target. The anode is a disc of tungsten or an alloy of tungsten and 10% rhenium. The anode rotates at a speed of 3,000–3,600 rpm or 9,000–1,0000 rpm. The rotor is made from copper cast or rod. The anode bearing must be properly lubricated by lead, gold, graphite, or silver [11].

Alpha, beta, and gamma ray radiations are the results of the disintegration of the nucleus of an unstable radioisotope before reaching a stable form (known as radioactive decay, which is the process in which a radioactive atom spontaneously gives off radiation in the form of energy or particles to reach a more stable state). For example, the decay chain that begins with uranium-238 (U-238) ends in lead-206 (Pb-206), after forming isotopes, such as uranium-234 (U-234), thorium-230 (Th-230), radium-226 (Ra-226), and radon-222 (Rn-222).

It is important to distinguish between radioactive material and the radiation it gives off.

Alpha particles are composite particles consisting of two protons and two neutrons tightly bound together. They are emitted from the nucleus of some radionuclides during a form of radioactive decay, called alpha-decay. An alpha-particle is identical to the nucleus of a normal (atomic mass four) helium atom (e.g., a doubly ionized helium atom). Alpha particles generally carry more energy than gamma or beta particles and deposit that energy very quickly while passing through tissue. Alpha particles can be stopped by a thin layer of light material, such as a sheet of paper, and cannot penetrate the outer, dead layer of skin. Therefore, they do not damage living tissue when outside the body. When alpha-emitting atoms are inhaled or swallowed, however, they are especially damaging because they transfer relatively large amounts of ionizing energy to living cells.

Beta particles are electrons ejected from the nucleus of a decaying atom. Although they can be stopped by a thin sheet of aluminum, beta particles can penetrate the dead skin layer, potentially causing burns. They can pose a serious direct or external radiation threat and can be lethal depending on the amount received. They also pose a serious internal radiation threat if beta-emitting atoms are ingested or inhaled.

Gamma rays represent high-energy electromagnetic radiation emitted by certain radionuclides when their nuclei transition from a higher to a lower energy state. These rays have high energy and a short wavelength. All gamma rays emitted from a given isotope have the same energy, a characteristic that enables scientists to identify which gamma emitters are present in a sample. Gamma rays penetrate tissue farther than do beta or alpha particles but leave a lower concentration of ions in their path to potentially cause cell damage. Gamma rays are very similar to X-rays.

As specified before, ionizing radiation is widely applied in medicine for diagnostic imaging and therapy. Many similarities exist between the equipment that use ionizing radiation in medicine: each have a X-ray or gamma source of radiation, a system that apply the radiation to the body, a detection system that collects the information after the radiation passes through the body (except for therapeutical equipment), an image display and storage system (except for therapeutical equipment), and a processing data system. Below is a short description of the major applications of ionizing radiation in therapy and diagnostic imaging.

Radiotherapy is a generic name for the use of X-ray and radioisotopes (e.g., ^{60}Cobalt) radiation for treatment. Radiation therapy (external beam or brachytherapy) is widely used to treat cancer [11]. External beam radiation therapy (EBRT) is performed with X-rays, gamma rays, electrons, or particle beams (protons are energized particles that have a positive charge) and uses large doses of radiation. For X-ray are used the photon or electron radiation accelerated, while proton therapy uses beams of high energy protons.

The dose rate of radiation therapy, defined as the intensity at which radiation is delivered to the surrounding medium, is usually expressed in grays per hour (Gy/h) and are categorized into the following types:

- Low-dose rate (LDR): < 2 Gy/h;
- Medium-dose rate (MDR): 2-12 Gy/h;
- High-dose rate (HDR): >12 Gy/h.

The electrical acceleration of charged particles is the basis of RF linear accelerators. These particles can range from electrons to ions. With electron linear accelerators (Linac) and betatrons (for photons), the level of the radiation beam can be 20 to 35 MeV (1 eV = 1.6×10^{-19} J). Photon therapy and electron treatment employ photons or electrons with various energies, depending on the treatment.

The accelerator operation begins with a modulator that generates high-voltage pulses delivered to the RF generator that work with microwave tubes (either a magnetron or a klystron), and operate in the s-band (as defined by IEEE is a frequency range from 2 to 4 GHz) with a frequency of 3 GHz. Magnetrons consist of some form of anode and cathode operated in a DC magnetic field normal to a DC electric field between the anode and the cathode. Because of the cross-fields between the anode and the cathode, electrons emitted from the cathode will move in curved paths. If the DC magnetic field is strong enough, the electrons will not arrive at the anode but return instead to the cathode. The modulator also supplies pulses to the electron gun. The RF generator is coupled to the accelerating structure with the isolator. The accelerating structure is also coupled to the RF generator by means of an automatic frequency control (AFC) servo. The detection for the AFC is supplied by two additional internal resonant cavities tuned above or below the resonant frequency. The electron beam strikes a target and becomes converted to X-rays. The X-rays pass through a primary collimator and then enter a conical flattering filter. The treatment head can also contain elements that modify the beam shape and an optical system that determines the field size and the skin-target distance. If the magnetron is employed as a source of RF energy, all electronic components as well as the acceleration structure and the treatment head are contained in a gantry that can rotate 360° around the rotation axis. The tumor center must be located at the isocenter, a point defined by the orthogonal intersection of the axis of rotation of the head and the therapy beam axis. The structure can generate electrons with energies of 6, 9, 12, 15, and 18 MeV.

Radiation with heavy particles involves the use of protons (the nuclei of the hydrogen atom; in other words, hydrogen atoms whose electrons have been removed), heavy ions, and alpha particles, as well as pi-mesons and neutrons. For biomedical purposes, the protons are used directly in biomedical radiation treatment, and after conversion into mesons, they can be used in neutron therapy. Proton beam therapy uses superconducting cyclotrons and synchrotrons to generate and accelerate protons to speed up to 180,000 km/s and energies of up to 250 MeV in a high energy-controlled beam that is delivered very precisely, through a nozzle, to the targeted tumor. The depth of penetration of the protons, which can be between 4–30 cm, is related to their energy. When the process starts, hydrogen atoms are separated into electrons and protons in a superconducting cyclotron. At maximum acceleration, the particles leave the cyclotron and are directed, with a constant speed and energy outside the cyclotron, through a vacuum tube with a preaccelerator to a synchrotron, where are accelerated. Here a radio frequency signal delivers a boost of energy that increases the energy of the protons. After leaving the synchrotron, the protons move through a beam transport system that directs them by deflector magnets to the beam delivery system (nozzle) to the tumor.

A gantry system is used to deliver beams from various directions. The gantry is a steel structure containing the end of the beam transport system and the scanning nozzle. The rotating gantry moves around the patient, delivering the beam at any desired angle. Together, the gantry and treatment table ensure precise positioning of the beam relative to tumor volume. To ensure precise dose delivery the mechanics of the gantry have to be able to keep the isocenter of rotation always within 1 mm under all rotation angles. This requires careful design of the mechanical structure since the overall weight can be several tens of tons.

With proton therapy, there is no radiation dose beyond the tumor, while X-rays continue to deposit radiation doses as they exit the patient body and the radiation could damage nearby healthy tissues, potentially causing side effects.

Brachytherapy is performed by placing radioactive material directly into the affected area giving a high dose rate. Most brachytherapy is performed by gamma rays in the photon energy range from 20 keV to 1.0 MeV. The gamma-ray sources are encapsulated in an inert nonradioactive metal that filters out low energy photons and alpha and beta particles. Common types of radiation sources used in brachytherapy are iodine, palladium, cesium, and iridium. Intravascular brachytherapy (IVB) is a minimally invasive technology that employs a controlled dose of ionizing radiation (gamma or beta radiation) to treat discreet lesions within a vessel or stent. In the IVB, the target (vessel wall) is located just within 1–5 mm from the radiation source. The system uses Strontium-90/Yttrium-90, which is a beta emitter with a maximum energy of 2.27 MeV. The long half-life (29 years) of the source facilitates treatment planning as the dose rate remains almost unchanged during the 6-month life cycle of the temporarily implanted therapy device [11].

Radiation therapy technology has rapidly advanced and now includes devices that use software and image-guided techniques to treat patients. Additionally, the development of methods, such as intraoperative radiation therapy (IORT), image-guided radiotherapy (IGRT), stereotactic radiosurgery (SRS), stereotactic body radiotherapy (SBRT), proton therapy, and the accompanying ability to modulate the energy, size, and shape of the radiation beam in real-time during delivery, has led to shorter fractionation schemes employing higher doses of radiation per fraction.

To ensure the safe use of these devices and treat patients accurately, quality assurance devices have been designed to validate dose calibration and delivery. Treatment planning software systems (TPSs) have been created to automate practices that can control the radiation beam that has become too complex to be done manually. Radiotherapy TPSs are used to plan the geometric, radiological, and dosimetry aspects of the therapy for optimization of treatment. Since absorbed dose distributions (beam model) cannot be measured directly in a patient, they are required to be calculated. With conventional simulation, is possible to identify the edge of a field and to estimate the entire volume being treated and the computer determines the isocenter of a field that will be targeted during the treatment. Modern TPSs packages fully integrate photon therapy, electron therapy, stereotaxic, brachytherapy, provision for simulation and image fusion, and intensity-modulated radiation therapy (IMRT). Numerous accessories, such as beam-limiting collimators, patient positioning systems, and patient motion tracking systems, are currently in use and have a substantial impact on the safe delivery of the radiation treatment beam.

Computerized tomography (CT), also known as computed axial tomography (CAT), is a noninvasive medical examination (procedure) that uses a rotating specialized X-ray source to pass X-rays through a patient's body for the production of cross-sectional images (slices) of a particular area of the body. These two-dimensional images can also be digitally combined to produce a single three-dimensional image. These images are used for a variety of diagnostic and therapeutic purposes.

CT scans can be performed on every region of the body for a variety of reasons (e.g., diagnostic, treatment planning, interventional, or screening) and work as follows:

- A motorized table moves the patient through a circular opening in the CT imaging system.
- While the patient is inside the opening, an X-ray source and a detector assembly within the system rotate around the patient. A single rotation typically

takes a second or less. During rotation, the X-ray source produces a narrow, fan-shaped beam of X-rays that passes through a section of the patient's body.

- Detectors in rows opposite the X-ray source register the X-rays that pass through the patient's body as a snapshot in the process of creating an image. Many different snapshots (at many angles through the patient) are collected during one complete rotation.
- For each rotation of the X-ray source and detector assembly, the image data are sent to a computer to reconstruct all of the individual snapshots into one or multiple cross-sectional images (slices) of the internal organs and tissues.

Over the past 30 years, CT technology has rapidly advanced and now includes such performance and safety features as rapid helical scanning, automatic tube-current modulation, various methods of incorporating multiple energy spectra acquisitions for added physical information in reconstructed images, the development of cone-beam X-ray sources with flat-panel detectors (cone-beam CT), and nonlinear iterative reconstruction algorithms for improved image quality or reduced radiation dose.

Cone-beam CT (CBCT) is a means of capturing and displaying volumetric X-ray data using a methodology that is similar to that of conventional CT. Conventional CT systems employ a fan-shaped beam that is scanned along the patient's length. These systems are routinely characterized by the acquisition slice thickness and the number of simultaneous channels of data acquired per gantry rotation [26].

In CBCT the concept of beam slice thickness and number of detector channels simultaneously acquired do not apply. A very broad beam is employed that is of sufficient breadth to entirely encompass the anatomy of interest without the need to scan along with the patient. A single rotation of the CT gantry is sufficient to collect the entire set of imaging data.

CBCT is now available either as an add-on feature for certain fluoroscopic systems or as a stand-alone, dedicated device. As dedicated systems, CBCT devices are finding widespread use in dentistry as well as with ENT and extremity imaging.

Certain types of medical X-ray imaging equipment that operate in a Digital Imaging and Communications in Medicine (DICOM), where a compliant environment can be configured to provide a means for capturing essential dose-related information as a distinct record. This DICOM feature, called Radiation Dose Structured Report (RDSR), reports the dosimetry output of X-ray devices (it does not report patient dose values).

A CT scanning system consists of an X-ray subsystem, a gantry, a patient table, and a controlling computer desk. A high-voltage (100–160 kV) X-ray generator supplies electric power to the X-ray rotating anode tube (120 kV at 200–500 mA), producing X-rays with an energy range of 30–120 keV. The gantry (which typically weighs about 1,800 kg) houses the X-ray tube, X-ray generator, filters, detector system, collimators, X-ray tube heat exchange (oil cooler), direct drive gantry motor, rotational frame, and rotation control mechanism. A separate power distribution cabinet consists of a high voltage generator, power converter AC to DC, line noise filter, control sensing, and other relevant parts; all CT systems consist of the following four major subsystems:

- Scanning system that takes suitable readings for a picture to be reconstructed and includes X-ray source and detectors. Currently, there are multislice computer tomographs (MSCT) up to 256-slice.
- A processing unit that converts these readings into intelligible picture information.
- Display parts that present this information in visual form and include other manipulative aids to assist in diagnostic.
- A storage unit that enables the information to be stored for future analysis.

The controlling computer desk, located outside of the CT operation room, consists of the control, processing, display, and storage units of the CT system [11].

Radiographic X-ray systems produce two-dimensional images of the body's internal structures. X-rays are produced by an X-ray tube, pass through the desired portion of the body, are partially absorbed by the body, and reach an image receptor. The varying intensities of X-rays that exit the body are reflective of the composition and densities of the body structures. Radiographic exams routinely consist of a small number of individual images. In a fluoroscopic procedure, a device passes X-rays through a patient's body for a brief length of time to capture a real-time moving image, which can be used to observe the movement of an object or substance in the body.

Historically, radiographic X-ray systems used X-ray film as the image receptor. Increasingly, the film is being replaced by digital image receptors that can be reused many times and offer many imaging performance improvements. Manufacturers have introduced devices consisting of a digital image receptor and associated software to allow an X-ray film system to be upgraded with a digital image receptor. Digital systems may take over functions from the existing X-ray system, such as user input of X-ray settings, initiating the exposure (exposure switch), and terminating an exposure once adequate radiation has been received at the image receptor.

The major components of a general-purpose X-ray analog system are the following: The X-ray tube, X-ray generator (50–150 kV at 0–600 mA), tube stand or support, intensifier screen, examination table, and control unit.

The digital X-ray machine is the modern development of the conventional X-ray technology that uses a digital-detector system instead of traditional photographic film.

A handheld X-ray system is a radiographic system that is portable and can be held by hand. This type of device can be used for dental procedures and imaging extremities. Handheld X-ray systems have unique concerns related to operator safety that can be addressed through shielding and labeling safety requirements.

Fluoroscopic X-ray systems employ the same basic concepts of image production as other radiographic X-ray systems. However, fluoroscopic systems produce these images repeatedly and in real ime, allowing visualization of the motion of internal structures of the body. Fluoroscopic X-ray systems can also be used to image materials and devices that are placed in the body during clinical procedures. These include, but are not limited to, imaging an ingested liquid as it passes through the digestive tract and monitoring the location of devices such as biopsy needles and instruments such as catheters, stents, blood clot filters, and other devices as they are moved through the vascular system. Fluoroscopic procedures can vary greatly in duration from a few seconds to minutes or even hours of X-ray exposure depending on the complexity of the procedure.

Mammographic radiographic units use low-dose X-rays to produce specialized images of the breast that provide information about breast morphology, normal anatomy, and gross pathology. The mammographic examination is used in comparison of the breast with ultrasound diagnostic imaging. The anode material for X-tube used in mammography is molybdenum (Mo) and the voltage applied to the X-ray tube is less (20–40 kV) than the high voltage used normally in the other radiographic procedures. Rhodium (Rh) is also used as a target material and filter.

In radiography and fluoroscopy mobile equipment the connection between the X-ray source and X-ray detector is made by C-arm (the name derived from the C-shaped arm used). Most C-arm units use an image intensifier that converts the X-rays into a visible image that is displayed on the C-arm monitor (Figure 5.9).

In diagnostic X-ray systems and their major components the effective dose leakage is limited at 1m from the source to 1 mSv in 1 hr and at 5 cm from any other components to 20 μSv in 1 hr. For mammographic equipment, the effective dose transmission through a mammographic image support system is limited at 5 cm to 1 μSv effective dose for each tube activation. In fluoroscopic equipment, entrance exposure rates are limited to an effective dose of 50 mSv/min (or 100 mSv/min with automatic exposure rate control). This limits the minimum possible distance between the X-ray tube and the patient's skin to 38, 30, 20, 19, or 10 cm depending on the device's design and intended use.

In a cabinet X-ray systems with X-ray tube installed in an enclosure, limits for radiation effective dose are at 5 cm is 5 μSv /hr under maximized operating conditions and door positions. Human access to the primary beam is also restricted and the following construction requirements are applied: two interlocks on each door with one resulting in physical disconnection of energy to the generator, use of key control, the presence of two independent X-rays on indicators, warning indicators and labels, and user instructions.

During a *nuclear medicine* procedure, such as a PET scan, a patient is given a small amount of a radioactive substance, called a radiopharmaceutical or radiotracer. A detector outside the body is then used to view an image of the radioactive

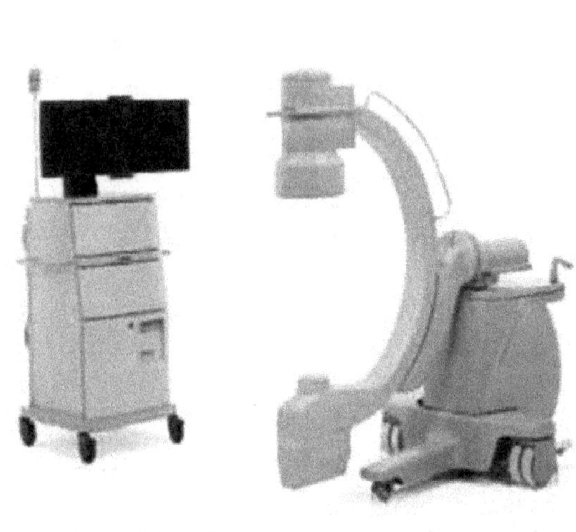

Figure 5.9 C-arm radiography mobile unit.

material as it moves through the body and to observe metabolic processes in the body. MRI and CT scans show what an organ looks like, while a PET scan can tell us how it is working. Some radiopharmaceutical is used for the treatment of specific illness. Once the radiopharmaceutical is given, the patient is usually asked to lie down on a table. A special camera that detects gamma radiation is placed over the patient's body to take pictures. A computer is used to show where the body concentrates the radioactive material. The radioactive materials usually leave the body within hours to months. PET scans (metabolic image) may be performed together with a CAT scan, that provides an image of the organ (anatomical image). PET scans provide a clear view of how the organs are working at the cellular level and if they have been damaged.

PET is based on the detection of coincidence events from electron-positron annihilation, which releases two 511-keV gamma rays moving in opposite directions. It is the detection of these gamma-rays that forms the basis of PET scanning. The main parts of a PET system include a gantry, a detector assembly (gamma-camera), coincidence circuits, a patient table, and a computer processing system. Four types of crystals are used in the PET cameras: bismuth germanate oxide (BGO), lutetium oxyorthosilicate (LSO), gadolinium oxyorthosilicate (GSO), and lutetium yttrium orthosilicate (LYSO). The performance standard for PET systems have been developed by the NEMA and the Society for Nuclear Medicine. These standards define common PET terminology, set experimental procedures, and provide tests for scanner evaluation [11].

Radiation sterilizers are used for the sterilization of gaseous, liquid, solid materials, homogenous and heterogeneous systems, and medical devices (syringes, needles, cannulas, etc.). In radiation sterilizers both nonionizing radiation and ionizing radiation are used. Ionizing radiation (X-ray and gamma-ray) with short wavelength and high intensity and a dose rate of 6.8 kGy/h destroy microorganisms. Cobalt-60 and Cesium-137 sources are an example of radioisotopes used in radiation sterilizers. The use of a radioisotope requires shielding (lead and stainless steel) to ensure the safety of the operators while in use and storage as these radioisotopes continuously emit gamma rays as they cannot be turned off. The maximum energy from Cobalt-60 is 1.33 MeV, which is not sufficient to cause nuclear transformations of any type in the product being sterilized. The product to be sterilized is exposed to radiation for 10–20 hours, depending on the strength of the source. Electrons with a high specific charge used in electron or photon therapy, are widely used also for sterilization purposes [11].

The *performance for diagnostic X-ray systems* and their major components (radiographic, fluoroscopic, CT equipment, etc.) are included in many IEC, ISO, and ASTM standards (see Table 5.22). These standards address aspects of radiation performance including control and an indication of technique factors, reproducibility of technique factors, the visual definition of the X-ray field, field alignment, source to skin distance, display of air kerma (kinetic energy released per unit mass = the radiation concentration delivered to a point) rates (AKR), computed tomography dose index (CTDI), information to be provided in the product labeling, and warning statements that must accompany the product.

Based on the existing standards and codes, each medical unit that conducts diagnostic and therapy based on ionizing radiation needs to develop *internal quality control* and assurance protocols to be used on the equipment and, in particular, the performance objectives and criteria. These protocols can be structured as follows:

- Performance objectives and criteria—generic;
- System description—custom;
- Acceptance tests and commissioning—largely generic;
- Quality control (QC) of equipment—largely generic;
- Documentation—generic;
- Table of QC tests—custom entries in a generic format.

These protocols focus on the following issues:

- Functionality;
- Reproducibility;
- Accuracy;
- Characterization and documentation;
- Data transfer and validation;
- Completeness.

System parts, such as linear accelerators, simulators, kV radiotherapy unit, isotopes therapy unit, multileaf collimators, and imaging devices, must establish daily, monthly, and annual quality control actions. Table 5.21 shows an example of a list of quality control tests for linear accelerator units [27].

For each test the necessary test equipment (which needs to be calibrated by an accredited laboratory), expected results, and accepted tolerance must specified. When establishing the tolerance, the aging of the equipment must be considered.

Ionizing radiation can penetrate the human body and the radiation energy can be absorbed in tissue and the person is irradiated. This has the potential to cause harmful effects to people, especially at high levels of exposure.

Because CT, fluoroscopy, and nuclear medicine procedures involve repeated or extended exposure to radiation, these types of examinations are associated with a higher radiation dose than conventional radiography [28] .

The *biological effects* of ionizing radiation on the body are described by effective dose measured in millisievert (mSv). The American College of Radiology recommends limiting lifetime diagnostic ionizing radiation exposure to 100 mSv. That is equal to 10,000 chest X-rays or up to 25 chest CTs. For example, the adult effective dose from a CT exam of the head is equivalent to the adult effective dose from roughly 100 chest X-rays [29]. The adult effective dose from a CT exam of the abdomen is roughly equivalent to the adult effective dose from roughly 400 chest X-rays.

The guiding principle of radiation safety is ALARA, which stands for as low as reasonably achievable. This principle means that even if it is a small dose, you should try to avoid it if receiving that dose has no direct benefit. To quantify this, there are three basic protective measures in radiation safety: time, distance, and shielding.

Table 5.21 Quality Control Tests for Kilovoltage Radiotherapy Units

Periodicity of Tests	*Quality Control Tests*
Daily	Door interlock/last person out functional
	Motion interlock functional
	Couch brakes functional
	Beam status indicators functional
	Patient audiovisual monitors functional
	Room radiation monitors functional
	Beam interrupt/counters functional
	Lasers/crosswires
	Optical distance indicator
	Optical back pointer
	Field size indicator
	Output constancy—photons
	Dynamic wedge factors
	Output constancy—electrons
Monthly	Emergency off functional
	Wedge, tray cone interlocks functional
	Accessories integrity and centering functional
	Gantry angle readouts
	Collimator angle readouts
	Couch position readouts
	Couch isocenter
	Couch angle
	Optical distance indicator
	Crosswire centering
	Light/radiation coincidence
	Field size indicator
	Relative dosimetry
	Central axis depth dose reproducibility
	Beam flatness
	Beam symmetry records
Annually	Reference dosimetry
	Relative output factor reproducibility
	Wedge transmission factor reproducibility
	Accessory transmission factor reproducibility
	Output reproducibility vs. gantry angle
	Beam symmetry reproducibility vs. gantry angle
	Monitor chamber linearity
	End monitor effect
	Collimator rotation isocenter
	Gantry rotation isocenter
	Couch rotation isocenter
	Coincidence of collimator, gantry, couch axes
	Coincidence of isocenters
	Couch deflection
	Independent quality control reviews

Time:

- Time refers to the amount of time spent near a radiation source;
- Minimize time near a radiation source to only as long as it takes to accomplish a task;
- First responders can use alarming dosimeters to help them minimize the amount of time they are in an area with elevated radiation levels.

Distance:

- Distance refers to how close you are to a radiation source;
- Maximize distance from a radioactive source as much as possible;
- If the distance from a radiation source is increased, the exposed dose will decrease.

Shielding:

- To shield yourself from a radiation source, put something between you and the source (a barrier).
- In a radiation emergency, officials may instruct you to get inside and put as many walls between you and the outside as possible. This is another way to use shielding.
- Protective clothing can shield first responders from alpha and beta particles, but will not protect them from gamma rays. Standing behind a wall or a fire truck can also serve as a shield.

When a radiation emergency occurs used personal protective equipment must be that can minimize the exposure:

- Respirators will help protect from inhalation hazards;
- Protective clothing helps keep radioactive material off skin and hair;
- Alarming dosimeters help manage stay time and track the accumulated doses in an area with elevated radiation levels.

At the entrance in the location, or in the work areas in which X-ray radiation sources are present, warning signs must be posted. A few examples of such signs are given in Figure 5.10.

A person exposed to radioactive material is not necessarily contaminated. For a person to be contaminated, radioactive material must be on or inside of their body. External contamination occurs when radioactive material comes into contact with a person's skin, hair, or clothing. It is important to remove the radioactive material as quickly as possible to lower the risk of harm and reduce the chance of spreading contamination. Internal contamination can occur when radioactive material is swallowed or breathed in. Internal contamination can also occur when radioactive material enters the body through an open wound. Different radioactive materials can accumulate in different body organs. In such situations decontamination must be made by medical personnel.

Figure 5.10 Warning signs for X-ray radiation.

Some causes of accidents associated with the work of X-ray equipment are

- Poor equipment configuration (e.g., unused beam ports not covered, interlock system is not engaged).
- Manipulation of equipment when energized (e.g., adjustment of samples or alignment of optics when the X-ray beam is on).
- Equipment failure (e.g., shutter failure, warning light failure).
- Inadequate training or violation of procedure (e.g., incorrect use of equipment, overriding interlocks).
- Installation of thick shielding walls around an X-ray source but ignoring the roof, X-rays can scatter off air molecules over shielding walls by radiation streaming and creating a radiation field known as sky shine.
- Whenever the voltage on a device can produce some X-rays even if the current is too low. This is called dark current and can cause unnecessary exposure that should be prevented.
- Use of inadequate materials (nonleaded glass windows, etc.) on the construction of a location consisting of an X-ray source of radiation.

Instruments used for radiation measurement fall into two broad categories [30]:

- Rate measuring instruments;
- Personal dose measuring instruments.

Rate measuring instruments measure the rate at which exposure is received (more commonly called the radiation intensity). Survey meters, audible alarms, and area monitors fall into this category.

These instruments present a radiation intensity reading relative to time, such as R/hr or mR/hr.

Dose measuring instruments are those that measure the total amount of exposure received during a measuring period. The dose-measuring instruments, or dosimeters, that are commonly used in industrial radiography are small devices that are designed to be worn by an individual to measure the exposure received by the individual.

The light meter is a portable unit designed to measure visible, UV, and near-UV light in a specified environment and is capable of reading any optical unit of energy or power level if the appropriate detector has been calibrated. The spectral range of the instrument is limited only by the choice of detector.

Steady-state measurements can be made from a steady-state source using the normal operation mode. Average measurements can be obtained from a flickering or modulated light source with the meter set in the fast function position. Flash measurements can be measured using the integrate function.

The ionizing radiation survey meter is useful for measuring radon decay products from air samples collected on filters. Wipe samples collected on a filter can also be counted with this detector and general area sampling can be done. Several types of ionizing radiation meters are commercially available.

The survey meter with the scintillation detector can be used to measure the presence of radon decay products in a dust sample. The barometric pressure should be noted for ionizing radiation chambers.

The survey meters that are used to measure radioactive isotopes radiation in the field typically consist of a detector and an analog or digital display. The detector consists of an anode and cathode in a gas-filled chamber where the gas becomes ionized whenever the counter is brought near radioactive substances. Depending on the voltage applied between the anode and the cathode, the detector may be considered an ion chamber, a proportional counter, or a Geiger-Müller (GM) detector. When used for gamma radiography, ion chamber detectors and GM meters are typically calibrated for the energy of the gamma radiation being used. Most often, gamma radiation from Cs-137 at 0.662 MeV provides the calibration. Only small errors occur when the radiographer uses Ir-192 (average energy about 0.34 MeV) or Co-60 (average energy about 1.25 MeV).

Pocket dosimeters are used to provide the wearer with an immediate reading of the exposure to X-rays and gamma rays and they are commonly worn in the pocket. The two types commonly used are the direct read pocket dosimeter and the digital electronic dosimeter. Audible alarms included in dosimeters emit a short beep when a predetermined exposure level from a gamma emitter has been received. Most audible alarms use a GM detector. Typical alarm rate meters will begin sounding in areas of 450-500 mR/h, but these audible alarms are not intended to be and should not be used as replacements for survey meters.

Personnel dosimetry film badges are commonly used to measure and record radiation exposure due to gamma rays, X-rays, and beta particles. The detector is a piece of radiation-sensitive film contained inside a film holder or badge. The film is packaged in a light-proof, vapor-proof envelope preventing light, moisture, or chemical vapors from affecting the film. The major advantages of a film badge as a personnel monitoring device are that it provides a permanent record, it can distinguish between different energies of photons, and it can measure doses due to different types of radiation. It is quite accurate for exposures greater than 100 millirem. Film badges need to be worn correctly so that the dose they receive accurately represents the dose the wearer receives. Whole-body badges are worn on the body between the neck and the waist, often on the belt or a shirt pocket. The clip-on badge is worn most often when performing X-ray or gamma radiography.

An alternative for a film badge is the thermoluminescent dosimeter (TLD), which is a phosphor, such as lithium fluoride (LiF) or calcium fluoride (CaF) in a solid crystal structure. A TLD is reusable (usually 3 months or less) and presents good linearity of response to dose, relative energy independence, and is sensitive to low doses. Instead of reading the optical density (blackness) of a film, as is done with film badges, the amount of light released versus the heating of the individual pieces of thermoluminescent material is measured.

There are four different but interrelated units for measuring radioactivity, exposure, absorbed dose, and dose equivalent, which are as follows:

- Radioactivity refers to the amount of ionizing radiation released when an element (such as uranium) spontaneously emits energy as a result of the radioactive decay (or disintegration) of an unstable atom. Whether it emits alpha or beta particles, gamma rays, X-rays, or neutrons, a quantity of radioactive material is expressed in terms of its radioactivity (or activity). Radioactivity is the term used to describe the rate at which radioactive material emits radiation, or how many atoms in the material decay (or disintegrate) in a given period of time. The units of measure for radioactivity are the becquerel (Bq), per International System (SI) unit, and curie (Ci), per British System (BS) unit. One Bq represents a rate of radioactive decay is equal to 1 disintegration per second and 1 Ci represents a rate of radioactive decay equal to 37 billion (37×10^9) disintegrations per second; 37 GBq = 1 curie (Ci).
- Exposure describes the amount of radiation traveling through the air. Many radiation monitors measure exposure. The units for exposure are the roentgen (R) and coulomb/kilogram (C/kg). One R is the amount of gamma or X-rays required to produce ions resulting in a charge of 258 μC/Kg of air under standard conditions.
- Absorbed dose describes the amount of radiation absorbed by an object or person (that is, the amount of energy that radioactive sources deposit in materials through which they pass). The units for absorbed dose are the radiation absorbed dose (rad) and gray (Gy) per SI. 1 Gy = 100 rad.
- Dose equivalent (or effective dose) combines the amount of radiation absorbed and the medical effects of that type of radiation. For beta and gamma radiation, the dose equivalent is the same as the absorbed dose. For alpha and neutron radiation, the dose equivalent is larger than the absorbed dose because these types of radiation are more damaging to the human body. In this case, the dose equivalent (in rems) is equal to the absorbed dose (in rads) multiplied by the quality factor of the type of radiation (see the 10 CFR 20.1004). Units for dose equivalent are the roentgen equivalent man (rem) and sievert (Sv) per SI. 1Sv is the dose equivalent equal to 1 J/Kg. The biological dose equivalents are commonly measured in 10^{-3} of rem (millirem or mrem). 1 Sv = 100 rem.

For practical purposes, 1R (exposure) = 1 rad (absorbed dose) = 1 rem or 1,000 mrem (dose equivalent) or 1R (exposure) = 0.01 Gy (absorbed dose) = 0.01 Sv (dose equivalent).

A measure given in Bq or Ci tells the radioactivity of a substance, while a measure in Sv or rem (or mrem) tells the amount of energy that a radioactive source deposits in living tissue.

International standards organizations develop standards for diagnostic and therapy X-ray, beta, and gamma radiation systems that address radiation performance and safety. The IEC publishes safety and essential performance standards for various types of such equipment. The IEC standards are consensus standards, with a large international group of stakeholders participating in their development including industry, academia, and end-users. IEC standards are currently used in the regulatory framework of the FDA, European Medical Device Authorities, and many other countries around the world. Therefore, medical device manufacturers are required to manufacture their devices so that they conform to relevant IEC standards.

IEC standards are usually reviewed every 5 years to determine if updates are necessary.

Table 5.22 lists examples of standards that refer to ionizing radiation.

5.3 Sound Waves

A sound wave is a disturbance that propagates through a medium (gases, liquids, solids) and time, usually with some transfer of energy. The excitation of a sound wave is a scalar (physical real number that has only magnitude) quantity. The propagating disturbance in the sound-conducting medium is in the form of alternate compressions and rarefactions (diminution in the density) of the medium, which is initially caused by the vibrating sound sources. This situation is specific for longitudinal waves, as are the sound waves. The pressure variations due to the propagating sound are superimposed on the ambient air pressure. The intensity of the sound is the amount of energy transmitted by a sinusoidal sound wave per unit of time through each unit area perpendicular to the direction of sound propagation. The sound intensity level is the magnitude of sound intensity, expressed in logarithmic units (decibels). A sound level of 0 dB is the threshold of hearing a sound of 1 kHz that has an intensity of 1 pW/m^2. The threshold of pain that generates damage in few minutes is 120 dB, representing an intensity of 1 W/m^2. The propagation speed of a sound through the dry air at 20°C is 343 m/s (1,100 ft/s) and in water or the body tissue is 1,540 m/s (4,800 ft/s). The sound waves have the following characteristics: frequency, period, wavelength, propagation speed (velocity), amplitude, and intensity. Frequency, period, amplitude, and intensity are determined by the sound source. Propagation speed is determined by the medium, and wavelength is determined by both the source and medium. Like light, the sound gets reflected at the surface of a solid or liquid and follows the same laws of reflection.

The sound waves can be divided into infrasound, audible sound, and ultrasound (see Table 5.1, Section B).

5.3.1 Acoustic Noise Exposure

Acoustical noise represents any unwanted sound or vibration that is audible to the human ear and/or tangible via touch. From the multitude of surrounding sound and vibration to which we are exposed, those that are within specifications for the task they perform and the noise level they create do not exceed safety standards and are

Table 5.22 Examples of Standards that Refer to Ionizing Radiation

Standard	*Description*
ASTM C1831	Standard Guide for Gamma Radiation Shielding Performance Testing
ASTM E181	Standard Test Methods for Detector Calibration and Analysis of Radionuclides
ASTM E666	Standard Practice for Calculating Absorbed Dose from Gamma or X Radiation
ASTM E1161	Standard Practice for Radiologic Examination of Semiconductors and Electronic Components
ASTM E1255	Standard Practice for Radioscopy
ASTM E1411	Standard Practice for Qualification of Radioscopic Systems
ASTM E1695	Standard Test Method for Measurement of CT System Performance
ASTM E1815	Standard Test Method for Classification of Film Systems for Industrial Radiography
ASTM E2738	Standard Practice for Digital Imaging and Communication in Non-destructive Evaluation (DICONDE) for CR Test Methods
ASTM F3094	Standard Test Method for Determining Protection Provided by X-ray Shielding Garments Used in Medical X-ray Fluoroscopy from Sources of Scattered X-Rays
ANSI N42.43	American National Standard Performance Criteria for Mobile and Transportable Radiation Monitors Used for Homeland Security
IEC 60601-1-3	Medical electrical equipment–Part 1-3: General requirements for basic safety and essential performance–Collateral Standard: Radiation protection in diagnostic X-ray equipment
IEC 60601-2-1	Medical electrical equipment–Part 2-1: Particular requirements for the basic safety and essential performance of electron accelerators in the range 1 MeV to 50 MeV
IEC 60601-2-11	Medical electrical equipment–Part 2-11: Particular requirements for the basic safety and essential performance of gamma beam therapy equipment
IEC 60601-2-28	Medical electrical equipment–Part 2-28: Particular requirements for the basic safety and essential performance of X-ray tube assemblies for medical diagnosis
IEC 6060102-44	Medical electrical equipment–Part 2-44: Particular requirements for the basic safety and essential performance of X-ray equipment for computed tomography
IEC 60601-2-54	Medical electrical equipment–Part 2-54: Particular requirements for the basic safety and essential performance of X-ray equipment for radiography and radioscopy
IEC 60601-2-63	Medical electrical equipment–Part 2-63: Particular requirements for the basic safety and essential performance of dental extra-oral X-ray equipment
IEC 60601-2-64	Medical electrical equipment–Part 2-64: Particular requirements for the basic safety and essential performance of light ion beam medical electrical equipment
IEC 60601-2-68	Electrical medical equipment–Part 2-68: Particular requirements for the basic safety and essential performance of X-ray-based image-guided radiotherapy equipment for use with electron accelerators, light ion beam therapy equipment and radionuclide beam therapy equipment
IEC 60731	Medical electrical equipment–Dosimeters with ionization chambers as used in radiotherapy
IEC 60846-2	Radiation protection instrumentation–Ambient and/or directional dose equivalent (rate) meters and/or monitors for beta, X, and gamma radiation–Part 2: High range beta and photon dose and dose rate portable instruments for emergency radiation protection purposes

Table 5.22 (continued)

Standard	*Description*
IEC 60976	Medical electrical equipment–Medical electron accelerators–Functional performance characteristics
IEC 61017	Radiation protection instrumentation–Transportable, mobile or installed equipment to measure photon radiation for environmental monitoring
IEC 61217	Radiotherapy equipment–Coordinates, movements, and scales
IEC 61322	Radiation protection instrumentation–Installed ambient dose equivalent rate meters, warning and monitoring assemblies for neutrons with energies from thermal to 20 MeV
IEC 61563	Radiation protection instrumentation–Equipment for measuring specific activity of gamma-emitting radionuclides in foodstuffs
IEC 62083	Medical electrical equipment–Requirements for the safety of radiotherapy treatment planning systems
IEC 62274	Medical electrical equipment–Safety of radiotherapy record and verify systems
IEC 62327	Radiation protection instrumentation–Hand-held instruments for the detection and identification of radionuclides and for the estimation of ambient dose equivalent rate from photon radiation
IEC 62401	Radiation protection instrumentation–Alarming Personal Radiation Devices (PRD) for detection of illicit trafficking of radioactive material
IEC 62484	Radiation protection instrumentation–Spectroscopy-based portal monitors used for the detection and identification of illicit trafficking of radioactive material
IEC 62618	Radiation protection instrumentation–Spectroscopy-based alarming Personal Radiation Detectors (SPRD) for the detection of illicit trafficking of radioactive material
IEC 62667	Medical electrical equipment–Light ion beam medical equipment - Performance characteristics
IEC 62963	Radiation protection instrumentation–X-ray CT inspection systems of bottled/canned liquids
IEC 62706	Radiation protection instrumentation–Recommended climatic, electromagnetic and mechanical performance requirements and methods of tests
IEC 62945	Radiation protection instrumentation–Measuring the imaging performance of X-ray CT security screening systems
IEC 62963	Radiation protection instrumentation–X-ray CT inspection systems of bottled/canned liquids
IEC/ TS 45B-832	Radiation protection instrumentation–Dosemeters for pulsed fields of ionizing radiation (working document)
ISO 3999	Radiation protection–Apparatus for industrial gamma radiography–Specifications for performance, design, and tests
ISO 4037 series	Radiological protection–X and gamma reference radiation for calibrating dosemeters and dose-rate meters and for determining their response as a function of photon energy
ISO 6980 series	Nuclear energy–Reference beta-particle radiation
ISO 7503 series	Measurement of radioactivity–Measurement and evaluation of surface contamination
ISO 7212	Enclosures for protection against ionizing radiation–Lead shielding units for 50 mm and 100 mm thick walls
ISO 8194	Radiation protection–Clothing for protection against radioactive contamination — Design, selection, testing and use
ISO 11665 series	Measurement of radioactivity in the environment–Air: radon-222
ISO 11929 series	Determination of the characteristic limits (decision threshold, detection limit and limits of the coverage interval) for measurements of ionizing radiation–Fundamentals and application

Table 5.22 (continued)

Standard	*Description*
ISO 12052	Health informatics–DICOM including workflow and data management
ISO 14146	Radiological protection–Criteria and performance limits for the periodic evaluation of dosimetry services
ISO 14152	Neutron radiation protection shielding–Design principles and considerations for the choice of appropriate materials
ISO 16645	Radiological protection–Medical electron accelerators–Requirements and recommendations for shielding design and evaluation
ISO 18589 series	Measurement of radioactivity in the environment–Soil
ISO 20785 series	Dosimetry for exposures to cosmic radiation in civilian aircraft
ISO 22188	Monitoring for inadvertent movement and illicit trafficking of radioactive material
ISO 27048	Radiation protection–Dose assessment for the monitoring of workers for internal radiation exposure
ISO/ASTM 52628	Standard practice for dosimetry in radiation processing
NEMA NU 2	Performance Measurements of PET

acceptable. Sounds at or below 70 dBA (decibels are on the A-weighted scale of a standard sound level meter at slow response) are generally considered safe. Those noises that cause discomfort or a health hazard become unwanted. A sound may be harmful if

- You have difficulty talking or hearing others talk over the sound;
- The sound makes your ears hurt;
- Your ears are ringing after hearing the sound;
- Other sounds seem muffled after you leave an area where there is a loud sound.

At work sites, means of protection against the effects of noise exposure will be provided when the sound levels exceed 85 dBA for 8 hours of exposure until 115 dBA up to 15 minutes or less. When the daily noise exposure is composed of two or more periods of noise exposure of different levels, their combined effect should be considered, rather than the individual effect of each. Such means of protection can be earplugs, earmuffs, or hearing bands. The FDA considers the maximum permissible average level of noise of 105 dB(A) (for 1 hour) and an overall maximum limit of 140 dB without ear protection.

In the United States for general industry, the standard 29 CFR 1910.95 "Occupational Noise Exposure" is designed to protect general industry workers, such as those working in the manufacturing, utilities, and service sectors establishing permissible noise exposures, requires the use of engineering and administrative controls and setting out the requirements of a hearing conservation program [31].

EU Directive 2002/49/EC [32] related to the assessment and management of environmental noise is the regulatory document that applies to noise exposure of humans to noise emitted by means of transport, road traffic, rail traffic, air traffic, and sites of industrial activity, in built-up areas, public parks, designated quiet areas, near schools, hospitals, and other noise-sensitive buildings and areas. The above directive does not apply to noise that is caused by the exposed person themself, noise from domestic activities, the noise created by neighbors, noise at workplaces, or noise inside means of transport or due to military activities in military areas. For these types of noises, specific requirements are used.

Hearing protection is used effectively by making sure that the protectors give enough protection; for example, at least to get below 85 dB at the ear (the limit level specified in occupational noise exposure documents). According to NIOSH Publication 89-126, the recommended exposure limit (REL) for occupational noise exposure is 85 dB, A-weighted, as an 8-hr time-weighted average (TWA) (85 dBA as an 8-hr TWA). Exposures at and above this level are considered hazardous. Recommended criteria documents provide the scientific basis for new occupational safety and health standards. These documents generally contain a critical review of the scientific and technical information available on the prevalence of hazards, the existence of safety and health risks, and the adequacy of control methods [33].

As a preventive means at least annually, audiometric testing (a graph of hearing threshold levels as a function of frequency as specified in ANSI S3.20-1995: audiogram) to users whose exposures equal or exceed an 8-hr time-weighted average of 85 dB should be conducted. Audiometric tests should be performed by a licensed or certified audiologist or otolaryngologist, with competence in administering audiometric examinations, obtaining valid audiograms, and properly using, maintaining, and checking calibration and proper functioning of the audiometers being used.

Acoustic noise exceeding a specified level affects not only the health of humans; them can also affects the measuring instruments (mechanical, optical, electrical, etc.) in different ways, depending on their sensing method and level of precision. Additionally, the electrical products due to some mechanical (belt, pump, incorrect soldering, cooling, loss of lubrication, out of balance, etc.) or electrical (coil, capacitor, motor, transformer, fan, etc.) parts can be generators of unwanted noises that accompanying the products during normal use. These noises can be an alarm signal warning that something is not in order and a product failure is approaching. Means to protect against these types of noises include, but are not limited to the use of the right design, use of approved components, and increasing the quality control in manufacturing. For identification of the noise sources and to evaluate the noise present in a given location, research-grade microphones, site survey equipment, and data analysis software to gather and analyze acoustic noise data should be used.

The best way to reduce the effects of acoustic noise is to eliminate the source of the noise. Some sources of acoustic noise are inherent to the environment, such as external wind or a building's HVAC equipment. There is a need to identify what is the best solution available for the equipment that needs protection from acoustic noises. Solutions such as acoustic curtains or baffles that can deflect acoustic energy that is transmitted directly from a nearby source may provide sufficient isolation for applications that are not severely sensitive to acoustic noise, but the amount of reduction provided by these solutions is minimal. Another option is to construct an acoustic enclosure using plywood, cardboard, foams, and various lining materials (resilient fiberglass with a resinous binder, rigid fiberglass board, open-cell acoustical foam, etc.) that absorb noise. When evaluating an acoustic enclosure, it is important to keep in mind that ergonomic requirements must be addressed.

Active noise cancellation (ANC), such as noise-cancellation headphones, has become a useful technology in the acoustic enclosure field. Active noise cancellation uses a noise-canceling system to reduce unwanted background noise. The system is based on microphones that listen to the sounds outside and inside of the earphone, an ANC chipset inverting the soundwaves, and a speaker inside the earphone canceling the outside sound by the neutralizing soundwaves. There are a few methods used for ANC:

- *Passive noise cancellation* uses well-designed ear cups to seal out unwanted noise. This is used for both over-ear headphones and in-ear earphones where the earbud itself will keep surrounding noise out.
- *Basic active noise cancellation* uses microphones and speakers to reduce background and surrounding noises. This is the most known type and has mostly been used in over-ear headphones. Technology has become so small and battery-efficient now that it can be used in true wireless in-ear earphones.
- *Adaptive active noise cancellation* uses microphones and speakers to automatically adjust the surrounding noise. This is the more sophisticated type of ANC where the level of noise cancelling digitally adapts to the surroundings.
- *Adjustable active noise cancellation* allows change in how much background noise is heard by manually adjusting noise cancellation levels. This is useful when full control is desired.
- *Transparency mode* lets the user easily tune back into the world around them, without switching off the music or taking earphones out of the ears.
- *Adjustable transparency mode* lets the user change how much of the outside world they want to pass through, without switching off the music.

For monitoring the sound level personal dosimeters and sound level meters are used.

Some sound level meters may have an octave or one-third octave band filter attached or integrated into the instrument. The filters are used to analyze the frequency content of the noise. They are also valuable for the calibration of audiometers and to determine the suitability of various types of noise control.

In normal operation, calibration of the instrument usually requires only checking. Prior to and after taking measurements, it is good practice to use a calibrator to check the ability of the personal dosimeter or sound level meter to correctly measure sound levels. As long as the sound level readout is within 0.2 dB of the known source, it is suggested that no adjustments to the calibration be made. If large fluctuations in the level occur (more than 1 dB) then either the calibrator or the instrument may have a problem.

Table 5.23 lists examples of standards dealing with acoustic noise exposure, measurement, and specifications.

5.3.2 Ultrasound

US is nonaudible sound waves in the frequency range of 20 kHz to 100 kHz generated by a transducer that usually consists of piezoelectric crystals (e.g., quartz),

Table 5.23 Standards for Acoustic Noise Exposure, Measurements, and Performances

Standard	*Description*
29 CFR 1910.95	Title 29, Part 1910 Occupational Safety and Health Standards, Subpart G, Standard 1910.9–Occupational Noise Exposure
40 CFR 211	Title 40-Protection of Environment, Ch.I, Subchapter G, Part 211–Product Noise Labeling
ANSI/ASA S1.25	American National Standard Specification for Personal Noise Dosimeters
ANSI/ASA S1.4 Part 3	American National Standard Specification for Sound Level Meters
ANSI/ASA S3.44 Part 1	American National Standard Determination of Occupational Noise Exposure and Estimation of Noise-Induced Hearing Impairment
ANSI/ ASA S3.6	American National Standard Specification for Audiometers
IEC 60500	Underwater acoustics–Hydrophones–Properties of hydrophones in the frequency range 1 Hz to 500 kHz
IEC 60565-1	Underwater acoustics–Hydrophones–Calibration of hydrophones–Part 1: Procedures for free-field calibration of hydrophones
IEC 61043	Electroacoustics–Instruments for the measurement of sound intensity–Measurements with pairs of pressure sensing microphones
IEC 61094	Measurement microphones
IEC 61842	Microphones and earphones for speech communications
IEC 62127	Ultrasonics–Hydrophones
NIOSH 98-126	Occupational Noise Exposure

ceramics, barium titanate (BTO), lithium sulfate (Li_2SO_4), or lead zirconate titanate (PZT) and converts an electrical signal into sound waves and vice versa.

The basic ultrasound process goes as follows. A signal generator (pulser) produces an electrical pulse and sends it to the transducer (probe), which changes the electrical pulse into a sound pulse and sends it into the target (body, material evaluated, material tested, etc.). The sound waves penetrate the target until a specific depth that depends on the used frequency (high frequency has small penetration) and is reflected to the transducer, which generates electrical signals from echo sound waves. Depending on the time it takes the sound waves to make the round trip into the target and back to the transducer, along with the intensity of the reflected waves, the incidence of the sound waves, and the phenomenon of absorption, a computer system that processes these electrical signals determines where on the display screen to make a dot and what shade of gray, from light to dark, it should be. The common display options are in amplitude mode (A-mode), B-mode, compounded B-mode, M-mode, or real-time mode.

A-mode is the simplest type of displaying ultrasound images. A single transducer scans a line through the target with the echoes plotted on screen as a function of depth. A one-dimensional presentation of the reflected sound wave is obtained in which echo amplitude (A) is displayed along the vertical axis and echo delay (depth) along the horizontal axis.

B-mode ultrasound consists of a linear array of transducers that simultaneously scan a plane through the target that can be viewed as a two-dimensional image on display. The image displayed is composed of bright dots representing the ultrasound echoes. The brightness of each dot is determined by the amplitude of the returned echo signal.

Compounded B-mode represents a method in which the information to be displayed is obtained from spatial (several different angles) application of ultrasound and combined to produce a single image.

M-mode is defined as time-motion display of the ultrasound wave along a chosen ultrasound line, using a very high sampling rate, which results in a high time resolution. This mode provides a mono-dimensional view of the target in movement (e.g., heart) so very rapid motions can be recorded, displayed, and measured.

Real-time mode represents a rapid succession of individual B-mode images using a phased array of detectors so that scans can be made electronically at a rate of 30 frames a second as a moving video display.

FDA specifies application areas for use of acoustic radiation in the following four categories of products: ultrasonic physical therapy, diagnostic imaging ultrasound, medical sound waves other than those used for therapy or diagnostic, and nonmedical sound waves.

Physical therapy ultrasonic products (diathermy products) use continuous or quasi-continuous ultrasound wave (with an output power of 0.5–15W of an acoustic frequency range of 0.5 MHz to 5 MHz), intended to deliver therapeutic heat to tissues. Ultrasonic diathermy devices are capable of heating deep tissue to a therapeutic temperature range of 40°–45°C for the treatment of pain, muscle spasms, and joint contractures. While ultrasonic diathermy devices are typically used only for physical therapy purposes, they can also be used in combination with radiation treatment protocols.

The 21 CFR 1050.10 is the performance standard for ultrasonic therapy products. The standard applies to generators operating above 16 kHz for physical therapy and indicates radiation parameters: average and temporal peak power and/or intensity, pulse duration, pulse repetition rate, effective radiating area, beam nonuniformity and spatial distributions, and so on.

For ultrasonic therapy devices, output power accuracy of +/- 20% and timer accuracy +/- 10% are required.

The IEC 60601-2-5 "Medical electrical equipment–Part 2-5: Particular requirements for the basic safety and essential performance of ultrasonic physiotherapy equipment," specifies requirements and tests for the safety of ultrasonic physiotherapy equipment, while the IEC 61689 specifies:

- Methods of measurement and characterization of the output;
- Characteristics to be specified by manufacturers;
- Guidelines for safety of the ultrasonic field generated;
- Methods to limit the likelihood of cavitation.

Diagnostic ultrasound imaging uses high-frequency sound waves (1–15 MHz) for real-time visualization of structures inside the body and detection of boundaries between different types of tissue. Diagnostic ultrasound systems can provide greyscale images of most soft tissues, such as the liver, heart, musculoskeletal structures, and other organs and structures. Diagnostic ultrasound in combination with doppler ultrasound can be used to visualize blood flow and calculate its velocity. Doppler fetal heart rate monitors are used to monitor the fetus's development and provide a reference for the calculation of the fetal growth rate. A bone sonometer is utilized to assess bone fragility (osteoporosis). Some ultrasound transducers are mounted on the tips of a catheter that can be inserted into blood vessels to obtain images of the insides of arteries.

US imaging is an example of diagnostic examination that does not involve exposure to ionizing radiation. Ultrasound imaging has a spatial resolution that depends on the size of the active aperture of the transducer, the center frequency, the bandwidth of the transducer, and the selected transmit focal depth. The commonly used focal depth-to-aperture ratio is five. The image resolution depends on the wavelength (short wavelength has good resolution). Air acts as a sound barrier and would result in poorer resolution. To prevent this, a lubricant, such as mineral oil is always placed on the skin. This provides a good connection between the transducer and the body.

For the ultrasonic diagnostic imaging equipment, the following information must be provided to the accompanying documents of the equipment:

- Maximum temporal-average power output (maximum power measured in W).
- Peak-negative acoustic pressure (p- measured in Pa) in the plane perpendicular to the beam-alignment axis containing the maximum pulse-pressure-squared integral (or maximum mean square acoustic pressure for continuous wave systems) in the whole ultrasonic field.
- Output beam intensity Iob measured in mW/cm^2.
- Spatial-peak temporal-average derived intensity (Ispta measured in mW/m^2) in the whole ultrasonic field.
- Nominal frequency *f* measured in Hz.
- Thermal and mechanical index.
- Pulse repetition rate (prr) for nonscanning modes or scan repetition rate (srr) for scanning modes. The scanning mode is a mode of operation that involves a sequence of ultrasonic pulses which give rise to ultrasonic scan lines that do not follow the same acoustic path.
- Distance (Ip) from the transducer output face to the point of maximum pulse-pressure squared integral (or maximum mean square acoustic pressure for continuous-wave system).

The IEC 60601-2-37 "Medical electrical equipment– Part 2-37: Particular requirements for the basic safety and essential performance of ultrasonic medical diagnostic and monitoring equipment," specifies requirements and tests for the ultrasonic medical diagnostic and monitoring equipment.

It is generally accepted that the most important adverse bioeffects due to acoustic output are cavitation and thermal heating. Regarding output and danger as a result of transmitted sound, the American Institute of Ultrasound in Medicine (AIUM) specifies that in the frequency range 500 kHz–10 MHz there is no confirmed significant adverse effects in tissue exposed to intensities below 100 mW/cm. For ultrasonic exposure times between 1–500s and when the intensity multiplied with exposure time gives a result less than 50 J/cm, these effects are not demonstrated even at higher intensities. The measurement of the acoustic output levels should be conducted with wide bandwidth calibrated hydrophones and force balances.

The ultrasound safety standards request for leakage currents measurement the use of a saline solution and a container in which to place the transducer.

Medical sound waves used in other applications other than therapy or diagnostic include high-intensity ultrasound devices for therapies other than diathermy and physical therapy. These include devices that use high-intensity ultrasound energy that are focused to ablate tissue (remove brain tumors, prostate tumors, uterine fibroids, etc.). The standard for high-intensity ultrasound devices is IEC 60601-2-62 "Medical electrical equipment–Part 2-62: Particular requirements for the basic safety and essential performance of high-intensity therapeutic ultrasound (HITU) equipment." Therapeutic ultrasound devices that use high-intensity focused ultrasound to ablate diseased tissue include risk mitigations to prevent ablating nontarget tissues.

The lithotripter is a medical equipment based on extracorporeally (ESWL) or intracorporeally induced focused pressure pulsed waves (shockwaves) used to treat some types of kidney stones (<1.5 cm) and stones in other organs (e.g., liver). The lithotripter sends shockwaves (pressure at focus until 150 Mpa) focused directly on the kidney stone, which are broken up into tiny pieces (<1 mm). The penetration depth in the body can attend 170 mm. The shockwaves are high-pressure sound waves that consist of a sharp peak in positive pressure followed by trailing negative waves that are sent to the target organ to fragment stones. X-ray or ultrasound imagining equipment is used to localize and monitor the stone during the treatment. A lithotripter consists of five basic components:

1. The pressure pulsed waves (shockwaves) can be generated in several ways as alectromagnetic, piezoelectric, and electrohydraulic;
2. The focusing system (acoustic lens; cylindrical, parabolic, or ellipsoid reflector) depends on the method used to generate the shockwaves;
3. Water is the typical medium used for the coupling mechanism;
4. The imaging/localization unit can be an X-ray fluoroscopy (working at 40–125 kV, 0.5–10 mA, or image intensifier size 30 cm) equipment or ultrasonic imaging (working with a 3.5 MHz convex sector transducer) equipment;
5. A motorized 3 axes patient table is used as a patient support device.

From a safety perspective, special attention needs to be focused on the high voltage present in electromagnetic shockwave generators and the values of the acoustic pressure field generated by lithotripsy equipment.

The standard for ESWL is IEC 60601-2-36 "Medical electrical equipment–Part 2-36: Particular requirements for the safety of equipment for extracorporeally induced lithotripsy."

The intracorporeal or endoscopic lithotripter is used to fragment stones in the body by employing a minimally invasive probe. The techniques used for intracorporeal lithotripters include ultrasonic (causing stone fragmentation by vibration in the 20–25 kHz range with a power of 100W), electrohydraulic, and pneumatic (converts the kinetic impact energy to a mechanical strain that fragments the stones). The pneumatic lithotripsy is accepted as the most powerful and faster method as all types of stones can be easily fragmented.

The ultrasound dental scaler is used to remove surface stains from teeth with high frequency (25–30 kHz) ultrasonic waves with an output power of 5–25W, associated with an air-controlled (pressure 450–600 kPa) water spray with input pressure of 26–60 psi.

A harmonic ultrasound surgical device is used for hemostatic cutting and/or coagulation of soft tissue. It performs these functions by using a frequency between 30–80 kHz, a harmonic maximum power output of 35W in continuous normal use, and a lower temperature than those used by electrosurgery or lasers.

Nonmedical ultrasound waves applications include a variety of products that use the sound waves for different nondestructive tests, measurements, and specific functions.

Ultrasonic cleaner is used for removal of impurity, or some other undesirable elements that spoils, corrupts, infects, and makes objects unfit by immersing them in a tank of liquid (heated at maximum 70°C) with the aid of high-frequency ultrasound waves in the range of 25–120 kHz and an appropriate cleaning solvent. This method of deep cleaning (not applicable for pure aluminum or very thin, soft metals) is employed in industry, laboratory, and medicine, where there is a need for clean, accessible, or inaccessible areas on devices or instruments. Ultrasound waves (AM with 100 Hz at a specific carrier frequency) from a generator (with power until 5 kW) are forwarded to transducers (at about 600 Vp-p), move through the cleaning solution, creating an effect called cavitation. The cavitation represents the formation of microscopic bubbles that rapidly expand and contract until implosion (when the pressure of the sound wave increases enough), releasing a very large amount of energy (high temperatures and pressures) in a tiny space. The cavitation bubbles act like microscopic scrubbing brushes within the solvent and as such remove the impurity and other undesirable elements from the object to be cleaned. The cleaning process is dependent on sound power, heat, frequency, time, and solvent type. The standard (technical report) IEC TR 60886 "Investigations on test procedures for ultrasonic cleaners" provides a reference for evaluating the various test procedures for checking the effectiveness of ultrasonic cleaners.

Insect and pest repellant devices use high-frequency sound waves (25–60 kHz), which humans are unable to hear, to repel insects and pests (mice, mosquitoes, flies, rats, cockroaches, ants, crickets, vermin, etc.) that are capable of hearing such frequencies, which irritate them and make them leave the indoor location.

An ultrasonic distance sensor (also referred to as a proximity sensor) is used in ultrasonic distance locators, level, and obstacle detection. These sensors generate and detect the echo of high-frequency sound waves (typically 30–80 kHz). The length of time between the sound pulse and of the echo from an object is used to calculate the distance of the object from the sensor (typically until 10–20m and varies by model). A large area of application of ultrasonic distance sensing is represented by indoor mobile robotics.

Sound navigation and ranging (sonar) is a device that uses ultrasonic waves to measure the distance, direction, and speed of underwater objects consisting of a transmitter and a detector. The transmitter produces and transmits ultrasonic waves. These waves travel through water and after striking the object on the seabed, get reflected and are sensed by the detector. The detector converts the ultrasonic waves into electrical signals, which are appropriately interpreted. The distance of the object that reflected the sound wave can be calculated by knowing the speed of sound in water and the time interval between transmission and reception of the ultrasound. The sonar technique is used to determine the depth of the sea and to locate objects such as underwater hills, valleys, submarines, icebergs, and sunken ships.

Sonic waves in *nondestructive testing* (NDT) uses high-frequency sound energy to conduct evaluations and make measurements. NDT can be used for flaw detection/evaluation, dimensional measurements, material characterization, and more.

This type of testing means any testing method that does not damage or destroy the sample. Usually, it consists of stimulating the sample with electricity, magnetism, electromagnetic radiation, or ultrasound and measuring the sample's response.

The NDT also has its limitations, which include:

- The surface must be accessible to transmit ultrasound;
- It normally requires a coupling medium to promote the transfer of sound energy into the test specimen;
- Materials that are rough, irregular in shape, very small, exceptionally thin, or not homogeneous are difficult to inspect;
- Cast iron and other coarse-grained materials are difficult to inspect due to low sound transmission and high signal to noise ratio;
- Linear defects oriented parallel to the sound beam may go undetected.

Reference standards are required for both equipment calibration and the characterization of flaws.

The ultrasound transducer used in nondestructive testing is usually made from lithium sulfate crystals because they are very efficient sound receivers.

Examples of ultrasonic nondestructive testing are [34]:

- Ultrasonic densitometry represents a method for determining density from the attenuation of ultrasound beams passing through a liquid or semisolid.
- Advanced ultrasonic backscatter technique (AUBT) is a technique developed for detecting damage from high-temperature hydrogen attacks (HTHAs). The technique is for use in pressure vessels and piping. The technique makes use of high-frequency, broadband US probes, and a digital oscilloscope. These allow it to provide both an A-scan display and frequency analysis.
- Phased array ultrasonic testing (PAUT) is a technique that utilizes a set of US transducers (up to 250 small elements). Each of the elements in a PAUT system is able to pulse individually. This is done with computer calculated timing, through a process known as phasing. This allows the system to steer focused beams through various angles and focal distances.
- Long-range ultrasonic testing (LRUT) is a method that allows testing large volumes of material from a single test point. This method works by fixing transducer rings uniformly around a pipe. These rings then generate a series of low-frequency guided waves. The waves can then propagate symmetrically along the pipe axis. This provides complete coverage of the pipe wall.

- Guided-wave ultrasonics (GWUT) is a technique that projects sound waves along pipe walls in order to detect corrosion or other damage. This method is performed by placing a ring of transducers around a pipe. The sound waves emitted from these transducers travel down both directions of the pipe. If they come in contact with corrosion or damage they will be reflected towards the transducers, which then collect the data automatically. GWUT is useful for examining a long length of pipe quickly and from a single location. This technique does not require contact with the pipe except for where the transducer ring is fitted. Guided-wave ultrasonics is an effective technique for inspecting pipes in difficult-to-reach locations, such as those in high locations, behind walls, or under insulation.
- An internal rotating inspection system (IRIS) is an ultrasonic technique used to detect corrosion in piping and tubing using an internally inserted probe into a flooded pipe that generates sound waves. The probe then moves through the pipe, scanning as it goes.
- Time-of-flight diffraction (TOFD) is a method used to look for flaws in welds. It uses the time of flight of an ultrasonic pulse to find the location of a reflector. To find the TOF, the method uses a pair of ultrasonic transducers. The transmitter emits low-frequency waves that propagate at an angle. They only diffract back to the receiver if they hit a defect.
- Dry-coupled ultrasonic testing (DCUT) is a method that does not require a liquid coupling to inspect metallic and nonmetallic material. DCUT transducers are capable of withstanding high voltages. DCUT is a versatile method that can be performed using flexible, contact, wheel, or remote transducers. Flexible transducers can be applied on an external or internal surface of a component to detect flaws or detect structural differences. Contact transducers are also used to detect flaws as well as thickness measurements. Wheel transducers allow inspectors to inspect long piping systems in a short amount of time. Remote transducers are advantageous because they can take thickness measurements at non-conventional angles (i.e., at angles that are not 90° to the surface). This eliminates the need to build supports or remove components to perform the inspection.
- Rapid ultrasonic gridding (RUG) is a method of performing ultrasonic thickness in which multiple ultrasonic thickness probes are utilized, simultaneously, to rapidly gather thickness measurements in a predefined or ad hoc space. Like other US methods, RUG captures raw A-mode data, which can be presented in B-mode or C-scan modes, or used to create visual representation as 3-D models. However, RUG is capable of capturing multiple A-scan data points at a much faster rate than traditional thickness measuring techniques.

Table 5.24 lists standards for products using sound waves describing the performance, testing, and safety.

Acoustic emission (AE) is the phenomenon of radiation of acoustic (elastic) waves in solids that occurs when a material undergoes irreversible changes (by a sudden redistribution of stress) in its internal structure; for example, as a result of crack formation or plastic deformation due to aging, temperature gradients, or external mechanical forces. When a piece of equipment is subjected to an external

Table 5.24 List of Standards Referring to Performance, Testing, and Safety of Products Using Sound Waves

Standard	*Description*
21 CFR 1050	Title 21, Chapter I - FDA, Subchapter J, Part 1050-Performance standard for sonic, infrasonic, and ultrasound radiation-emitting products
ASTM A1038	Standard Test Method for Portable Hardness Testing by the Ultrasonic Contact Impedance Method
ASTM E114	Standard Practice for Ultrasonic Pulse-Echo Straight-Beam Contact Testing
ASTM E213	Standard Practice for Ultrasonic Testing of Metal Pipe and Tubing
ASTM E543	Standard Specification for Agencies Performing Non-destructive Testing
ASTM E1324	Standard Guide for Measuring Some Electronic Characteristics of Ultrasonic Testing Instruments
ASTM E2375	Standard Practice for Ultrasonic Testing of Wrought Products
IEC 60601-2-5	Medical electrical equipment–Part 2-5: Particular requirements for the basic safety and essential performance of ultrasonic physiotherapy equipment
IEC 60601-2-36	Medical electrical equipment–Part 2-36: Particular requirements for the safety of equipment for extracorporeally induced lithotripsy
IEC 60601-2-37	Medical electrical equipment–Part 2-37: Particular requirements for the basic safety and essential performance of ultrasonic medical diagnostic and monitoring equipment
IEC 60601-2-62	Medical electrical equipment–Part 2-62: Particular requirements for the basic safety and essential performance of HITU equipment
IEC 61157	Standard means for the reporting of the acoustic output of medical diagnostic ultrasonic equipment
IEC 61161	Ultrasonics–Power measurement–Radiation force balances and performance requirements
IEC TS 61206	Ultrasonics–Continuous-wave Doppler systems–Test procedures
IEC 61266	Ultrasonics–Hand-held probe Doppler fetal heartbeat detectors–Performance requirements and methods of measurement and reporting
IEC 61689	Ultrasonics–Physiotherapy systems–Field specifications and methods of measurement in the frequency range 0.5 MHz to 5 MHz
IEC 61828	Ultrasonics–Transducers–Definitions and measurement methods regarding focusing for the transmitted fields
IEC 61846	Ultrasonics–Pressure pulse lithotripters–Characteristics of fields
IEC 61847	Ultrasonics–Surgical systems–Measurement and declaration of the basic output characteristics
IEC 62127-1	Ultrasonics–Hydrophones–Part 1: Measurement and characterization of medical ultrasonic fields up to 40 MHz
IEC 62359	Ultrasonics–Field characterization–Test methods for the determination of thermal and mechanical indices related to medical diagnostic ultrasonic fields
IEC 62555	Ultrasonics–Power measurement–HITU transducers and systems
ISO 12710	Non-destructive testing–Ultrasonic inspection–Evaluation of electronic characteristics of instruments;
ISO 16810	Non-destructive testing. Ultrasonic testing. General principles
ISO 18081	Non-destructive testing — Acoustic emission testing— Leak detection by means of acoustic emission
ISO 22232-1	Non-destructive testing — Characterization and verification of ultrasonic test equipment — Part 1: Instruments
ISO 24647	Non-destructive testing—Standard test method for robotic ultrasonic testing (document in work)
ISO 29821	Condition monitoring and diagnostics of machines. Ultrasound. General guidelines, procedures and validation

Table 5.24 (continued)

Standard	*Description*
IEC TR 60854	Methods of measuring the performance of ultrasonic pulse-echo diagnostic equipment
IEC TR 60886	Investigations on test procedures for ultrasonic cleaners
IEC TR 60782	Measurements of ultrasonic magneto-strictive transducers
IEC TR 61088	Characteristics and measurements of ultrasonic piezoceramic transducers
IEC TR 62799	Models for evaluation of thermal hazard in medical diagnostic ultrasonic fields
IEC TS 61206	Ultrasonics–Continuous-wave Doppler systems–Test procedures
IEC TS 63081	Ultrasonics–Methods for the characterization of the ultrasonic properties of materials

stimulus, such as a change in pressure, load, or temperature, this triggers the release of energy in the form of stress waves, which propagate to the surface and are recorded by sensors. Acoustic emissions can come from natural sources, such as earthquakes or rock bursts, or from the equipment itself, such as melting, twinning, and phase transformations in metals. Detection and analysis of AE signals can provide information on the origin and importance of discontinuities in a material.

AE testing is a powerful method for examining the behavior of materials deforming under stress and is most often used in a dynamic test environment, meaning that it is used to monitor for crack detection in pressure equipment when the equipment is experiencing an increase in stress. The AE NDT technique is based on the detection and conversion of high-frequency elastic waves into electrical signals. This is accomplished by directly coupling piezoelectric transducers (that respond to movement in a range of 30 kHz to 1 MHz) on the surface of the structure under test and loading the structure. Sensors are coupled to the structure and the output of each sensor (during structure loading) is amplified through a low-noise preamplifier, filtered to remove any extraneous noise, and further processed by suitable electronic equipment for measurement, display, and storage. For materials with high attenuation, such as plastic composites, lower frequencies may be used to better distinguish AE signals.

The technique of AE uses one or more sensors to listen to a wide range of events that may take place inside a solid material. Depending on the source of this high-frequency sound, there are broadly three application areas:

- Structural testing and surveillance (the source of AE is probably a defect that radiates elastic waves as it grows);
- Process monitoring and control (the acoustic signature of processes is monitored);
- Materials characterization (as an additional diagnostic technique for the study of, for instance, fracture, because it gives unique dynamic information on defect growth).

Due to its versatility, acoustic emission testing (AET) has many applications within the industry, such as assessing structural integrity, detecting flaws, testing for leaks, or monitoring weld quality. Because of the diverse number of situations it can be applied to, it sees extensive use in several areas including the detection of active corrosion in the bottom of above-ground storage tanks, detecting creep damage in high-energy piping (HEP) systems, pressure vessel inspection, and leak detection [34].

AET is different than other NDT techniques in two major ways:

1. Instead of supplying energy to the object under examination, AET listens for the energy released by the object naturally. AE tests can be, and often are, performed on structures while they are in operation since this provides adequate loading for propagating defects and triggering acoustic emissions.
2. AET deals with dynamic processes in a material. This is particularly useful because only active features are highlighted during the examination. Thus, it is possible to discern between developing and stagnant defects. However, one should be aware that flaws can go undetected if the loading is not high enough to cause an acoustic event that can be detected by the system.

References

[1] 21 CFR 1000.3(k), "Title 21–Food and Drugs- Chapter I-Food and Drug Administration Department of Health and Human Services- Subchapter J - Radiological Health_Part 1000–General - Subpart A - General Provisions- Sec. 1000.3 Definitions," FDA, 2020.

[2] NASA's Imagine the Universe, https://imagine.gsfc.nasa.gov/science/toolbox/emspectrum1.html.

[3] ICNIPR 2020, "Guidelines for Limiting Exposure to Electromagnetic Fields (EMFs) (100 KHz to 300 GHz)," http://www.icnirp.org/cms/upload/publications/ICNIRPemfgdl.pdf.

[4] EU Directive 1999/519/EC, "The Limitation of Exposure of the General Public to Electromagnetic Fields (0 Hz to 300 GHz)," Brussels, 1999.

[5] IEC 60601-2-33, "Medical Electrical Equipment - Part 2-33: Particular Requirements for the Basic Safety and Essential Performance of Magnetic Resonance Equipment for Medical Diagnosis," Geneva, 2010.

[6] Perez, R., *Design of Medical Electronic Devices*, San Diego, CA:Academic Press, 2002.

[7] ASTM F2503, "Standard Practice for Marking Medical Devices and Other Items for Safety in the Magnetic Resonance Environment," *ASTM International*, West Conshohocken, PA, 2005.

[8] Golio, M., *The RF and Microwave Handbook*, Boca Raton, FL: CRC Press, 2001.

[9] EU Directive 2014/53/EU, "Relating to the Making Available on the Market of Radio Equipment (RED)," Brussels, 2014.

[10] "Guide on CE Marking and RED for Low-Power Wireless Transmitters," Elite Electronic Engineering Inc., Downers Grove: IL, 2019.

[11] Khandpur, R. S., Compendium of Biomedical Instrumentation, Hoboken, NJ: John Wiley & Sons, 2020.

[12] IEC 60601-2-2, "Medical Electrical Equipment–Part 2-2: Particular Requirements for the Basic Safety and Essential Performance of High Frequency Surgical Equipment and High Frequency Surgical Accessories," Geneva, 2017.

[13] Wi-Fi Alliance, https://www.wi-fi.org/discover-wi-fi.

[14] NAVAIR-NAWCWD TP 8347, *Electronic Warfare and Radar Systems Engineering Handbook*, Naval Air Warfare Center Weapons Division Point Mugu, CA, 2012.

[15] https://economictimes.indiatimes.com/definition/doppler-effect.

[16] Directive 2006/25/EC, "Artificial Optical Radiation Directive," Brussels, 2006.

[17] IEC 62471, "Photobiological Safety of Lamps and Lamp Systems," Geneva, 2006.

[18] IEC 60598-1, "Luminaires - Part 1: General Requirements and Tests," Geneva, 2020.

[19] Schubert, E. F., *Light-Emitting Diodes*, Cambridge, MA: Cambridge University Press, 2006.

[20] ANSES (The French Agency for Food, Environmental and Occupational Health & Safety), https://www.anses.fr/en/content/anses-request-based-opinions-and-reports.

[21] AMA (American Medical Association), https://www.ama-assn.org/practice-management.

[22] Bjorn, L. O., Photobiology, New York: Springer, 2008.

[23] FDA-CDRH Form 3632, "Guide for Preparing Product Reports for Lasers and Product Containing Lasers," 2020.

[24] FDA, "Technical Electronic Product Radiation Safety Standards Committee," October 2016.

[25] CDC (Center for Disease Control and Prevention), "Radiation Safety," 2015, https://www.cdc.gov/nceh/radiation/safety.html.

[26] FDA, Computed Tomography, https://www.fda.gov/radiation-emitting-products/medical-x-ray-imaging/computed-tomography-ct.

[27] Dunscombe, P., et al., "Development of Quality Control Standards for Radiation Therapy Equipment in Canada," *Journal of Applied Clinical Medical Physics*, Vol. 8, No. 1, 2007.

[28] FDA, "Initiative to Reduce Unnecessary Radiation Exposure from Medical Imaging," 2010.

[29] Mettler, F. A., et al., "Effective Doses in Radiology and Diagnostic Nuclear Medicine: A Catalog," *Radiology*, Vol. 248, July 2008.

[30] *Handbook of International Electrical Safety Practices*, Princeton Energy Resources International, Scrivener and Wiley, 2010.

[31] OSHA, "Technical Manual Section III Chapter 5, TED-01-00-015," Washington DC, 2013, https://www.osha.gov/dts/osta/otm/new_noise/index.html#standards.

[32] Directive 2002/49/EC, "Assessment and Management of Environmental Noise," Brussels, 2002.

[33] NIOSH (National Institute for Occupational Safety and Health), "Occupational Noise Exposure," Publication 98-126, 1998.

[34] Inspectioneering, "Ultrasonic Testing," https://inspectioneering.com/tag/ultrasonics.

Selected Bibliography

http://www.fda.gov/MedicalDevices/DeviceRegulationandGuidance/GuidanceDocuments/ucm094345.htm.

http://www.fda.gov/RadiationemittingProducts/RadiationEmittingProductsandProcedures/Medical Imaging/Medical X-Rays/ucm315011.htm.

http://www.fda.gov/radiation-emittingproducts/radiationemittingproductsandprocedures/medicalimaging/medicalx-rays/.

https://www.icnirp.org/en/publications/index.html.

ITU-International Telecommunication Union: Recommendation ITU-R V.431-7.

https://www.tnuda.org.il/en/table-use-frequencies.

https://www.lightingeurope.org.

"Radiological Safety Training for Radiation-Producing (X-Ray) Devices," DOE Handbook, DOE-HDBK-1109-97, August 1997, Reaffirmation with Errata, July 2002.

CHAPTER 6

Safety for Hazardous Locations

A hazardous area is an area in which an explosive atmosphere is present or may be expected to be present in quantities that require special precautions for the construction, installation, and use of electrical products [1].

The areas are known by different names, such as ex-areas, hazardous locations, hazardous areas, or explosive atmospheres, and relate to areas where flammable liquids, vapors, gases, or combustible dusts are likely to occur and to cause a fire or explosion.

Explosions happen when a convergence of fuel (gas, vapor, mist, dust), oxygen (air) and an ignition source (spark, heat, flame, etc.) exists and can occur from materials processed, used, and released by electrical equipment, the present surrounding of electrical equipment, or used in the construction of electrical equipment.

The information in this chapter is intended only as a guide, which cannot be exhaustive, on the principles of the complex domain of hazardous locations (HazLoc), hazardous materials (HazMat), and electrical equipment (apparatus) used in potential HazLoc in the presence of HazMat.

The U. S. NEC and the Canadian Electrical Code (CEC) define hazardous locations as areas where fire or explosion may exist due to flammable gases, flammable liquids -produces vapors, combustible dusts, or ignitable fibers/flying [2]. The Occupational Safety and Health Administration (OSHA) provides a similar definition: areas where flammable liquids, gases or vapors, or combustible dusts exist in sufficient quantities to produce an explosion or fire.

Explosion protection is an essential part of the overall risk management to be conducted for industrial plants and products to ensure safety in industrial processes using or producing hazardous materials. The basic principles of explosion protection have been applied in industry for over 100 years. The NEC addressed the installation of equipment for use in hazardous locations in 1920 and first published standard to address equipment for use in these area dates from 1930.

Within the field of analysis of hazardous location activities, three main aspects are involved:

1. The study of the areas contaminated by an explosive atmosphere—which even today seems to be not sufficiently developed and studied—involves many variations and specific circumstances occurring in daily practice.
2. Rapid changes in the design and construction of the electrical equipment used in these locations due to advances in technology should be recognized.

3. Keeping in mind points 1 and 2 above, the lack of consistency in follow-up and complete documentation regarding the installation, operation, maintenance, and regular inspection of such equipment. This occurs despite efforts to include rules within electrical codes to cover all related aspects, issuing regulation after regulation, amendments, interpretations.

In short, explosion protection must take into account the physical equipment itself, all possible conditions in which this equipment will exist, and then create or modify relevant regulations to ensure consistency, continuity, and adaptability.

The modern-day development of industry has meant an increased need to use equipment in ex-areas. Such equipment is termed "ex-equipment." Note that even electrical or nonelectrical equipment can become a source of ignition in these volatile areas.

Ex-equipment in such areas include, but are not limited to [3] automotive refueling stations or gas stations; oil refineries, rigs, and oil processing plants; chemical processing industries; printing industries, paper and textile industries; hospital operating theaters; aircraft refueling and hangars; surface coating industries; underground coal mines; sewage treatment plants; gas pipelines and distribution centers; grain handling and storage; woodworking areas; sugar refineries; and metal surface grinding, especially aluminum dusts and particles.

The risks to the lives and health of people or loss of property when using the equipment that operates in the presence of explosive atmospheres are high. It is known that sometimes when a high risk is identified it may be too late to take action before accidents occur. We learn from these accidents and they invariably act as feedback that results in changes in the operating systems and equipment, having the continuous purpose to minimize the likelihood of a repeat event.

It is important here to understand that there are no short procedures to be used as a process of acceptance of electrical equipment designated for hazardous locations. The rules of certification or listings must be followed. However, the fact that there are no shortcuts available to speed up the approval process should be considered as a positive step toward having safer equipment.

To mitigate hazards generated by these risks, the principle of explosion protection was elaborated on and codified in international standards, such as the IEC 60079 series, and conformity assessment best practices, such as the ISO/IEC Guide 67. They are also at the basis of product certification systems, such as the IEC System for Certification to Standards relating to Equipment for Use in Explosive Atmospheres (IECEx). The European Union published directives, known as ATEX, related to safe design of ex-equipment and for safe use in such atmospheres.

Most national regulatory frameworks require that conformity assessment be conducted by independent, third-party inspection bodies. This is a prerequisite for safety in a sector where hazards are substantial and may involve many casualties.

When referring to hazardous (flammable and explosive) locations, the following must be considered:

- Flammable and explosive atmospheres (flammable and combustible materials, source of ignition);
- Equipment and protection systems (types, means of protection, components, construction, documentation, and marking);

- Services (installation, maintenance, repair, inspection of hazardous area, and testing of the equipment).

6.1 Flammable and Explosive Environments

An explosive atmosphere is one with flammable gases, vapors, mists, and/or dust suspended in the air. Electrical and mechanical equipment may ignite such an atmosphere if appropriate measures are not taken.

For an explosion to take place, flammable substances (in a specific concentration), air, and a source of ignition need to simultaneously be present.

An explosive atmosphere (environment) is a mixture with air, under atmospheric conditions (pressures 80 kPa to 110 kPa, temperatures -20°C to +60°C, and oxygen content normally 21%) [4] of some flammable or combustible substances (in the form of gas, vapor, dust, fibers, flyings, liquids, and solids), which, after ignition, permit self-sustaining propagation [1], and when it explodes, causes damages.

A flammable atmosphere (environment) is a condition of flammable and combustible substances (in the form of gas, vapor, dust, fibers, flyings, liquids, and solids) that will ignite at the flash point and continue to burn in the air if exposed to a source of ignition. (Note that the terms "flammable" and "explosive" are used interchangeably since unconfined vapors mixed in the air will burn while confined vapors will produce an explosion.)

There are two primary systems used in the world to classify hazardous areas: the class/division/group system (used only in the United States and Canada) and the zone system (used internationally). While these systems have many similarities, they have significant differences that impact the manufacturers and users around the world, making it very difficult for a product or installation to comply with both systems at the same time.

In 1996, the United States adopted the IEC Zone system and this system became the internationally accepted system with national deviations in different counties. In the United States and Canada, both systems are used.

The division system classification of hazardous atmospheres areas as specified in NEC Article 500 is presented in Table 6.1. Under this scheme, an area classified as Class I, Division 1 would represent the most hazardous area within that class [2].

In the IEC Zone classification system presented in Table 6.2, the U.S. Class I Divisions 1 and 2 are divided into three zones: Zones 0, 1, and 2 for gas explosive atmosphere, as illustrated in Figure 6.1 [5]. However, the U.S. Class II Divisions 1 and 2 are divided into three other zones: Zones 20, 21, and 22 for dust explosive atmosphere, as illustrated in Figure 6.2 [5]. The Division 1 of U.S. Class I is related to Zone 0 and Zone 1, while the Division 1 of the U.S. Class II correspond with Zone 20 and Zone 21. The U.S. Division 2 of Class I corresponds with Zone 2, while Division 2 of Class II corresponds with Zone 22. The U.S. Class III Divisions 1 and 2 do not correspond with zone classification system and are included in Zones 20, 21, and 22 (see Table 6.3). The Zone classification system was adopted by the United States in Articles 505 and 506 of the NEC [6] and Section 18 plus Annex J in the CEC.

Table 6.1 U.S. Class and Division Classification of Hazardous Atmospheres Depending on Type and Presence of Hazard

Hazardous Location	*Hazard*	*Hazard Type*	*Division*	*Description of Division (Presence of Hazard)*	*Group of HazMat*	*Examples of*
Class I	Gases, vapors, and liquids that can burn or explode	G	1	Ignitable flammable hazards can exist all the time (continuous or intermittent) under normal conditions*	A B C D	Acetylene Hydrogen Ethylene Propane
			2	Ignitable flammable hazards are not likely to exist under normal conditions (exist only in abnormal conditions)	A B C D	Acetylene Hydrogen Ethylene Propane
Class II	Combustible dust that can burn or explode	D	1	Ignitable flammable hazards can exist all the time (continuous or intermittent) under normal conditions	E F G	Conductive dust (metal: aluminum or magnesium) Nonconductive dust (e.g., coal, flour, sugar, grain, wood, plastic)
			2	Ignitable flammable hazards are not likely to exist under normal conditions (exist only in abnormal conditions)	E F G	Nonconductive dust (e.g., coal, flour, sugar, grain, wood, plastic)
Class III	Ignitable fibers, flyings that can burn or explode	D	1	Ignitable flammable hazards are used or manufactured	No	Ignitable fibers, flyings (e.g., cotton lint, flax, rayon)
			2	Ignitable flammable hazards are handled or stored		

*Normal conditions refer to operating condition.

Class I, Division 2 is essentially the equivalent of Zone 2. This zone is an area in which an explosive gas-air mixture is not likely to occur and, if it does occur, will exist only for a short time. Zone 0 defines those locations in which an explosive gas-air mixture is present continuously or for long periods during normal operation. An example of this would be the vapor space of a closed, but vented, process vessel or storage tank. Zone 1 includes the remainder of what would otherwise have been classified as Class I, Division 1. However, by isolating the worst hazards with a Zone 0 classification, designers are free to use less restrictive and less costly systems in the areas defined as Zone 1.

The zone system Group II of hazardous materials is subdivided according to the nature of the explosive gas atmosphere:

- IIA, a typical gas is propane or gases and vapors of equivalent hazard, corresponding with Group D in the U.S. class/division system;

Table 6.2 IEC Zone Classification System of Hazardous Atmospheres Depending on Type and Presence of Hazard

Hazardous Location	*Hazard*	*Hazard Type*	*Zone*	*Description of Zone (Presence of Hazard)*	*Group of HazMat*	*Examples of HazMat*
Mines	Firedamp that can burn or explode	N/A	N/A	N/A	I	Methane
Class I	Gases, vapors, and liquids that can burn or explode	G	0	Ignitable flammable hazards can exist all the time (continuous) under normal conditions*	IIC	Acetylene Hydrogen
					IIB	Ethylene
					IIA	Propane
			1	Ignitable flammable hazards are likely to exist (intermittent) under normal conditions	IIC	Acetylene Hydrogen
					IIB	Ethylene
					IIA	Propane
			2	Ignitable flammable hazards are not likely to exist under normal conditions (exist only in abnormal conditions)	IIC	Acetylene Hydrogen
					IIB	Ethylene
					IIA	Propane

- IIB, a typical gas is ethylene or gases and vapors of equivalent hazard, corresponding with Group C in the U.S. class/division system;
- IIC, a typical gas is hydrogen or gases and vapors of equivalent hazard, corresponding with Groups A and B in the U.S. class/division system.

Note that this subdivision is based on the maximum experimental safe gap (MESG) or the minimum ignition current ratio (MIC ratio) of the explosive gas atmosphere in which the equipment may be installed [7]. Equipment marked to be used in IIB gas atmosphere is suitable also for an atmosphere in which Group IIA gas is present. Similarly, equipment marked IIC is suitable also for an atmosphere in which Group IIA or Group IIB gas is present.

Zone system Group III of hazardous materials is subdivided according to the nature of the explosive dust atmosphere:

- IIIA, combustible flyings, corresponding with Class III in the U.S. class/division system;
- IIIB, nonconductive dust, corresponding with Groups F and G in the U.S. class/division system;
- IIIC, conductive dust, corresponding with Group E in the U.S. class/division system only for Division 1.

Note that equipment marked to be used in IIIB dust atmosphere is also suitable for an atmosphere in which Group IIIA dust is present. Similarly, equipment

Table 6.2 (continued)

Hazardous Location	*Hazard*	*Hazard Type*	*Zone*	*Description of Zone (Presence of Hazard)*	*Group of HazMat*	*Examples of HazMat*
Class II	Combustible dust or ignitable fibers, flyings that can burn or explode	D	20	Ignitable flammable hazards can exist all the time (continuous) under normal conditions	IIIC	Conductive dust (metal: aluminum or magnesium)
					IIIB	Nonconductive dust (e.g., coal, flour, sugar, grain, wood, plastic)
					IIIA	Ignitable fibers, flyings (e.g., cotton lint, flax, rayon)
			21	Ignitable flammable hazards are likely to exist (intermittent) under normal conditions	IIIC	Conductive dust (metal: aluminum or magnesium)
					IIIB	Nonconductive dust (e.g., coal, flour, sugar, grain, wood, plastic)
					IIIA	Ignitable fibers, flyings (e.g., cotton lint, flax, rayon)
			22	Ignitable flammable hazards are not likely to exist under normal conditions (exist only in abnormal conditions)	IIIC	Conductive dust (metal: aluminum or magnesium)
					IIIB	Nonconductive dust (e.g., coal, flour, sugar, grain, wood, plastic)
					IIIA	Ignitable fibers, flyings (e.g., cotton lint, flax, rayon)

* Normal conditions refer to operating condition.

† In most modern plants this area is Zone 2 and only small areas are Zone 1. Zone 0 is restricted to a few places such as the insides of storage tanks where there is normally no need for electrical equipment.

marked IIIC is also suitable for an atmosphere in which Group IIIA or Group IIIB dust is present.

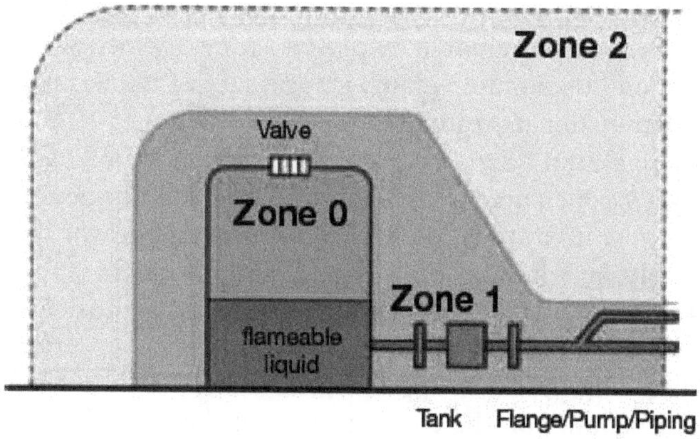

Figure 6.1 Example of the zone classification of explosive gas atmospheres to IEC 60079-10-1. (Courtesy of Eaton.)

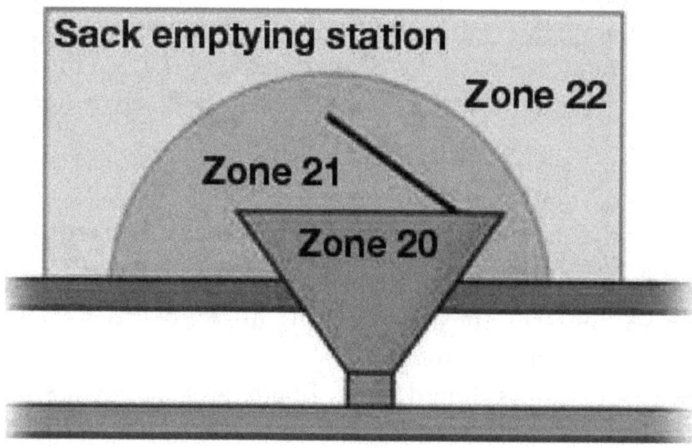

Figure 6.2 Example of the zone classification of explosive dust atmospheres to IEC 60079-10-2. (Courtesy of Eaton.)

Table 6.3 Comparison between NEC Class/Division System and IEC Zone Classification System of Hazardous Area

US Class/Division System NEC 500		*Zone Classification System (NEC 505, 506; CEC 18; ATEX; IECEx)*	*Hazardous Atmosphere*
Class I	Division 1	Zone 0	Gas
		Zone 1	
	Division 2	Zone 2	
Class II	Division 1	Zone 20	Dust
		Zone 21	
	Division 2	Zone 22	
Class III	Division 1	Do not have correspondent in Zone system; are included in Zones 20, 21, and 22	
	Division 2		

An important parameter in a hazardous area is represented by the flammable (explosive) limits/flammable (explosive) range, which is the domain between the minimum and maximum volume percentages of the hazardous material in air that forms a flammable mixture [8].

The minimum gas or vapor concentration in air, under atmospheric conditions, below which an explosive atmosphere will not be formed (when an ignition is present, the mixture is too lean to burn), is the lower flammable limit (LFL) or lower exploasive limit (LEL). The maximum gas or vapor concentration in air under atmospheric conditions, above which an explosive atmosphere will not be formed (when an ignition is present, the mixture is too rich to burn), is the upper flammable limit (UFL) or upper explosive limit (UEL). These limits are substance-specific depending on the concentration of the hazardous substances with the air, air temperature, and atmospheric pressure (see Table 6.4). Note that under other atmospheric conditions than those specified above, the explosion limits change. For example, as the content of oxygen increases, the upper explosive limit is raised.

If an atmosphere monitoring device for a specific location reading is 25% of LEL for a flammable substance (e.g., 1% hydrogen) this indicates that the flammable substance-air mixture is a quarter of the way to the lowest flammable substance concentration that can burn. It is important to never enter a >25% LEL atmosphere.

6.1.1 Flammable and Combustible Substances: Hazards

Flammable and combustible substances are components which allow explosions when present in a specific concentration in an explosive or flammable atmosphere that characterizes the hazard location.

The gases, vapors, liquid, dust, fibers, and flyings are the most typical flammable and combustible substances. Knowing the characteristics of these substances as well as the means of monitoring concentrations aid in preventing critical situations caused by fire or explosion. Special attention needs to be paid to the storage, use, dispensing, mixing, transporting, and handling of flammable and combustible substances.

Flammable gas is a gas which, at 20°C (68°F) or less and with a standard atmospheric pressure of 1 atm (101.3 kPa or 14.7 psi), forms a flammable mixture with the air at a concentration of 13% by volume or less, or which, at ambient

Table 6.4 Examples of LEL and UEL for Gases

Substance	*LEL**	*Most Easily Ignitable Range**	*UEL**
Acetylene	2.5%	6%–12%	100%
Butane	1.86%		8,41%
Hydrogen	4%	14%–34%	75%
Methane	5%	8%–9%	16%
Propane	2%		10%
Silane	1.5%		98%

* These limits specify the content by volume of the flammable gas in the mixture with air at 20°C and atmospheric pressure.

temperature and pressure, forms a range of flammable mixtures with air wider than 12% by volume, regardless of the lower limit.

Gases can be classified into three groups: oxidizers (e.g., air, O_2, NO_2; non-flammable), inert gases (e.g., CO_2, Ne, He, Xe; not combustible), and flammable gases (can be explosive when mixed with air or oxygen in specific proportions, usually in a confined volume). Flammable gases may be diluted with inert gas to the point where the mixture is not flammable.

Examples of flammable gases are hydrogen, butane, methane, ethylene, acetylene, ammonia, ethane, propane, propylene, and silane. These are substances that, in contact with water, emit flammable gases.

Flammable gases are typically divided into two primary categories: fuel gases and industrial gases. Fuel gases include natural gas (methane) and liquified petroleum gas (propane). Industrial gases include all flammable gases not usually used as fuels (acetylene, hydrogen, etc.). Chlorine trifluoride is the most flammable gas.

Table 6.4 lists examples of LEL and UEL for some flammable gases [9]. Ventilation, natural or mechanical, must be sufficient to limit the concentration of flammable gases or vapors to a maximum level of 25% of their LEL/LFL.

Leaking fuel gases may form ignitable mixtures with the surrounding air and lead to fire or explosions. For this reason, many of them are odorized (have added a smell), which means that leaks can be more readily identified.

The ISO/IEC 80079-20-1 standard "Explosive atmospheres–Part 20-1: Material characteristics for gas and vapor classification–Test methods and data" provides guidance on classification of gases and vapors and tabulated values of chemical and engineering properties of substances helpful in the selection of equipment to be used in hazardous areas.

The flammability of gases can be determined by tests or by calculation in accordance with methods from the ISO 10156 "Gases and gas mixtures–Determination of fire potential and oxidizing ability for the selection of cylinder valve outlets" standard, which deals with specific methods for determining whether or not a gas or gas mixture is flammable in the air and whether a gas or gas mixture is more or less oxidizing than the air under atmospheric conditions.

Flammable and combustible liquids can burn and are classified, or grouped, as either flammable or combustible by their flashpoints. Flammable liquids will ignite and burn easily at normal working temperatures. Combustible liquids have the ability to burn at temperatures that are usually above working temperatures.

Some common terms and definitions to be aware of include the following [8]:

- Boiling point is the temperature at which a liquid's vapor pressure is equal to the atmospheric pressure. Liquids with low boiling points are very volatile.
- Flash point is the minimum temperature at which a liquid gives off a sufficient concentration of vapor to form a vapor-air mixture that will ignite near the surface of the liquid and propagate a flame away from the ignition source (flash fire is not continuous combustion). The flash point of flammable and combustible liquids may be altered by the presence of impurity or additive. This may have an effect on the classification or determination of the category/packing group for the material. For these reasons, always check a product's classification using the product safety data sheet. If there is doubt

as to the flash point and thus the classification of the materials (e.g., waste liquids or mixtures), a flash-point test should be performed as specified in the ISO 2719, ISO 3679, ASTM D93–20, and AS2106 standards.

- Vapor pressure is the measure of the pressure created by a liquid's vapor at a specific temperature. Flammable or combustible liquids with a high vapor pressure at room temperature are more hazardous than liquids with lower vapor pressures because they will produce more flammable vapor without heating.
- Vapor density is the weight of the volume of pure vapor or gas (with no air present) compared to the weight of an equal volume of dry air at the same temperature and pressure. A vapor density figure less than one indicates the vapor is lighter than air. A figure greater than one indicates the vapor is heavier than air.

Flammable liquids give off vapor that can easily be ignited at normal working temperatures, having a flash point below 37.8°C (100°F) while combustible liquids are liquids having a flash point at or above 37.8°C (100°F) and below 93.3°C (200°F). A combustible liquid does not catch fire as easily as a flammable liquid.

The vapors produced by flammable liquids often present the most serious hazard because they can easily ignite or explode under certain conditions. These vapors burn readily and intensely or can travel some distance to a source of ignition and flash back.

Flammable and combustible liquids are regulated in the United States under the CFR Title 29, Section 1910.106. The NFPA 30 "Flammable and Combustible Liquids Code" provides requirements for the safe storage, handling, and use of flammable and combustible liquids. The provisions of this code should be considered necessary to provide a reasonable level of protection against loss of life and property from fire and explosion, reflecting the state of the art prevalent at the time the code was issued.

Flammable and combustible liquids are classified under the Globally Harmonized System of Classification and Labelling of Chemicals (GHS) [10], based on the flash point. A flammable liquid is classified into one of four categories for this class, which is shown in Table 6.5.

A flammable solid is a solid, other than a blasting agent or explosive, that has an ignition temperature below 100°C (212°F), is capable of causing a fire through friction or retained heat from manufacturing or processing, absorption of moisture,

Table 6.5 Classification of Flammable Liquids

Category	*Criteria*
1	Flash point <23°C (73.4°F) and initial boiling point < 35°C (95°F)
2	Flash point <23°C (73.4°F) and initial boiling point > 35°C (95°F)
3	Flash point >23°C (73.4°F) and < 60°C (140°F)
4	Flash point >60°C (140°F) and < 93°C (199°F)

From: [10].

or spontaneous chemical change, or can be ignited readily and, when ignited, burns so vigorously and persistently that it creates a serious hazard.

Readily combustible solids are powdered, granular, or pasty chemicals that are dangerous if they can be easily ignited by brief contact with an ignition source, such as a burning match, and if the flame spreads rapidly.

Flammable solids are divided into two categories depending on the burning rate test and burning time.

The first category is represented by powdered, granular, or pasty chemicals when the burning time is less than 45 seconds or the burning rate is more than 2.2 mm/s, and the second category is represented by powders of metals or metal alloys when they can be ignited and the fire spreads over the whole length of the sample in 10 minutes or less.

Flammable solids are more hazardous when widely dispersed in a confined space (e.g., finely divided metal powders). When aluminum becomes dust (or heated to a high temperature) it can readily ignite with a fierce flame that cannot be extinguished with a typical fire extinguisher. Sodium, calcium, potassium, and calcium carbide can be considered flammable solids that will emit a flammable gas when wet and react violently with water and steam. Nitrocellulose and magnesium are other flammable solids.

Proper ventilation is extremely important when handling larger quantities of flammable solids.

6.1.2 Sources of Ignition

In order to prevent the ignition of a hazardous explosive atmosphere, it is necessary to be aware of all possible ignition sources that may occur and to ensure that these ignition sources cannot become effective by applying explosion-protection measures.

Potential sources of ignition include [11]:

1. Hot surfaces (e.g., surfaces heated by coils, resistors, lamps, brakes, or hot bearings), including incandescent materials;
2. Flames and hot gases (e.g., exhausts from internal combustion engines or particles that are formed by the switching sparks of power switches eroding material from the switch contacts), including hot particles and welding;
3. Mechanically generated sparks (e.g. ,when casings or enclosures are struck, tools such as a rusty hammer and chisel in contact with light alloys or the metal fork of a forklift truck);
4. Arcing, flashes, and sparking produced by electrical equipment (e.g., in contactors, circuit breakers, commutators, switching sparks, sparks at collectors or slip rings);
5. Stray electric currents, cathodic corrosion protection (e.g., electric rails and other earthed voltage supplies for electric corrosion protection of equipment and reverse currents from generators, caused by a short circuit to exposed conductive part/ground fault, induction);

6. Static electricity (e.g., transmission belts made from plastic materials, enclosures of portable devices, synthetic clothing material, separation processes when rolling out paper or plastic foil, plastic pipe systems, pumping or pouring an electric isolating fluid);
7. Lightning (e.g., atmospheric weather disturbances and large currents flowing from where lightning strikes can produce sparks in the vicinity of the point of impact);
8. RF electromagnetic waves from 10 kHz to 3 THz (e.g., wireless signals, industrial high-frequency generators for heating, drying, or cutting);
9. Electromagnetic waves from 300 GHz to 3 PHz;
10. Ionizing radiation (e.g., X-ray apparatus, radioactive material, absorption of energy leads to heating up);
11. Optical radiation (especially in the infrared range), including lasers;
12. Ultrasonic (e.g., absorption of energy in solid/liquid materials leads to heating up);
13. Adiabatic compression and shock waves (e.g., sudden opening of valves);
14. Exothermic chemical reactions (e.g., chemical reaction leads to heating up), including self-ignition of dusts.

Equipment can have its own potential source of ignition, if, when operated as intended (including malfunctions) in a potentially explosive atmosphere, it is capable of igniting the latter unless specific safety measures are taken. Therefore, equipment must ensure the required level of protection. To ensure this required level of protection, various techniques can be applied (e.g., intrinsic safety, pressurization, increased safety, and encapsulation) (see Table 6.8).

6.1.3 Standards and Codes

Over time IEC, ISO, UL, CENELEC, and other standards developers have issued many standards to which equipment must be constructed and subsequently tested and assessed by conformity and certification bodies worldwide. The list below is not exhaustive, but it gives the most commonly used standards.

Table 6.6 represents a list of IEC, ISO, EN, and other standards related to explosive atmosphere organized around basic concepts and methodology, determination of characteristics of explosive atmospheres, design requirements for electrical equipment, design requirements for nonelectrical equipment, production of the equipment, material characteristics, classification of areas, electrical installations design, selection, erection, inspection, and maintenance, and equipment repair, overhaul, and reclamation.

Table 6.7 lists the UL standards related to hazardous (classified) locations, not including the UL 60079 series of standards equivalent with IEC 60079 series of standards with U.S. national deviations.

The U.S. NEC (NFPA 70) in Articles 500-510 and Canadian CEC (Section 18 and Annex J) specify requirements applicable to design, use, installation, and servicing of electrical equipment intended for explosive atmospheres.

The current set of standards and codes of design, production, installation, use, inspection, maintenance, and repair for electrical equipment in explosive

Table 6.6 IEC, ISO, EN, and Other Standards Related to Explosive Atmospheres

Standard for Explosive Atmospheres	*Description*
Basic Concepts and Methodology	
EN 1127-1	Explosive atmospheres. Explosion prevention and protection. Basic concepts and methodology
EN 13237	Potentially explosive atmospheres. Terms and definitions for equipment and protective systems intended for use in potentially explosive atmospheres
EN 15233	Methodology for functional safety assessment of protective systems for potentially explosive atmospheres
EN 15198	Methodology for the risk assessment of nonelectrical equipment and components for intended use in potentially explosive atmospheres
FM 3600	Electrical Equipment for Use in Hazardous (Classified) Locations
IEC 60079-44	Explosive atmospheres. Personal Competence
Determination of Characteristics of Explosive Atmospheres	
EN 14034-1	Determination of explosion characteristics of dust clouds. Determination of the maximum explosion pressure pmax of dust clouds
EN 14034-3	Determination of explosion characteristics of dust clouds. Determination of the lower explosion limit LEL of dust clouds
EN 14034-2	Determination of explosion characteristics of dust clouds. Determination of the maximum rate of explosion pressure rise (dp/dt)max of dust clouds
EN 14034-4	Determination of explosion characteristics of dust clouds. Determination of the limiting oxygen concentration LOC of dust clouds
EN 15188	Determination of the spontaneous ignition behaviour of dust accumulations
EN 1839	Determination of the explosion limits and the limiting oxygen concentration (LOC) for flammable gases and vapours
EN 14522	Determination of the auto ignition temperature of gases and vapors
EN 15794	Determination of explosion points of flammable liquids
EN 15967	Determination of maximum explosion pressure and the maximum rate of pressure rise of gases and vapours
Design Requirements for Electrical Equipment	
IEC 60079-0	Explosive atmospheres. Equipment. General requirements
IEC 60079-1	Explosive atmospheres. Equipment protection by flameproof enclosures "d"
IEC 60079-2	Explosive atmospheres. Equipment protection by pressurized enclosure "p"
IEC 60079-5	Explosive atmospheres. Equipment protection by powder filling "q"
IEC 60079-6	Explosive atmospheres – Equipment protection by liquid immersion "o"
IEC 60079-7	Explosive atmospheres. Equipment protection by increased safety "e"
IEC 60079-11	Explosive atmospheres. Equipment protection by intrinsic safety "I"
IEC 60079-13	Explosive atmospheres. Equipment protection by pressurized room "p" and artificially ventilated room "v"
IEC 60079-15	Explosive atmospheres. Equipment protection by type of protection "n"
IEC TR 60079-16	Electrical apparatus for explosive gas atmospheres. Artificial ventilation for the protection of analyzer(s) houses
IEC 60079-18	Explosive atmospheres. Equipment protection by encapsulation "m"
IEC 60079-25	Explosive atmospheres. Intrinsically safe electrical systems
IEC 60079-26	Explosive atmospheres. Equipment with Equipment Protection Level (EPL) Ga
IEC 60079-28	Explosive atmospheres. Protection of equipment and transmission systems using optical radiation

Table 6.6 (continued)

Standard for Explosive Atmospheres	*Description*
IEC 60079-29-1	Explosive atmospheres. Gas detectors–Performance requirements of detectors for flammable gases
IEC 60079-29-2	Explosive atmospheres. Gas detectors. Selection, installation, use and maintenance of detectors for flammable gases and oxygen
IEC 60079-29-3	Explosive atmospheres. Gas detectors–Guidance on functional safety of fixed gas detection systems
IEC 60079-29-4	Explosive atmospheres. Gas detectors. Performance requirements of open path detectors for flammable gases
IEC 60079-30-1	Explosive atmospheres. Electrical resistance trace heating. General and testing requirements
IEC 60079-30-2	Explosive atmospheres. Electrical resistance trace heating. Application guide for design, installation and maintenance
IEC 60079-31	Explosive atmospheres. Equipment dust ignition protection by enclosure "t"
IEC TR 60079-32-1	Explosive atmospheres. Electrostatic hazards, guidance
IEC 60079-33	Explosive atmospheres. Equipment protection by special protection "s"
IEC TS 60079-39	Explosive atmospheres. Intrinsically safe systems with electronically controlled spark duration limitation
IEC TS 60079-40	Explosive atmospheres. Requirements for process sealing between flammable process fluids and electrical systems
IEC TS 60079-42	Explosive atmospheres. Electrical safety devices for the control of potential ignition sources for Ex-Equipment
IEC TS 60079-43	Explosive atmospheres. Equipment in adverse service conditions
IEC 60079-45	Explosive atmospheres. Electrical Ignition Systems for Internal Combustion Engines (Document in work)
IEC TS 60079-47	Explosive atmospheres. Equipment protection by 2-Wire Intrinsically Safe Ethernet concept (2-WISE) (Document in work)
EN 14986	Design of fans working in potentially explosive atmospheres
EN 14994	Gas explosion venting protective systems
EN 14797	Explosion venting devices
EN 14460	Explosion resistant equipment
EN 14491	Dust explosion venting protective systems
EN 15947-4	Pyrotechnic articles. Fireworks, Categories F1, F2, and F3. Test methods
ISO 16852	Flame arresters. Performance requirements, test methods and limits for use
EN 15089	Explosion isolation systems
EN 16447	Explosion isolation flap valves
EN 50495	Safety devices required for the safe functioning of equipment with respect to explosion risks
Design Requirements for Non-Electrical Equipment	
ISO/IEC 80079-36	Explosive atmospheres. Non-electrical equipment for explosive atmospheres. Basic method and requirements
ISO/IEC 80079-37	Explosive atmospheres. Non-electrical equipment for explosive atmospheres. Non-electrical type of protection constructional safety "c," control of ignition sources "b," liquid immersion "k"
Production of Equipment	
ISO/IEC 80079-34	Explosive atmospheres. Application of quality management systems for Ex Product manufacture
IEC TS 60079-46	Explosive atmospheres. Equipment assemblies

Table 6.6 (continued)

Standard for Explosive Atmospheres	*Description*
Material Characteristics	
ISO/IEC 80079-20-1	Explosive atmospheres — Part 20-1: Material characteristics for gas and vapour classification — Test methods and data
DIN IEC 60079-20-2	Explosive atmospheres — Part 20-2: Material characteristics — Combustible dusts test methods (Draft document)
Classification of Areas	
IEC 60079-10-1	Explosive atmospheres. Classification of areas – Explosive gas atmospheres
IEC 60079-10-2	Explosive atmospheres. Classification of areas. Explosive dust atmospheres
Electrical Installations Design, Selection, Erection, Inspection, and Maintenance	
IEC 60079-14	Explosive atmospheres. Electrical installations design, selection, and erection
IEC 60079-17	Explosive atmospheres. Electrical installations inspection and maintenance
Equipment Repair, Overhaul and Reclamation	
IEC 60079-19	Explosive atmospheres. Equipment repair, overhaul, and reclamation

atmosphere form a solid base for manufacturers and evaluation bodies. The trend of globalization and harmonization of the IEC 60079 series-based requirements is hard and useful work on that will provide safer equipment, but many areas still remain where differences in applications and acceptance of these requirements exist.

6.2 Equipment and Type of Protection

A few basic steps are needed to protect against an explosion:

1. Limiting the concentration of flammable substances to mix with atmospheric oxygen to the point where there is no danger of an explosive mixture forming, increasing air circulation by natural or artificial ventilation, or by using enclosures filled with an inert substance.
2. Concentration monitoring of the level of hazardous substances by means of a gas detection system, which will sound off an alarm or switch off the system.
3. Using electrical equipment with types of protection corresponding to the necessary level of protection. For this is necessary to know the frequency and duration of the occurrence of a hazardous explosive atmosphere and the key explosion related figures for the flammable materials (e.g., temperature classes, gas and dust ignition temperatures, or explosion subgroups), as well as the local ambient conditions.
4. Provide the means to limit the impact of an explosion and/or to reduce it to a nonhazardous level by an explosion-resistant design to withstand an explosion inside an explosion relief by deploying bursting discs or explosion flaps that open in a safe direction, and explosion suppression and

Table 6.7 UL Standards Related to Hazardous (Classified) Locations

UL Standard	*Designation*
674	Standard for Electric Motors and Generators for Use in Division 1 Hazardous (Classified) Locations (Ed. 5)
698A	Standard for Industrial Control Panels Relating to Hazardous (Classified) Locations (Ed. 4)
783	Standard for Electric Flashlights and Lanterns for Use in Hazardous (Classified) Locations (Ed. 6)
823	Standard for Electric Heaters For Use in Hazardous (Classified) Locations (Ed. 9)
844	Standard for Luminaires for Use in Hazardous (Classified) Locations (Ed. 13)
913	Standard for Intrinsically Safe Apparatus and Associated Apparatus for Use in Class I, II, III, Division 1, Hazardous (Classified) Locations (Ed. 8)
1203	Standard for Explosion-Proof and Dust-Ignition-Proof Electrical Equipment for Use in Hazardous (Classified) Locations (Ed. 5)
1389	Standard for Plant Oil Extraction Equipment for Installation and Use in Ordinary (Unclassified) Locations and Hazardous (Classified) Locations (Ed.1)
1836	Outline of Investigation for Electric Motors and Generators for Use in Class I, Division 2, Class I, Zone 2, Class II, Division 2 and Zone 22 Hazardous (Classified) Locations (Ed. 5)
2029	Gas/Vapor-Blocked Cable Classified of Use in Class I Hazardous (Classified) Locations (Ed. 4)
2062	Enclosures for Use in Hazardous (Classified) Locations (Ed. 2)
2225	Standard for Cables and Cable-Fittings for Use In Hazardous (Classified) Locations (Ed. 4)
2874	Outline of Investigation for Electric Motors for Use in Hazardous (Classified) Locations – Protection by Pressurized Atmosphere Maintained above the UFL (Ed. 2)
4300	Outline of Investigation for Electrically Heated Insulated Covers for Compressed Gas Cylinders for Use in Hazardous (Classified) Locations (Ed. 1)
120002	Certificate Standard for Ex Equipment for Hazardous (Classified) Locations (Ed. 1)
121201	Nonincendive Electrical Equipment for Use in Class I and II, Division 2 and Class III, Divisions 1 and 2 Hazardous (Classified) Locations (Ed. 9)
121203	Portable Electronic Products Suitable for Use in Class I and II, Division 2, Class I, Zone 2 and Class III, Division 2 and 2 Hazardous (Classified) Locations (Ed. 2)
122001	General Requirements for Electrical Ignition Systems for Internal Combustion Engines in Class I, Division 2 or Zone 2, Hazardous (Classified) Locations (Ed. 1)

preventing propagation of the explosion by rapidly injecting extinguishing agents into containers and plant.

Equipment for explosive atmospheres is a general term including apparatus, fittings, devices, components, and the like used as a part of, or in connection with, an electrical installation in an explosive atmosphere [4]. An ex-component is a part of electrical equipment or a module, marked with the symbol U, which is not intended to be used alone and requires additional consideration when incorporated into equipment or systems for use in explosive atmospheres. Ex-components include an empty enclosure and components or assemblies of components for use with equipment that complies with the requirements of one or more of the types of protection (e.g., a type e earth terminal, ammeter, heater or indicator, a type d or m switch component or thermostat, a type i supply, sensor or indicator, or a type *d* or *t* push button switch, a type *t* limit switch, or indicating lamp).

When equipment is classified according to the zone system, it needs to refer to the equipment category within each equipment-group (I or II), determining the requisite level of protection to be ensured.

Equipment group I means equipment intended for use in underground parts of mines and in surface installations of such mines exposed to mine gas (firedamp) and/or combustible dust, comprising equipment categories M1 and M2.

The equipment in category M1 must ensure a very high level of protection and is required to remain functional, even in the event of rare incidents relating to equipment. If a failure of one means of protection occurs, at least an independent second means must provide the requested level of protection, or this level is assured in the event of two faults occurring independently of each other. Additionally, the equipment must be so constructed that no dust can penetrate it and the opening of equipment parts that may access sources of ignition is possible only under nonactive or intrinsically safe conditions or be fitted with appropriate additional interlocking systems.

Equipment category M2 must ensure a high level of protection and need to be deenergized in the event of an explosive atmosphere. The requested level of protection must ensure that sources of ignition do not become active in the case of more severe operating conditions, in particular those arising from rough handling and changing environmental conditions. Additionally, the equipment must be so constructed that the opening of equipment parts that may access sources of ignition is possible only under nonactive conditions or be fitted with appropriate additional interlocking systems. Where it is not possible to render equipment nonactive, the manufacturer must include in the equipment label a safety warning such as, "do not open while energized."

For M1 and M2 categories, zoning is not employed.

This group of equipment is covered by the standard EN 1127-2 "Explosive atmospheres–Explosion prevention and protection–Part 2: Basic concepts and methodology for mining."

Equipment group II means equipment intended for use in other locations that can be exposed to explosive atmospheres, comprising of equipment categories 1, 2, and 3.

Equipment category 1 is intended for use in areas in which explosive atmospheres caused by mixtures of air and gases, vapors, or mists (or by air/dust, fibers, or flyings mixtures) are present continuously, for long periods or frequently, and can be installed in Zones 0, 1, and 2.

This category of equipment must provide a very high level of protection and is required to remain functional, even in the event of rare incidents relating to equipment. If a failure of one means of protection occurs, at least an independent second means shall provide the requested level of protection, or this level is assured in the event of two faults occurring independently of each other. Additionally, the equipment must be constructed so that no dust can penetrate it and the opening of equipment parts that may access sources of ignition is possible only under nonactive or intrinsically safe conditions. If the intrinsically safe condition is not possible to reach by the equipment, an interlocking system can be added to ensure the appropriate condition. The dust can enter or escape from the equipment only at specifically designated points. This requirement must also be met by cable entries and connecting parts.

Equipment category 2 is intended for use in areas in which explosive atmospheres caused by gases, vapors, or mists (or by air/dust, fibers or flyings mixtures) are likely to occur occasionally and can be installed in Zones 1 and 2. This equipment must ensure a high level of protection preventing ignition sources arising, even in the event of frequently occurring disturbances or equipment operating faults. Additionally, the equipment must be so constructed that the opening of equipment parts that may access sources of ignition is possible only under nonactive conditions or be fitted with appropriate additional interlocking systems. Where it is not possible to render equipment nonactive, the manufacturer must include in the equipment label a safety warning such as, "Do not open while energized."

Equipment category 3 is intended for use in areas in which explosive atmospheres caused by gases, vapors, mists, or air/dust, fibers, or flyings mixtures are unlikely to occur or, if they do occur, are likely to do so only infrequently and for a short period only and can be installed in Zone 2 only. This equipment must ensure a lower level of protection as to prevent foreseeable ignition sources that can occur during normal operation.

This type of protection represents the specific measures applied to equipment to avoid ignition of a surrounding explosive atmosphere.

Table 6.8 summarizes the classes of equipment according to the IECEx Scheme and European Union ATEX Directive 2014/34/EU, the type of protection provided, and the applicable Standards for operation in an explosive environment.

The above methods of protection prevent or delay the diffusion of a hazardous substance into an electrical equipment where possible ignition could occur.

Table 6.9 presents the protection principles and applications of a few of these methods.

The Ingress Protection (IP) Code per the IEC 60529 standard, the IK Code per the IEC 62262 standard, and the NEMA Enclosures Type Number for protection provided by enclosures, depending on the use of environmental conditions, was presented in the first volume of this book series in Section 7.7.

6.3 Components and Construction

Certain components that are considered infallible are designed to not fail in a way that could be dangerous or detrimental in a hazardous location.

The concept of infallible components and assemblies may be applied to main transformers, capacitors, damping winding, current limiting resistors, safety shunts, and others (see Table 6.10). When the basic component is the one most probable to fail in safe mode, then a redundant assembly of two or more such devices may be considered infallible.

The limitation of energy within electrical circuits now brings us to the concept of intrinsic safety for electrical equipment used in hazardous locations.

6.3.1 Electrical

Intrinsically safe circuits [12], as applied to hazardous (classified) locations, is a design engineering concept related to explosion protection applied to electrical equipment. Intrinsically safe equipment and wiring must not be capable of releasing any type of energy (electrical, thermal, etc.) under any normal or fault condition that can generate the ignition of a hazardous mixture in its surroundings. An intrinsically safe (IS) circuit can be considered an energy-limited circuit in any foreseeable fault conditions and even in multiple failure conditions. The basic design of an intrinsic safety barrier uses zener diodes to limit voltage, resistors to limit current and a fuse. IS circuits are therefore inherently safe as evidenced by the fact that they are only technique accepted for Zone 0 (high-risk) hazardous areas.

The intrinsically safe circuit type of protection i (with levels ia, ib, and ic) is applicable to electrical equipment in which the electrical circuits themselves are incapable of causing an explosion in the surrounding explosive atmosphere.

The standard that deals with this type of protection for electrical equipment is the IEC 60079-11.

The requirements applicable to the equipment cover also an associated electrical equipment that contains both intrinsically safe circuits and nonintrinsically safe circuits and is constructed so that the nonintrinsically safe circuits cannot adversely affect the intrinsically safe circuits; for example, a recorder that is not itself in an explosive atmosphere, but is connected to a thermocouple situated within an explosive atmosphere where only the recorder input circuit is intrinsically safe.

All relevant requirements of the IEC 60079-11 standard also apply for the following parts of the equipment considered to be simple apparatus:

- Passive components; for example, switches, junction boxes, resistors, and simple semiconductor devices;
- Sources of stored energy consisting of single components in simple circuits with well-defined parameters; for example, capacitors or inductors;
- Sources of generated energy; for example, thermocouples and photocells, which do not generate more than 1.2V, 100 mA, 20 mJ, and 25 mW;
- The above parts must not contain any means of increasing the available voltage or current; for example, DC-DC converters, and must not achieve safety by the inclusion of voltage and/or current-limiting and/or suppression device.

Some significant requirements from the IEC 60079-11 standard include [12]:

- The enclosure for equipment in Group I needs to have a degree of protection of IP54 in accordance with the IEC 60529.
- Where separate intrinsically safe circuits are being considered, the clearance distance between bare conducting parts of external connection facilities must be at least 6 mm between the separate intrinsically safe circuits and at least 3 mm from earthed parts.
- The failure of a component that is encapsulated or hermetically sealed, for example a semiconductor, in which internal clearances and distances through encapsulant are not defined, is to be considered as a single fault.
- Solid insulation is considered to be prefabricated; for example, sheet, sleeving, or elastomeric insulation on wiring. Varnish and similar coatings are not considered to be solid insulation.

Table 6.8 Type of Protection and Standards Applicable to Different Types of Equipment Operating in an Explosive Environment

Explosive Environment				*Equipment Type*				*Protection*			
				ATEX Directive 2014/34/EU		*IECEx Scheme (IEC 60079-0)*					
Presence of	*Type*	*Zone*	*Div.*	*Equipment Group*	*Category†*	*Hazard-Group*	*EPL‡*	*Level*	*Type**	*Provided by*	*Standard*
Gases, vapors, liquids (Class I)	G	0	1	II	1	II A,B,C	Ga	Very high	d, da	Flameproof enclosure	IEC 60079-1
									ia	Intrinsic safety	IEC 60079-11; UL 913; CSA 157
									ma	Encapsulation	IEC 60079-18
									Ga	E with PL§	IEC 60079-26
									op is	ISOR¶	IEC 60079-28
		1	1	II	2	II A,B,C	Gb	High	d, db	Flameproof enclosure	IEC 60079-1
									px	Pressurized enclosure	IEC 60079-2; NFPA 496
									py		
									q, qb	Powder filling	IEC 60079-5
									o, ob	Liquid (oil) immersion	IEC 60079-6
									e, eb	Increased safety	IEC 60079-7
									ib	Intrinsic safety	IEC 60079-11
									p/v	PE/V#	IEC60079-13
									mb	Encapsulation	IEC 60079-18
									i	Intrinsic system	IEC 60079-25
									op is	ISOR¶	IEC 60079-28
									op pr	S/IOR**	IEC 60079-28
									op sh		
		2	2	II	3	II A,B,C	Gc	Enhanced	d, dc	Flameproof enclosure	IEC 60079-1
									pz	Pressurized enclosure	IEC 60079-2
									o, oc	Liquid (oil) immersion	IEC 60079-6
									ec	Increased safety	IEC 60079-7
									ic	Intrinsic safety	IEC 60079-11
									nA	Nonsparking	IEC 60079-15
									nC	Hermetically sealed	IEC 60079-15
									nR	RBE††	IEC 60079-15
									mc	Encapsulation	IEC 60079-18
									op is	ISOR¶	IEC 60079-28
									op pr	S/IOR**	IEC 60079-28
									op sh		

Dust (Class II), fibers, flyings (Class III)	D	20	1	II	1	III A,B,C	Da	Very high	ia	Intrinsic safety	IEC 60079-11
									ma	Encapsulation	IEC 60079-18
									op is	ISOR¶	IEC 60079-28
									ta	By enclosure	IEC 60079-31
		21	1	II	2	III A,B,C	Db	High	p	Pressurized enclosure	IEC 60079-2
									ib	Intrinsic safety	IEC 60079-11
									mb	Encapsulation	IEC 60079-18
									op is	ISOR¶	IEC 60079-28
									op pr	S/IOR**	IEC 60079-28
									op sh		
									tb	By enclosure	IEC 60079-31
		22	2	II	3	III A,B,C	Dc	Enhanced	p	Pressurized enclosure	IEC 60079-2
									ic	Intrinsic safety	IEC 60079-11
									mc	Encapsulation	IEC 60079-18
									op is	ISOR¶	IEC 60079-28
									op pr	S/IOR**	IEC 60079-28
									op sh		
									tc	By enclosure	IEC 60079-31
Methane, coal dust, firedamp		Mining		I	M1	I	Ma	Very high	ia	Intrinsic safety	IEC 60079-11; UL 913; CSA 157
									ma	Encapsulation	IEC 60079-18
		Mining		I	M2	I	Mb	High	d, da	Flameproof enclosure	IEC 60079-1
									e, eb	Increased safety	IEC 60079-7
									ib	Intrinsic safety	IEC 60079-11
									mb	Encapsulation	IEC 60079-18
									px	Pressurized enclosure	IEC 60079-2; NFPA 496
									q	Powder filling	IEC 60079-5

*Equipment may be specified with more than one level of protection, provided by different parameters for each level of protection.

†The correspondence between zone and category is according to the ATEX "workplace" Directive 1999/92/EC: category 1–zone 0 or 20; category 2–zone

or 21; category 3–zone 2 or 22.

‡EPL = equipment protection level.

§E with PL = equipment with protection level.

¶ISOR = inherent safe optical radiation.

#PE/V = pressurized enclosure/ventilation.

**S/IOR = safe/interlocked optical radiation.

††RBE = restricted breathing equipment.

Table 6.9 Protection Principle for Equipment Used in an Explosive Atmosphere

Protective Method	*Type*	*Protection Principle*
Flameproof enclosure	d	The parts which can ignite an explosive atmosphere that can withstand the pressure developed during an internal explosion of an explosive mixture and prevents the transmission of the explosion to the explosive atmosphere surrounding the enclosure. Direct cable entries are made with the aid of flexible seal rings or seal materials. Used for switchgear, motors, light fittings, control stations, indicating equipment, control systems, transformers, heating equipment, slip rings, collectors, adjustable resistors, fuses or lamps, and friction brakes.
Pressurization	p	The interior of the enclosure is kept at a pressure slightly higher than the atmosphere surrounding the enclosure. Any hazardous substance that may be present in the atmosphere cannot enter the enclosure and be ignited. Used for control cabinets, analyzer units, analytical instruments, switchgear, and large motors.
Purging	P	The interior of the enclosure is pressurized and a flow of air or inert gas (helium, neon, argon, krypton, xenon, or radon) is maintained at a level adequate to reduce the concentration of a hazardous substance that may be present. Used in power equipment.
Oil immersion	o	The electrical components are immersed in inert oil that extinguish any spark or flame and controls surface temperature to a safe level. Used for transformers, starting resistors, and gear.
Hermetic sealing	nC	Devices are sealed within an envelope by fusion (e.g., soldering, brazing, welding, or the fusion of glass to metal) against the entrance of an external atmosphere. The electrical contact is sealed within a glass tube to prevent sparks. Used for all electrical equipment for zone 2.
Encapsulation (potting)	m	A molding material is used to envelop the source of electrical energy and prevent any spark from coming in contact with the hazardous substance. Used for switchgears with small capacity, control and signaling units, display units, sensors, and PCBs.
Restricted breathing	nR	This is a form of sealing the enclosure with gaskets in such a way that the ingress of gases is restricted. Used for all electrical equipment for zone 2, measuring and monitoring instrumentation and information systems and equipment, complex machinery, and large machines.
Intrinsic safety	i	Limitation of the spark's energy and temperature of the surface of the equipment at values that will not ignite the surrounding hazardous atmosphere. Used for instrumentation technology, fieldbus technology, sensors, and actuators.
Increased safety	e	The equipment must not arc, spark, or ignition capable hot surfaces. Used for luminaires, AC motors, terminal and junction boxes, control stations for installing Ex components (with a different type of protection), and squirrel-cage motors.
Powder filling	q	Filling the enclosure with a finely grained powder, an arc within the enclosure is unable and the housings or enclosures is designed to prevent escape of the powder in operating position if the covers are removed. Only dry silica glass or hard glass particles with no metallic mixtures are approved as filling. Used in capacitors, electronic assembly groups, or transformers.
By enclosure (housing)	t	The enclosure is sealed so tight, that no combustible dust can enter. The surface temperature of the external enclosure is limited. Used in various equipment where during normal operation sparks, electric arcs, or hot surfaces occur.

- The insulation between an intrinsically safe circuit and the earthed frame of the electrical equipment or parts that may be earthed must withstand an

Table 6.10 Infallible Component Parameters of Concern

Infallible Component	*Parameters of Concern*
Mains transformers	Wire sizes, segregation, production tests, fuse in primary current limiting resistor in output.
Capacitors	High reliability; at least two capacitors mounted in series not allowed use of electrolytic or tantalum.
Current limiting resistor	Wire wound that may only fail in an open circuit mode vitreous enameled; carbon resistor not acceptable.
Safety shunts	Mode of failure only short circuit; at least two components in parallel.

r.m.s. AC test voltage of twice the voltage of the intrinsically safe circuit or 500V r.m.s., whichever is greater.

- The insulation between an intrinsically safe circuit and a nonintrinsically safe circuit must be capable of withstanding an r.m.s. AC test voltage of 2U + 1,00 V, with a minimum of 1,500V r.m.s., where U is the sum of the r.m.s. values of the voltages of the intrinsically safe circuit and the nonintrinsically safe circuit.
- Where the coil of a relay is connected to an intrinsically safe circuit, the contacts in normal operation must not switch more than the nominal value of 5A r.m.s., 250V r.m.s., or 100 VA.
- Fuses for levels of protection ia and ib that may carry current when located in explosive atmospheres must be encapsulated and the compound must not enter the fuse interior.
- Cells and batteries complying with the UL1642 or the IEC 62133 or other relevant safety standards are permitted to be connected in parallel in intrinsically safe equipment.
- Soldering points and welds inside an item of equipment must be protected with insulating enamel.
- The use of three series blocking diodes in circuits of level of protection ia is permitted.
- Semiconductors and controllable semiconductor devices must be permitted in series current limiting circuits in level of protection ib or ic equipment.
- Where a fault can lead to a subsequent fault or faults, then the primary and subsequent faults must be considered to be a single fault.
- The insulated conductor must be covered with at least two layers of insulation, or a single layer of solid insulation of thickness greater than 0.5 mm between adjacent conductors.
- Blocking capacitors must be of a high reliability solid dielectric type. Electrolytic or tantalum capacitors must not be used.
- The assembly must be protected against access in order to prevent repair or replacement of any components on which safety depends either by encapsulation or by an enclosure that forms a nonrecoverable unit. The entire assembly must form a single entity.

- The main type tests and verifications are the assessment of intrinsically safe circuits (assessment using reference curves and tables, assessment of simply circuits, permitted reduction of effective capacitance when protected by a series resistance), measurement of creepage distances, clearances, and separation distances through casting compound and through solid insulation, encapsulation (adherence and temperature), fieldbus intrinsically safe concept (FISCO) (IEC 60079-11), spark ignition, temperature dielectric strength, determination of parameters of loosely specified components, tests for cells and batteries (electrolyte leakage, spark ignition and surface temperature, and battery container pressure), mechanical (casting compound, determination of the acceptability of fuses requiring encapsulation, partitions), equipment containing piezoelectric devices, diode safety barriers and safety shunts, cable pull, transformer, optical isolators (thermal conditioning, dielectric and carbonization, overload test at the receiver side, overload test at the transmitter side, thermal conditioning and dielectric strength, carbonization, dielectric, and short-circuit), and current carrying capacity of infallible printed circuit board connections.
- The recommended routine and verifications tests are routine tests for diode safety barriers, and routine tests for infallible transformers.

6.3.2 Mechanical

Traditionally, protection from an explosion in hazardous environments has been accomplished by either using explosion-proof equipment that can contain an explosion inside an enclosure, or pressurization or purging, which isolates the explosive hazard (gas, dust) from the electrical equipment.

There are two main classes of enclosures used in hazardous locations:

1. Explosion-proof enclosures: According to NEC, this kind of enclosure must be capable of withstanding an explosion of a specified gas or vapor that may occur within it and operating at such an external temperature that a surrounding flammable gas or vapor atmosphere will not be ignited thereby.
2. Dust ignition-proof enclosures: Such enclosure should exclude the entrance inside of an ignitable quantity of dusts and do not permit to arcs, sparks, or heat generated or liberated inside the enclosures to cause ignition of exterior dust accumulations on the enclosure or of atmospheric dust suspensions in the vicinity of the enclosure.

These enclosures are designed to meet dust ignition-proof requirements for indoor use, and with some modifications these may be used for outdoor applications.

In the United States for hazardous (classified) location, NEMA has established types of designations for enclosures as presented in Table 6.11.

The electrical equipment included within one or more categories must be connected using means that must respect the level of degree of protection provided by the enclosures. For this purpose, conduit and cable seals are used. The sealing of each conduit must be performed in a manner that prevents the propagation of

Table 6.11 NEMA Enclosures for Use in Hazardous Areas

NEMA Enclosure Type Number	*Standard*	*Location in which the Enclosure Can Be Used*	*Description*
7	NEC - NFPA 70	Class I, Division 1, Groups A, B, C, or D	Designed to hold an internal explosion without causing an external hazard.
8	NEC - NFPA 70	Class I, Division 1, Groups A, B, C, and D	Designed to prevent combustion through the use of oil-immersed equipment.
9	NEC - NFPA 70	Class II, Division 1, Groups E, F, or G	Designed to prevent the ignition of combustible dust.
10	30 CFR Part 18	Mines	Designed to hold an internal explosion without causing an external hazard.

flames and explosive pressure from the interior of an enclosure into the external conduit system.

In situations when the available energy is insufficient to ignite the hazardous substances that are or may be present, the special enclosures described above are not required.

Materials used in the construction of enclosures of electrical equipment must not contain in total, by mass, more than are specified in Table 6.12 [4]:

- Plastic materials must have a temperature index (TI) or relative temperature index (RTI) and mechanical of at least 20K greater than the maximum service temperature of the enclosure or the part of the enclosure.
- The elastomers and materials used for cementing shall have a continuous operating temperature (COT) range that includes a minimum temperature that is below, or equal to, the minimum service temperature and a maximum temperature that is at least 20K above the maximum service temperature.

Table 6.12 Content of Aluminum, Magnesium, Titanium, and Zirconium in the Construction of Enclosures of Electrical Equipment Used in Explosive Atmospheres

Equipment	*EPL*	*Content, by Mass, of Aluminum, Magnesium, Titanium, and Zirconium*	*Content, by Mass, of Magnesium, Titanium, and Zirconium*
Group I	Ma	15 %	7.5 %
	Mb		
Group II	Ga	10 %	7.5 %
	Gb	N/A	7.5 %
	Gc	*	*
Group III	Da	N/A	7.5 %
	Db	N/A	7.5 %
	Dc	*	*

*No requirements except for fan impellors, fan hoods, and ventilating screens, which must comply with the requirements for EPL Gb.

- If plastic with a surface area exceeding 500 mm^2 is employed as a covering on a conductive material, the plastic must have one or more of the following characteristics:
  - Surface resistance;
  - A breakdown voltage ≤4 kV measured across the thickness of the insulating material;
  - A thickness ≥ 8 mm of the external insulation on metal parts.
- Fastening screws for enclosures of materials containing light metals may be made of light metal or nonmetallic material if the material of the fastener is compatible with that of the enclosure, and must be capable of being released or removed only with the aid of a tool.
- Bushings used as connection facilities and may be subjected to a torque during connection or disconnection must be mounted in such a way that all parts are secured against turning and need to pass a torque test.
- PE conductor connection facilities must allow for the effective connection of at least one conductor with a cross-sectional area given in Table 6.13 [6].
- Equipotential bonding connection facilities on the outside of electrical equipment must provide effective connection of a conductor with a cross-sectional area of at least 4 mm^2.
- Connection facilities must be effectively protected against corrosion. Special precautions must be taken if one of the parts in contact consists of a material containing light metal, for example by using an intermediate part made of steel when making a connection to a material containing light metals.
- Entry into the equipment must either be by a plain or threaded hole located in the wall of the enclosure or an adaptor plate designed to be fitted in or on the walls of the enclosure.
- Blanking elements, intended to close unused openings in the enclosure walls of electrical equipment, must satisfy the requirements of the specific type of protection concerned. The blanking element must only be removable with the aid of a tool.
- The degree of protection (IP) of ventilation openings must be at least IP20 on the air inlet side and IP10 on the air outlet side.
- The mechanical stability of PCBs must be sufficient to prevent reduction of creepage distances and air clearances. PCB can be protected by at least two layers of an adherent insulating coating.

Table 6.13 Minimum Cross-Sectional Area of PE Conductors

Cross-Sectional Area of Phase Conductors, S mm²	*Cross-Sectional Area of the Corresponding PE Conductor, Sp mm²*
$S \leq 16$	S
$16 < S \leq 35$	16
$S > 35$	0,5 S

- The rotating parts of a fan must be enclosed by a fan hood that is not considered to be part of the enclosure of any electrical equipment used in the fan (e.g., the electrical motor).
- Lubricants and seals used in bearings must be suitable for the maximum temperature of the bearings.
- Switchgear must not have contacts immersed in flammable dielectric.
- Where switchgear includes a disconnector, it must disconnect all poles. The switchgear must be designed so that either the position of the disconnector contacts is visible, or their open position is reliably indicated.
- Connectors and terminals must maintain contact pressure on and resist loosening of the earth connection by vibration.
- Plugs and components remaining energized when not engaged with a socket outlet are not permitted.
- Tests for metallic enclosures, metallic parts of enclosures, and glass parts of enclosures must be performed on four samples in the following order:
  - Tests for resistance to impact;
  - Drop test, if applicable;
  - Tests for degrees of IP;
  - Any other tests required by a specific standard;
  - Any other test specific to the type of protection concerned.

6.4 Installation in Hazardous Locations

When systems are installed in potentially explosive areas, a great number of precautionary measures must be taken.

It is important that personnel installing electrical equipment in a hazardous location know when explosions are likely to happen and how to prevent it. A joint effort by the manufacturers of explosion-proof electrical equipment and the constructors and operators of industrial plants can help ensure the safe operation of electrical equipment in hazardous locations [13].

The personnel involved in installation of equipment in a potential explosive atmosphere need to have the following skills:

- Basic knowledge of basic principles of explosive atmospheres;
- Basic awareness to enter hazardous area site;
  - Basic knowledge of the classification of hazardous areas;
  - Capability to perform detailed inspection of electrical installations;
  - Knowledge of testing of electrical installations;
    - Well trained in installation of explosion protected equipment and wiring systems;
    - Knowledge of maintenance of equipment in hazardous areas;
    - Knowledge of overhaul and repair of explosion protected equipment.

Specific details related to the installation of electrical equipment in a hazardous location are described as follows [13]:

1. In a proper grounding system, all grounds are tied together at only one grounding point.
2. The ground path must have less than 1Ω of resistance from the furthest barrier to the main grounding stud. The CEC recommends redundant grounding conductors.
3. Intrinsically safe apparatus, enclosures, and raceways, if made of metal, must be connected to the equipment grounding conductor (green or yellow/green wire), which must be a minimum of 12 AWG. The manufacturer must provide details on any grounding or equipotential bonding required for the installation.
4. In all situations, installation, maintenance, and troubleshooting, it is important to install the ground first and disconnect the ground last.
5. IS conductors must be separated from all other wiring by placing them in separate conduits or by a separation of minimum 5.5mm of air space. Within an enclosure the conductors can be separated by a grounded metal or insulated partition.
6. IS wiring may be light blue when no other conductors colored light blue are used.
7. The barriers normally are installed in a dust- and moisture-free NEMA enclosure located in the nonhazardous area.
8. When barriers are installed in explosion-proof enclosures, which are located in the hazardous area, explosion-proof seals are required on the enclosure.
9. Once a year the barriers should be checked to ensure that the connections are tight, the ground wiring has less 1Ω of resistance, and the barriers are free from moisture and dirt.
10. The enclosure must be as close as possible to the hazardous area to minimize cable runs and increase capacitance of the circuit.
11. A flameproof d enclosure marked for Class I, Division 2, requires the installation of a flameproof conduit seal or flameproof cable gland at each entry to maintain the type of protection flameproof d.
12. The resistance between coating and either the point of bond (in the case of equipment for fixed installations) or the farthest point of potential contact with the enclosure (in the case of portable equipment) must not exceed 10^9W.
13. The warning markings must be located so that it is likely to be readily visible after installation.

Any equipment must be permitted to be installed in any hazardous location in which the maximum surface temperature is not exceeded.

The temperature classes are used to designate the maximum operating temperatures on the surface of the equipment, which should not exceed the ignition temperature of the surrounding atmosphere. Ignition temperature is the minimum temperature required, at normal atmospheric pressure in the absence of a spark or flame, to set a fire or cause self-sustained combustion independently of the heating or heated element (see Table 6.14) [6].

The IEC 60079-14 and the IEC 60079-17 standards contain the specific requirements for electrical installations within hazardous areas, where the hazard may be caused by flammable gases, vapors, mists, dusts, fibers, or flyings.

Information concerning the installation of equipment utilizing optical emissions technology (such as laser equipment) that could potentially become an ignition source in hazardous (classified) locations are provided in the IEC 60079-28 standard.

6.5 Documentation and Marking

The manufacturer must prepare documents that give full and correct specifications of the explosion safety aspects of the electrical equipment. This documentation must include instructions, providing the following particulars at minimum [4]:

- Manufacturer name and address.
- The information with which the electrical equipment is marked (with explanations), except for the serial number.
- Ratings of the unit.
- Information to facilitate maintenance (e.g., address of the importer, service provider).
- List of the standards, including the edition, with which the equipment is declared to comply.

Table 6.14 Temperature Class of Surfaces

IECEx, ATEX, US + Canada (Classified)	*US + Canada Division*	*Maximum Surface Temperature °C*
T1	T1	450
T2	T2	300
-	T2A	280
-	T2B	260
-	T2C	230
-	T2D	215
T3	T3	200
-	T3A	180
-	T3B	165
-	T3C	160
T4	T4	135
-	T4A	120
T5	T5	100
T6	T6	85

- The environmental conditions for which the equipment is suitable.
- The ambient pollution degree and overvoltage category, if applicable.
- Explanation of symbols, cautions, and warnings.
- Instructions for:
  - Putting into service;
  - Use;
  - Assembling and dismantling;
  - Maintenance, overhaul, and repair;
  - Installation;
  - Adjustment;
  - Possible misuse which might occur.
- Technical data:
- Classification detail;
- Electrical parameters for the complete unit:
  - Power output data, such as Uo (peak AC or DC), which represent maximum output voltage in a protected circuit that can appear under open-circuit conditions at the connection facilities of the apparatus at any applied voltage up to the maximum voltage, including Um and Ui; Io (peak AC or DC), which represents maximum current in a protected circuit that can be taken from the connection facilities of the apparatus; Po, which represents maximum electrical power in a protected circuit that can be taken from the apparatus.
  - Maximum external values of Co, representing maximum external capacitance in a protected circuit that can be connected to the connection facilities of the apparatus without invalidating the type of protection; Lo, representing maximum external inductance in a protected circuit that can be connected to the connection facilities of the apparatus without invalidating the type of protection; and/or the permissible Lo/Ro ratio, which can be connected to the connection facilities of the equipment without invalidating the type of protection.
  - Power input data, such as Ui, representing maximum input voltage (peak AC or DC) that can be applied to the connection facilities for protected circuits without invalidating the type of protection; Ii, representing maximum input current (peak AC or DC), which can be applied to the connection facilities for protected circuits without invalidating the type of protection; Pi, representing maximum input power in a protected circuit that can be dissipated within an apparatus when it is connected to an external source without invalidating the type of protection.
  - Maximum internal values of Ci (total equivalent internal capacitance), Li (total equivalent internal inductance) and the Li/Ri ratio, which is considered as appearing at the connection facilities.
  - The maximum value of Um that may be applied to terminals of nonprotected circuits or associated equipment without invalidating the type of protection.

- Pressure parameters for the complete unit.
- Maximum surface temperatures and other limit values.
- The designation of the surfaces of any enclosure only in circumstances where this is relevant to type of protection.

Note that a control drawing is a recommended form of consolidating connection information and special requirements for installation and use.

- Other information:
  - Where necessary, training instructions;
  - Where applicable, any special conditions of use;
  - Where necessary, the essential characteristics of tools that may be fitted to the equipment;
  - Where applicable, any special conditions that are assumed in determining the type of protection, for example that the voltage is to be supplied from a protective transformer or through a diode safety barrier;
  - Where applicable, certification details;
  - Type of detection equipment, its approvals, installation location(s), alarm and shutdown criteria, and calibration frequency must be documented when combustible gas detectors are used as a protection technique.

The electrical equipment for use in explosive gas atmospheres is identified based on designations that address:

- The likelihood that the explosive gas atmosphere is present when equipment is operating (Zone 0, Zone 1, and Zone 2);
- The ignition-related properties of the explosive gas atmosphere (Group IIA, Group IIB, and Group IIC in each Zone 0, 1, and 2);
- The maximum surface temperature of the equipment under normal operating conditions (T1 – T6 in each Zone 0, 1, and 2);
- The protection methods used by equipment to prevent ignition of the surrounding atmosphere (for Zone 0: da, Ga, ia, ma, op is; for Zone 1: db, eb, i, ib, mb, ob, op is, op pr, op sh, pxb, pyb, p/v, q; for Zone 2: dc, ec, ic, mc, nA, nC, nR, oc, op is, op pr, op sh, pzc; see Table 6.7).

Figure 6.3 provides an example of how these designations are incorporated into typical equipment markings.

Figure 6.4 shows an example of a label for electrical equipment intended to be used in an explosive atmosphere. The marking refers to all situations:

- Zone system (IECEx scheme and ATEX directive);
- Use of the equipment in gas hazardous atmosphere;
- Use of equipment in dust hazardous atmosphere;
- United States division system.

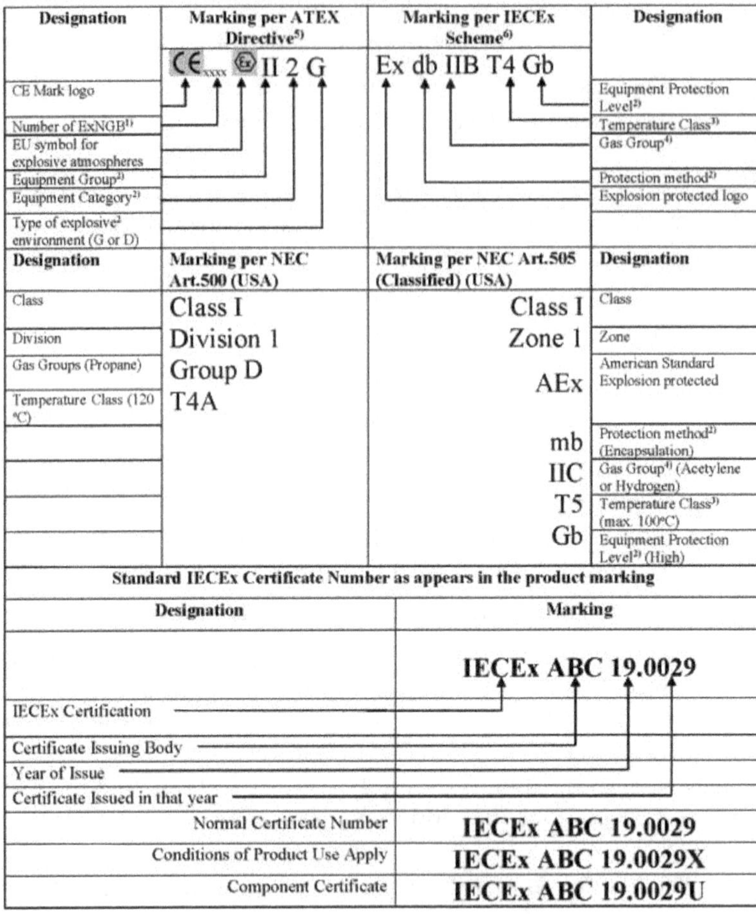

1. ExNGB = European ATEX Notified Bodies Group.
2. See Table 6.8.
3. For Zone 20, 21 and 22 (Explosive Environment Type D–presence of combustible dusts or ignitable fibers and flyings) equipment must be marked with value of maximum operating temperature in place of temperature class (see Table 6.14).
4. See Table 6.2. When the electrical equipment is for use only in a particular gas, the chemical formula or the name of the gas in parentheses.
5. This marking refers to equipment Group II Category 2 intended for use in areas in which explosive atmospheres caused by gases, vapors, mists (G) are likely to occur occasionally.
6. This marking refers to equipment with a high level of protection against explosion (EPL Gb) in an explosive atmosphere caused by ethylene gas (IIB) by means of flameproof enclosure (db) in conformity with IEC 60079-1 standard. The maximum surface temperature can be at 135° C (T4).

Figure 6.3 Examples of Incorporation of Designations of Explosive Gas Atmosphere into Equipment Markings

A real label should only include the requested marking depending on the type of conformity, applicable system, and type of hazardous atmosphere.

Electrical equipment designed for use in a normal ambient temperature range of –20°C to +40°C does not require marking of the ambient temperature range. However, electrical equipment designed for use in other than this normal am-

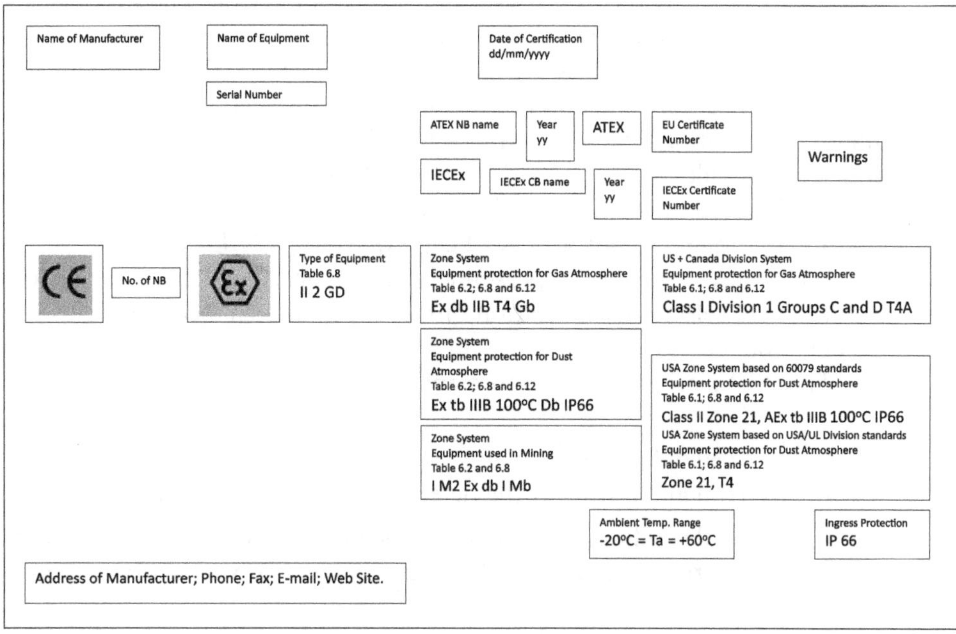

Warnings: DO NOT OPEN WHEN ENERGIZED; DO NOT OPEN WHEN AN EXPLOSIVE GAS/DUST ATMOSPHERE IS PRESENT

NB = Notified Body; CB = Certification Body

II 2 GD = Equipment Group **II** (surface industry); Equipment category **2**; For gas **(G)** or dust **(D)** atmosphere; To be installed in **Zone 1 and Zone 21.**

Ex d IIC T4 Gb = Equipment for Explosive Atmosphere **Ex**; Protection provided by Flameproof enclosure **db** (IEC 60079-1); Gas Group **IIB Ethylene**; Temperature Class **T4 max. 135°C**; Equipment Protection Level (EPL) **High Gb in Zone 1.**

Ex tb IIIB 100°C Db IP66 = Equipment for Explosive Atmosphere **Ex**; Protection provided by Enclosure **tb** (IEC 60079-31); Dust Group **IIIB Non-Conductive Dust** (e.g. coal, flour, sugar); Temperature Max. **100°C**; Equipment Protection Level (EPL) **High Db in Zone 21;** Ingress Protection (IEC 60529) **IP66.**

Class II Zone 21, AEx tb IIIB 100°C IP66 = Ignitable dust intermittent atmosphere under normal conditions **Class II, Zone 21**; American Standard Explosion protected **AEx**; Protection provided by Enclosure **tb** (IEC 60079-31); Dust Group **IIIB Non-Conductive Dust** (e.g. coal, flour, sugar); Temperature Max. **100°C**; Ingress Protection (IEC 60529) **IP66.**

Zone 21, T4 = Classification Zone **Zone 21**; Temperature Class **T4 max. 135°C.**

I M2 Ex d I Mb = Equipment Group **I (Mining)**; Equipment category **M2**; Equipment for Explosive Atmosphere **Ex**; Protection provided by Flameproof enclosure **db** (IEC 60079-1); Gas Group **I Methane**; Equipment Protection Level (EPL) **High Mb in Mining** (De-energized when Ex Atmosphere is recognized).

IP66 = Degree of Protection of the equipment against ingress of solid foreign objects and water as per IEC 60529 standard

Figure 6.4 Example of markings in electrical equipment for use in an explosive atmosphere.

bient temperature range is considered to be a special case and needs to be marked accordingly (e.g., –30°C ≤ Ta ≤ +50°C).

6.6 IECEx and ATEX

The IECEx system is one of the four conformity assessment systems operated by the IEC [3].

IECEx as a global certification scheme became operational in 1996, allowing a unitary certification of electrical equipment and services intended for use in explosive atmospheres. The objective of the IECEx scheme is the worldwide acceptance of one certification based on one standard and one mark for these activities, facilitating international trade in this sector. IECEx has the administration structure based on the rules, operational documents (OD), and decision sheets, and works in partnership with the United Nations Economic Commission for Europe (UNECE) and International Organization of Legal Metrology (OILM).

IECEx covers products, personnel competencies, and services, such as equipment design, selection, installation, inspection, maintenance, repair, overhaul, and reclamation.

The IECEx scheme applies to manufacturers of:

- Electrical apparatus for explosive gas atmospheres;
- Electrical apparatus for the detection and measurement of flammable gases;
- Electrical products, such as switches, outlets and outlet boxes, circuit breakers, electric motors, and lighting used in hazardous environments.

The IECEx system includes the following four separate international certification elements:

1. IECEx certified equipment scheme;
2. IECEx certified services scheme;
3. IECEx conformity mark licensing system;
4. IECEx certified competent persons scheme.

At present the status of IECEx is as follows [3]:

- Thirty-five countries participate at IECEx and the issued certificates of conformity are accepted at a national level in all these countries;
- Fifty-nine organizations from 29 countries are approved to operate within the IECEx certified equipment scheme and to issue IECEx test reports (ExTRs), IECEx quality assessment reports (QARs), and the certificate of conformity (CoC);
- Sixteen organizations from 12 counties can issue certificates of conformity covering service facilities for the repair and overhaul of equipment for use in explosive atmospheres;
- Thirteen organizations from 9 countries can issue IECEx conformity mark licenses covering the use of the IECEx conformity mark;
- Sixteen organizations from 10 countries can issue IECEx certificates of personnel competencies.

For a certified equipment scheme, an IECEx CoC is issued by an IECEx approved certification body, ExCB (similar to a European ATEX notified bodies group, which in fact also operate as an ExCB), and is made publicly available (at the IECEx website) to provide proof that:

- An IECEx ExTR has been issued by an IECEx test laboratory (ExTL) to cover the samples tested (compliance with all parts of the relevant IEC standards, e.g., IEC 60079 series, is mandatory; these standards are listed on the IECEx certificate).
- A QAR has been issued and is current to cover the ongoing surveillance of certified manufacturing (by factory inspection at the manufacturer's premises). A QAR is valid for 3 years. At the end of this period a full audit is required.

For certified services schemes an IECEx CoC-service facilities is issued by an IECEx approved certification body, ExCB, and is made publicly available (at the IECEx website) to provide proof that:

- A facilities assessment report (FAR) was issued and is current to cover the ongoing surveillance of certified services (by factory inspection at service facility). Reassessment is required every 3 years.
- An assessment of facility staff competencies was conducted with acceptable results and is current. Reassessment is required every 3 years.

For a certified competent persons scheme an IECEx certificate of personnel competence (CoPC) is issued by an IECEx approved certification body, ExCB, and is made publicly available (at the IECEx website) to provide proof that:

- A personnel competence assessment report (PCAR) was issued according to competency unit requirements and is current.
- An assessment of applicant's competencies was conducted with acceptable results and is current. Reassessment is required every 3 years.

For all schemes, self-certification is not permitted, also the use of manufacturer data or unwitnessed testing results for certification purposes is not accepted.

The IECEx system is not a regulatory framework. Instead, it is a true certification scheme, where IECEx ExCB and IECEx ExTL are visited and assessed by IECEx expert assessors for qualification. For these organizations the assessment regime consists of an initial peer assessment prior to acceptance for entry to the IECEx system followed by ongoing surveillance assessment audits (the frequency of these audits is established by IECEx management), and after 5 years, a full reassessment audit [3].

The atmospheres explosible (ATEX) directive 2014/34/EU is a European directive and falls within the scope of CE Marking [14]. It applies to mechanical and electrical equipment and protective systems to be used in potentially explosive atmospheres, defining the essential health and safety requirements (ESHR) and conformity assessment procedures, to be applied before products are placed on the EU market, replacing the previous directive 94/9/EC.

The ATEX EU directive is primarily concerned with trade and the manufacture and sale of ex-equipment (electrical and nonelectrical equipment with a potential ignition source). The responsibility for compliance is left with the manufacturer.

The ATEX directive is in direct relation with 1999/92/EC directive, which is primarily concerned with use of ex-equipment, the safety of workers and applies to the classification of hazardous areas, and the correct selection, installation, inspection, and maintenance of ex-equipment. The responsibility for compliance is left with the employers and workers.

An explosive atmosphere means a mixture with air, under atmospheric conditions, of flammable substances in the form of gases, vapors, mists, or dusts in which, after ignition has occurred, combustion spreads to the entire unburned mixture.

A potentially explosive atmosphere means an atmosphere that could become explosive due to local and operational conditions.

The ATEX directive covers [14]:

- Equipment (fixed or mobile machines and instrumentation used for the generation, transfer, storage, measurement, control and conversion of energy, and/or for the processing of material and capable of causing an explosion through their own potential sources of ignition) or a protective system (devices other than components of equipment that are intended to prevent explosion, stop incipient explosions immediately, and/or to limit the effective range of an explosion that are separately made available on the market for use as autonomous systems), intended for use in potentially explosive atmospheres;
- Safety, controlling, or regulating devices that are intended to assist safe operation of equipment and/or protective system for use outside potentially explosive atmospheres but are required for or contributing to the safe functioning of equipment and protective systems with respect to the risks of explosion;
- A component, intended to be incorporated into equipment and protective systems (any item essential to the safe functioning of equipment and protective systems but with no autonomous function).

Note that the ATEX directive 2014/34/EU does not regulate the process of installation.

Some examples of the equipment and protective systems that fall under the scope of the ATEX directive are as follows: paint spray booths, filter units and vented silo bins, gas turbines, steam turbines, gasoline pumps, cables, rotating mechanical seals, bucket elevators, forklift trucks, transportable, pressurized cabins, automatically lubricating systems, electrical trace heating systems, motor protection for category 3 motors, Wi-Fi access points, refrigerators, and storage cabinets for volatile substances.

The technical documentation must contain the following elements [15]:

- Description of the equipment or protective system;
- Design and manufacturing drawings and layouts;
- Descriptions to explain the drawings and functioning of the product;
- Proof that the product complies with harmonized standards, completely or partially;

- If harmonized standards have not been complied with or have been partially applied, details of the steps taken to conform to the directive's requirements must be made clear;
- Calculations and examinations;
- Test reports;
- Certificate from the notified body.

The marking must contain some elements.

All equipment and protective systems must be marked legibly and indelibly with the following minimum particulars [15]:

- Name, registered trade name or registered trademark, and address of the manufacturer;
- CE Marking (as in Annex II to Regulation (EC) No 765/2008);
- Designation of series or type;
- Year of construction;
- Specific marking of explosion protection («epsilon-x», or «the hexagon») followed by the symbol of the equipment group and category;
- For equipment group II, the letter G (concerning explosive atmospheres caused by gases, vapors, or mists);
- The letter D (concerning explosive atmospheres caused by dust, fibers, or flyings).

Note that importers are required to put their names and addresses on the products.

An ATEX certificate does not confirm compliance with any particular standard. Compliance to standards, CENELEC or others, as harmonized (e.g., the IEC 60079 series) are not a mandatory requirement; however, they are often used to assess products and most manufacturers use relevant standards to demonstrate compliance with the essential health and safety requirements (EHSR) of Annex II of the ATEX directive.

Bodies offering ATEX certification are known as the European ATEX Notified Bodies Group (ExNBG). These third-party bodies are appointed by member states and the directive 2014/34/EU introduced a requirement for an ExNBG to have national accreditation.

ATEX ExNBG issue certificates called EU type examination certificates. These are not the same as an IECEx CoC and have more in common with the IECEx test report. The certificate does not directly relate to subsequently manufactured items, but only to the type or sample that was examined.

EU type examination certificates indicate that the product complies with EHSR requirements standards that have been used to support the justification of compliance. The EHSRs will be listed in the EU type examination certificate, but complete compliance with those standards is not guaranteed.

The assessment and testing/certification process of ATEX as with EU directives is risk-based, meaning that for lower-risk areas ATEX allows a full manufacturer's declaration of conformity without involvement of an NB. Other higher-risk areas may require the involvement of an NB. The legislation covers both electrical and nonelectrical products.

In the case of equipment intended for use in a Zone 0 (Category 1) or for electrical equipment intended for use in Zone 1 (Category 2) the manufacturer needs to engage a European ATEX notified bodies group for examination/testing of the samples and a European ATEX notified bodies group for the factory inspection. The ExNBG conduct regular audits of manufacturers in respect of all Category 1 equipment and electrical equipment of Category 2. There is no audit in respect of nonelectrical Category 2 equipment or in respect of all Category 3 equipment.

The declaration of conformity (DoC) made on the sole responsibility of the manufacturer is a self-declaration confirming that they comply with EHSRs and are in possession of the type of examination certificate (which covers the assessment of samples) as well as a quality assessment notification (QAN) which covers the factory inspection (current manufacturing of products).

Harmonized CEN/CENELEC standards EN specified in the EU Official Journal as applicable for ATEX directive have been developed specifically to allow a presumption of conformity with the EHSR's. The word presumption means that full compliance is not automatic when applying to CEN/CENELEC standards.

ExNBG issues [15]:

- EU type examination certificates: These document the evaluation and testing of the equipment to the applicable EN 60079 standards.
- Ex QAN: These document the suitability of the manufacturer's QA system as related to ISO/IEC 80079-34. The preparation of these follow an identical process to that for IECEx QARs.

Interrelations between IECEx and ATEX:

- The ATEX directive requires that the essential health and safety requirements are satisfied, for this compliance with a relevant standard can be considered as fulfilling these requirements.
- ATEX does not require compliance with the relevant IEC standard(s).
- In most cases an IECEx ExTR can be accepted on issuing an ATEX EC type examination certificate where the technical requirements in IEC and EN standards are identical (normally this is the case).
- In all cases an IECEx QAR can sustain an ATEX QAN.
- In many cases European ExCBs will issue IECEx certification and ATEX documentation at the same time for specific equipment.
- ATEX is a regulatory framework (it is not a certification scheme) applied within the EU and refers only to equipment.
- IECEx scheme covers the equipment and services (e.g., repair and overhaul) industries.

- Products with IECEx certificate accepted in several countries. Alternatively, a test report (ExTR) is accepted by any IECEx scheme member country and can use the ExTR to issue locally accepted certification.
- Test results and assessment to IEC standards can be used for both IECEx and ATEX.
- ATEX documentation alone is not sufficient to satisfy IECEx certification requirements.
- Some countries, such as Hong Kong, Taiwan, Vietnam, and Indonesia, do not accept the IECEx test report, but will accept the ATEX test report.
- For workplace requirements, IECEx recommends to refer to national regulations while ATEX contains special requirements for workers and management.

References

[1] IEC 60050-426, "International Electrotechnical Vocabulary (IEV)–Part 426: Explosive Atmospheres" Geneva, Switzerland, 2020.

[2] NEC–NFPA 70, Article 500, *National Electrical Code*, Quincy, MA: National Fire Protection Association, 2017.

[3] IECEx, https://www.iecex.com/information/about-iecex/.

[4] IEC 60079-0, "Explosive Atmospheres–Part 0: General Requirements," Geneva, Switzerland, 2017.

[5] CEAG, *Principle of Explosion Protection*, Houston, TX: Cooper Industries Ltd, 2012.

[6] NEC–NFPA 70, Article 505-506, *National Electrical Code*, Quincy, MA: National Fire Protection Association, 2017.

[7] ISO/ IEC 80079-20-1, "Explosive Atmospheres–Part 20-1: Material Characteristics for Gas and Vapor Classification–Test Methods and Data," Geneva, Switzerland, 2017.

[8] National Fire Protection Association, *NFPA 30: Flammable and Combustible Liquids Code*, Batterymarch, PA, 2003

[9] Engineering ToolBox, (2003), Gases - Explosion and Flammability Concentration Limits, https://www.engineeringtoolbox.com/explosive-concentration-limits-d_423.html.

[10] United Nations, "Globally Harmonized System of Classification and Labelling of Chemicals (GHS)," New York, Geneva, 2019.

[11] EN 1127-1, "Explosive Atmospheres. Explosion Prevention and Protection. Basic Concepts and Methodology," CENELEC, Brussels, 2019.

[12] IEC 60079-11, "Explosive Atmospheres. Part 11: Equipment Protection by Intrinsic Safety I," IEC, Geneva, Switzerland, 2011.

[13] IEC 60079-14, "Explosive Atmospheres–Part14: Electrical Installations Design, Selection and Erection," IEC, Geneva, Switzerland, 2013.

[14] Directive 2014/34/EU -in EUR-Lex, http://eur-lex.europa.eu/legal-content/EN/TXT/?qid=1460102657898&uri=CELEX:32014L0034.

[15] European Commission's website on Equipment for potentially explosive atmospheres (ATEX), http://ec.europa.eu/growth/sectors/mechanical-engineering/atex/.

Selected Bibliography

Bottrill, G., D. Cheyne, and G. Vijayaraghavan, Practical Electrical Equipment and Installations in Hazardous Areas, Newnes, 2005.

CEN/TC 305, "Potentially Explosive Atmospheres–Explosion Prevention and Protection," https://standards.cen.eu/dyn/www/f?p=204:7:0::::FSP_ORG_ID:6286&cs=1DBA2E1F81140C9AE9D8C65E805416228.

Eaton, "Principle of Ex-Protection," 2017.

Groh, H., *Explosion Protection*, Amsterdam: Elsevier, 2004.

IML Group PLC, http://www.hazardexonthenet.net/.

McMillan, A., Electrical Installations in Hazardous Areas, Oxford, UK: Butterworth-Heinemann, 1998.

National Fire Protection Association, *NFPA 45: Laboratories Using Chemicals*, Batterymarch, PA: NFPA, 2004.

OSHA, "General Industry Safety and Health Standards 29CFR 1910.106," Washington DC: Bureau of National Affairs, 2007.

CHAPTER 7

Practical Aspects Related to Global Market Access

7.1 Introduction

In Chapter 2 of Volume I of our book, *Electrical Product Compliance and Safety Engineering*, under "International Regulations and Global Market Access," we discussed regional regulations, testing and certification bodies (the history of such a body, LICE, can be found in [1]), and certification marking requirements. We emphasized the need for raising the level of credibility by involving independent testing and certification bodies during the approval process of the electrical equipment intended to be placed in the global market.

Although there are many obstacles to overcome before we reach this final objective—access to the global market—we consider that access is not as difficult as actually being able to stay in it.

In this chapter, our intention is to provide more details in the actual world when those remaining players of the game will be faced with even more challenges. We will try to make some connections of the existing quality management systems in a way that may reduce the involved risks when some equipment will arrive on the market and may be able to stay there in good conditions for the life of it.

Based on our experiences, resulting from defining the equipment, designing it, performing internal testing and evaluation, managing quality systems, introducing electrical equipment in manufacturing, obtaining the necessary final approvals, and implementing measures of continuous improvement for each step during those activities, we will share few practical ideas that may assist compliance professionals to better understand global market access.

In today's marketplace, where each piece of equipment will be scrutinized even more than before, we are trying to build confidence that once in the market, a manufacturer will be able to stay in business by having the necessary tools and insight to withstand the whirlwind pace of global business development.

In our opinion, the international market, often referred to as the global market, will be very cautious going forward, and will undoubtedly not show the same flexibility that many companies took advantage of in the past. Governments, regulatory agencies, international organizations, and manufacturers will certainly be very cautious in giving the green light to projects and products, and under existing financial and economic pressures may well only issue short-term decisions that will characterize a very volatile environment for accessing the market. This will make it necessary for all players to become as flexible as possible to accommodate the regulations for access. Several changes (political/war/unrest/globalization) have affected Eastern Europe, Africa, and even the North and South American markets.

Manufacturers will have to understand the changes very quickly, and to be able to stay in the game, they must be focused on their main objective: product safety and customer satisfaction. They must adapt their policies to be able to clearly describe their processes. Otherwise, they will suffer the Deming prognosis, "If you can't describe what you are doing, you do not know what you are doing," and they will, so to speak, be eliminated from the game (market) before the end of the match.

To develop new markets, manufacturers need to follow these steps for evaluation of new foreign markets and entry strategies to penetrate them:

- Identify local regulatory requirements and approval strategies;
- Identify local partners/representatives;
- Develop customer profiles;
- Highlight the advantages of using your product;
- Enlist expert opinions about product performances;
- Coordinate the strategic execution of manufacturing plans (time, capacity, service, etc.).

7.2 Required Documentation

Before gathering the necessary documentation, a manufacturer needs to be aware of the following aspects:

- The latest valid regulations in the targeted markets;
- Specific national requirements and/or deviations from the requirements of the relevant standards;
- Contents of the documentation that must be submitted;
- The expected cost of the necessary official approvals.

The documentation should start to be gathered well before even the product definition document is issued and approved, and must be continuously updated during the design stage, internal testing, implementation in production, and marketing of the equipment.

The regulatory engineer will have the duty of gathering all the design solutions and to include them within the technical construction file (TCF) of an electrical product.

The TCF will include:

- All the parameters expected to be met;
- Testing according to the relevant standards;
- Critical design information;
- List of critical components (LCCs) that are a significant part of the equipment (these components are discussed in detail in Chapter 8 of Volume I of this book series);
- Marking according to specific market.

The next step, which would typically occur at the final stage, would be the decision of obtaining the necessary proof of compliance, and therefore should be a celebration—not a time of stress—for any compliance and/or approval department.

If the process was properly managed and the documentation was done properly, then it is expected that the hardware and the software teams will be in the position to fully satisfy each requirement.

The marketing team has an important role and must not make the design team and/or the compliance team face last-minute marketing changes. Any late changes may compromise the speed to market in a very dramatic way, and, we observed that equipment may be shelved in the last minute of accessing the market, for a while or forever, due to very late changes.

The challenges that may appear during the final process of obtaining the proof of compliance may be generated mainly by the following factors:

- Availability of the critical components;
- Insufficient knowledge of the performance and regulatory requirements;
- Electromagnetic compatibility testing (complex systems may be subject to some last-minute failures);
- Lack of attention to details.

In order to eliminate the first factor, the availability of critical components, we insist on having time for each critical component that defines performance and/or product safety compliance of the equipment for at least two alternate components. This will save time and money for the manufacturer.

The second factor is strictly related to determining the grid for compliance. We strongly believe that the manufacturing/design team knows the most about their equipment, and should be able to fully disclose all the potential hazards during the evaluation. By not doing so, this will have repurcussions later, when the customer realizes the lack of quality control. The approval process of equipment is not a hide-and-seek game with the chosen testing laboratory, and the testing and evaluation should be a straightforward process on the part of the testing laboratory. Engineering judgments should be issued according to engineering principles that are not subject to interpretation. When choosing a testing laboratory, we strongly recommend doing a thorough investigation regarding the prices for evaluation, keeping in mind the reputation and history of each organization, in order to save further headaches and maximize your budget.

The third factor may be the most difficult. In EMC compliance, even if parts of a system are found fully compliant during the design stage, the final assembly will not necessarily be a compliant system.

Understanding this requires thorough knowledge of the intimate interaction of the parts of the system (interconnections, grounding, bonding, etc.).

Regarding the fourth factor, it is critical to ensure that everything—from the name and model of the equipment and to the information included on the Marking—must be consistent and accurate. In many cases, border authorities have rejected batches of imported equipment simply due to discrepancies between the Marking label and the accompanying documentation.

At the point, the manufacturer will have the technical construction file along with the acceptance test results for the intended market. The manufacturer will make the latest requested modifications required from the evaluation, within the installation manual, safety instructions, user's manual, perhaps even within the marking label(s), and will draw all the declarations that have to be signed, (declaration of conformity, declaration regarding the country of origin, declaration regarding the routine tests, RoHS, WEEE, etc.), and finally will ensure that all the necessary steps to assure the quality of the equipment will be provided, such as routine tests in production and performance tests. These will be performed with calibrated test equipment, after which another step will prepare the equipment for shipping. This step represents another concern to ensure that the customer will receive the shipped equipment in perfect condition even if the shipping is continental or intercontinental and requires different transportation standards to be met (e.g., International Safe Transit Association (ISTA).

Once the equipment is shipped, there may be a sense of relief and even excitement, from the sales department down through the rest of the teams. Once it leaves the company, the equipment will remain for the rest of its life a symbol of the quality supplied by that manufacturer. It is a well-known aspect of business that a good job is rarely recognized, but just one error will lead to a nonstop chain of complaints. This means that the customer must be satisfied, and to do this, protective measures should be in place, some of which are in the hands of the testing laboratory that evaluated the equipment, but generally speaking, the overall responsibility falls on the manufacturer. The following sections will discuss the steps that must be taken to ensure that your equipment is compliant and ready to go into the market.

7.3 Specific Labeling and Marking

As required by local imposed laws and regulations, and following the local and national electrical codes to be accepted in a market, in most of the cases, an electrical product must be approved and thus must bear a sign that it was evaluated and meets the applicable requirements of the intended market.

The signs that are affixed to the approved equipment are labels and/or a specific marking. These act almost like a business card for the equipment and thus, include the minimum information that the end user will need to use the equipment.

It is common to be faced at home or work with the situation of having to find a spare part for a piece of equipment or even consumables for it. Despite the trend to standardize each industry, there is absolutely no certainty that we will be able to find interchangeable parts for equipment even from the same manufacturer.

In order to protect customers, and also for traceability, local authorities having jurisdiction (AHJ) issue the requirements for labeling and or marking. Note that the labeling system is less stringent than the marking system.

Labels are generally issued by an inspection authority who inspected all or a few samples of a batch of equipment based on some locally issued regulations. This type of approval is called special inspection or field inspection and it limits the number of pieces of equipment that are approved. Labels are issued by the mandated organization based on a summary of tests that the randomly chosen samples will be subjected to.

Generally speaking, the special inspection type of approval is suitable for custom-built equipment for special applications, equipment that is manufactured on a nonrepetitive basis, equipment already installed or ready for use on-site and awaiting acceptance by the AHJ, and equipment with a production run with a limited number of units per model, per year.

The above clearly provides an open door for manufacturers to reach the market very fast, and at the same time, we observed that it is used by smaller companies with a low volume of equipment. However, not all equipment is suitable to be accepted via this procedure and thus, labeling based on a field (special) inspection has a limited field of application. Therefore, equipment and/or components that involve a more significant product safety review and documentation are excluded from acceptance through this approval process.

The field (special inspection) must not be considered a certification (listing) service, as it is not meant to be used for wiring devices, medical-electrical equipment, hazardous locations equipment, manlifts, elevators, climb assists, high-voltage equipment, or components that need further evaluation as part of the end product. The manufacturer receives a limited number of labels that will be affixed to the approved batch only. Along with the labels, it is the duty of the party who applied for the field (special) inspection to keep the field inspection report, which will show the number of approved pieces and the serial number of the issued labels for the inspected equipment. Random testing is involved during the field/special inspection, along with an attentive construction and components review. Only equipment that includes the approved type of components within the mains circuits will be considered for this type of acceptance.

The field/special inspection process involves several steps, including:

- Request for inspection;
- Equipment evaluation:
- Writing and submitting a field evaluation report to the manufacturer.

To prepare for the evaluation, the manufacturer or the company that wants to place the equipment on the market will need to provide the inspection body with documentation, such as electrical schematics, mechanical drawings, bill of materials, including a detailed specification of all the critical components, manual(s), nameplates, related certificates, and any available client test data related to the involved equipment.

The certification/listing process represents the highest level of acceptance and it brings with it the highest degree of confidence in the approved equipment. The evaluation of the certified/listed equipment involves type tests as well as a thorough surveillance process during the lifetime of the product. The certification agency will be the observing part by those scheduled periodical inspections (follow-up service) at the manufacturing locations and will keep the manufacturing documentation. This ensures that all the critical components accepted during the type testing and evaluation remain the same and all the quality management procedures are respected during the manufacturing process.

The marking of the certified/listed equipment is printed by companies approved to do this type of label and in turn, they are under the surveillance and control of the agency in charge of the certification.

The contents and the mark included on the label represents the manufacturer and the testing agency. It is a symbol of recognition that the equipment on which the label is affixed represents fully evaluated equipment based on one or more standards for product safety.

During the lifetime of a product, no environmental conditions or any other factor may influence the durability and legibility of the affixed labels. At any time, the affixed marking must be legible and provide all relevant information to identify the equipment, the manufacturer, and the certification body that approved the product. Once the equipment is approved, one of the conditions, as we stated above, is to use an approved label supplier. Approved label suppliers use durable materials that are indelible, making it less likely that the text on the label will disappear over time.

It is not surprising that the marking process is very complex and labels are becoming a complete story in of themselves of the destination and capabilities of the equipment.

It is typical that certain markings are preferred by customers themselves, and some will look in North America for UL listed equipment, while other customers will look for CSA or TUV certified equipment; in the European Union, the specific CE marking is essentially the entrance ticket into member countries. These markings, accompanied by a certification mark from a well-recognized certification body, represent a level of security, and thus make the equipment more competitive in the market. The same evaluation criteria, from the point of view of consumers, applies to other markets worldwide.

This is not only a marketing aspect. Over time, NRTLs were able to build confidence in their certification services and thus, each testing laboratory and certification body currently has a good name in a certain field. The authors will not detail any related aspect to this to not endorse any testing or certification organization. We have complete respect for all the test engineers who are involved in the process of keeping consumers safe and sound.

Marks and symbols are put together to reflect the fact that the equipment on which that label was affixed was found compliant with certain requirements. Generally speaking, these labels are in fact providing value added for the product on which the labels are affixed.

Marketing departments should take advantage of the markings for this reason, and consider keeping a complex marking on their equipment even if is not mandatory.

We discussed the contents of marking for certified equipment within Volume 1 (Sections 2.5 and 12.1) of this book series.

7.4 Accompanying Documents and Languages

As noted previously, a manufacturer, whether established inside or outside of a market, is ultimately responsible for the proper use of their equipment. And thus, besides the proper marking, the manufacturer is fully responsible for providing all the information regarding the proper use of that equipment (all the documents necessary for the border authorities, electrical inspection agencies, and the end user).

A manufacturer is defined within the EU directives as, "the natural or legal person with responsibility for the design, manufacture, packaging, and labeling of a device before it is placed on the market under its name." Typically, a manufacturer established outside of the country where the equipment is going may appoint an authorized representative (e.g., a distributor that is established in that country to act on their behalf). We have to emphasize here that placing a product on the market is a legal procedure that must be followed in accordance with all relevant market requirements, despite who may be the legal entity that places the equipment on the market (a manufacturer or a representative/distributor).

The complete list of documents that must accompany an electrical product depends on the type of equipment and of the approval type that it requires. The list also depends on to who will be the end user, and who will install and service the equipment in case of failure.

Then, the end user will need to ensure that the equipment is installed and operated in a completely safe manner, and that no undue hazard to persons or property will likely be generated.

The accompanying documents are usually required and verified when the equipment is crossing a border. To avoid unwanted headaches, it is strongly recommended to have all of the required documentation in the language of the intended market. It will make the acceptance process smooth and may eliminate any cause of misunderstanding. Although English is widely accepted, some EU directives require having the main documents (e.g., declaration of conformity, safety instructions, information to demonstrate the conformity with directives, contact details for manufacturer and importer) in the language of the EU countries where the equipment is going. The manufacturer and/or their representative must have available the product documentation and documentation that offer the proof of approval as specified in Table 7.1.

We discussed the contents of accompanying documents in Volume I (Sections 12.5 and 12.6) of this book series.

7.5 Approvals for the Intended Market: Internal Nonaccredited Testing versus an ISO/IEC 17025 Accreditation

The process of approval for the intended market will generate the proof of approval we discussed earlier. The complexity of the approval process depends on the type of

Table 7.1 Documents That Need to Accompany a Product to Access a Specific Market

Approval Documents	*Product Specifications*	*Marking*	*Manuals*
Method of approval as it is required by the local legislation. Test and evaluation reports or certification reports. WEEE documents (for EU). European authorized representative (if applicable). Customs documents.	Identification of product. Ratings. Environmental condition for installation and use. Means of supply (electrical, water, air, etc,) as applicable. Relevant parameters that will offer the proof that the equipment is suitable for use on the intended distribution system. Special environmental conditions (as applicable).	Product labels per applicable standards. Certification marks (if applicable). Additional local marking (as applicable).	User manual. Installation manual. Service manual (optional).

the equipment and as already stated, this may be an electrical inspection procedure continuing with full testing and evaluation of the equipment.

We will not discuss here a product that must be tested for certification, which has a very clear path that gives the right to apply a specific certification mark from a certification body on the product.

There are circumstances where it is acceptable for the testing and evaluation to be performed by the manufacturer; as we know, declarations of conformity for the acceptance of equipment on the EU market are voluntary declarations. Thus, a complete test and evaluation report of essential requirements will be sufficient to offer proof of compliance with the imposed legislation (directive). A similar methodology is accepted by the FDA when requesting compliance with the recognized standards.

The manufacturer may have numerous choices in performing the required test and evaluation of their equipment. We will discuss here only two of them, as the rest are very straightforward. One is to go to a testing house and submit a request for testing (obtaining a certification mark, or simply a compliance with a specific standard test report, which will bring the manufacturer some peace of mind). In fact, the idea of testing and evaluation consists of performing a conformity assessment with one or more standards that cover the performances and product safety aspects that must be fulfilled for the subject application.

The starting point is to determine which standard applies to the product. It is not a secret that the majority of manufacturers will choose one standard only. It is a matter of choice to fully understand the difference between a simple conformity assessment with one standard and to test and evaluate from the point of view of performance and product safety to determine that the equipment is qualitative and safe. In our engineering judgment, there may be a huge difference between a piece of equipment that is designed in conformity with a standard and a piece of equipment that is qualitative and safe.

The quality and safety of a product represent a characteristic similar to a bulletproof vest for some electrical equipment. How this is determined and who is the most important involved part is very easy to state: the design team and the manufacturer of that equipment.

Over time, we have seen several instances where electrical equipment that is fully conformant with applicable clauses of a specific standard was not qualitative and safe. In our opinion, when a manufacturer is fully involved in the process of testing and evaluation of their product, the chances of having a performant and safer product rise considerably.

What are the ways to perform an evaluation in-house?

In order to provide a comprehensive answer, we have to review all the ways that may be chosen to achieve the same result: documentation that shows the product is an approved type.

We already know that the NCBs (e.g.,VDE, TUV, DEKRA, BSI, SGS, Intertek, NEMKO, CSA, UL, Japan Electrical Safety and Environment Technology Laboratories (JET), China Quality Certification Center (CQC), IMQ) have a large testing laboratory network. Almost all are able to perform the any type of testing that any manufacturer will ever need. We also know that the services obtained from these testing laboratories will not always be in line with the budgets of all manufacturers even though they will provide a level of credibility at the end of the evaluation.

We strongly suggest that manufacturers become fully familiar with as many applicable standards for their equipment related to but not limited to enclosures, materials, components, wiring, and all the foreseeable hazards that may occur during the lifetime of the equipment. This will ensure that the equipment meets the necessary criteria for operating safely in the field.

We want to stress again that the standards represent only the minimum requirements for any equipment, and thus all the responsibility and liability falls to the manufacturer despite any conformity assessment to which the equipment will be subjected. It also bears repeating that the design team and the original manufacturer have the best knowledge of their equipment.

To be very clear, we do not consider that the standards are not thorough enough in covering any specific category of equipment. It is just a fact that as technology evolves, it will require new measures to ensure safety and compliance. As well, cost reduction is often a factor in quality assurance even if not disclosed directly, components often change, the reliability of components supplied by the same manufacturer can vary, and thus the end application changes the equipment characteristics.

In today's ever-changing world of industry, under great pressure coming from many directions, each manufacturer tends to trim as much as they can at every stage in order to place their equipment on the market at a competitive price. We recognize that the performance and safety characteristics of this equipment will be somewhat sacrificed and public safety may also be compromised. Manufacturers will produce equipment that will be safe most of the time, using lower-performing components, and because of this the price may go down a bit. The equipment will start its journey into the market with less assurance that the end user will enjoy it for the duration that it was supposed to work. We have seen time and again cases where expensive equipment is abandoned due to reliability and safety issues that the end user encountered way before the equipment met its life expectation. Expensive equipment manufactured by companies that eventually go bankrupt (not coincidentally), and the product ends up in a warehouse waiting to be recycled.

Manufacturers who demonstrate competence will be much more likely to stay in business. They will ensure that their quality management system integrates follow-up services as a tool for quality maintenance and improvement, they will study and analyze the evolution of their equipment in the field, and they will mitigate any possible damages that may appear.

A typical pattern should be developed where the manufacturer will start to test and evaluate their own equipment.

Below are several scenarios to better understand the steps of testing and evaluation. We state first that in real life, the process is done out of order. The equipment is designed, manufactured, and then presented for the final exam of conformity with a standard.

Most of the time the exam will be failed during the first try (we will not dwell on the causes of the failures), but these failures will generate lots of frustration, and more than once we have found that the initial approvals budget will be exceeded, and the most important element (in our opinion)—the time to market—will not be met.

There is a solution that will help any manufacturer who starts their own testing laboratory. Knowing the standards that are applicable for the equipment intended to be placed on the market, a manufacturer will procure the test equipment taking care to have all the equipment calibrated.

Then the equipment will start to be developed on paper. From that moment, the regulatory approvals engineer(s) become an active participant, and the evaluation process will start with choosing the critical components in a manner that everything at the end will not present any issues when the equipment is ready for the type tests (all the tests that apply to the equipment under test (EUT)) described within the chosen standards.

During the design stages the regulatory engineer will evaluate and test printed circuit boards offering live feedback to the design team. The design team will keep under strict control the existence of the correct printed circuit board layout including creepage distances, air clearances, and location of the components to better meet the next level of testing from the EMC and product safety point of view. Additionally, they will take care of the review of each critical component, and last but not least, will ask for alternate components for those which are considered critical in order to ensure the continuity of manufacturing if any shortage should appear on the supply chain side.

Once the boards are assembled, preliminary testing programs are already designed and will be performed. Some of the tests on the enclosures can be made way before the equipment is ready by using dummy load from the point of view of the weight and mechanical strength.

Assuming that everything has gone according to plan up, the type tests will be a success.

There are several legitimate questions related to this process of evaluation and testing. If performed in-house, does it offer the necessary confidence to place the equipment on the market? Is it sufficient for acceptance as a valid testing and evaluation of the equipment?

The regulatory engineer will be in charge of providing comprehensive answers. He or she will review the intended market requirements, study the legislation, local rules, regulations applicable to the equipment, and the local electrical code that is relevant to the targeted market.

In short, having a regulatory engineer is an invaluable resource.

It is also useful to have a testing laboratory in-house. For some markets, detailed in-house testing will be satisfactory; a complete test and evaluation report that is included within a technical construction file of the equipment will be sufficient.

We know the frustration of discovering that for other markets, an in-house test report will not be acceptable. Then what can be done?

Based our experience, two avenues can be approached at the same time:

1. To work in conjunction with an NRTL. Currently, almost all NRTLs are working closely with manufacturers on building programs that their clients are allowed to perform in-house witnessed (by the NRTL) testing. In time, NRTL will eventually offer a gradual recognition.

   The procedure is as follows: First, the NRTL agrees to assess the capabilities of the manufacturer by observing their in-house tests. Then, they expand the program at the manufacturer's request to a more relaxed program in which the NRTL will accept the test results as valid based on minimal but effective supervision. This supervision is done by auditing the tests that the manufacturer performs on an annual basis, and, at the same time, auditing the quality management system that the manufacturer has in place, which offers proof of a healthy testing environment. This type of umbrella represents an ideal solution for high-volume manufacturers who want to ensure that the time-to-market requirements are fully met. In our opinion, such a solution does not represent the most economical route for a low-volume manufacturer who may be faced with hefty fees for the services rendered by the NRTL.
2. To obtain official recognition (accreditation) of competence based on the ISO/IEC 17025 standard in performing the testing in-house, the test and evaluation report must be issued according to the specific product standards.

   Official recognition of competence in testing may be obtained by applying for accreditation for the standards that the manufacturer is intending to use when evaluating the equipment manufactured on their premises, regardless of how many manufacturing locations they may have.

The laboratory accreditation is given by a body (e.g., A2LA, UKAS, NAVLAP) that has the mandate to offer accreditation.

To fully understand what means to become accredited, we have to understand what accreditation means. Accreditation is a formal, third-party recognition that an organization is competent to perform specific tasks, the work for which they are accredited. Accreditation can be voluntary or mandated by the government. It is a continuous quality improvement process to demonstrate that internationally and/or nationally prescribed standards have been met.

Accreditation bodies are established in many economies with the primary purpose of ensuring that conformity assessment bodies and testing laboratories are subject to oversight by an authoritative body. We will not provide details about which organizations are appointing the accreditors. However, we found that in Canada, the Standard Council of Canada was appointed by the federal government. Other accreditors were established as not-for-profit organizations or had been around for a long time and evolved based on common law principles.

As an example, UKAS in the United Kingdom is a not-for-profit company limited by guarantee. It is independent of the government and operates under a memorandum of understanding with the government through the secretary of state for business, energy, and industrial strategy. UKAS was appointed as the National Accreditation Body by the Accreditation Regulations 2009 (SI No 3155/2009) and the EU Regulation (EC) 765/2008 [3].

Generally speaking, some of the accreditors are appointed by local governments, and for certain markets, may reflect the local political orientation.

When looking for an accreditor of an internal testing laboratory, it is important to research their history as an accreditation organization and to consider if they have a good reputation and are an independent organization that does not directly report to any government agency, which may generate a conflict of interest or political interference.

We contacted several accredited testing labs and asked their opinions about that accreditor. Finally, due to the size of our laboratory and the involved fees, and taking into account all the aspects mentioned above, we decided on a well-recognized accreditor with extensive experience in this business. After almost 20 years with this accreditor, A2LA, we can state that their long history in this business, their transparency in operations, and their excellent customer service verifies their standing as a reliable company. This accreditor has built confidence in our testing and we were rewarded with the work that this accreditor did under the International Laboratory Accreditation Cooperation (ILAC) umbrella, resulting in the worldwide acceptance of our test reports.

According to our experience as participants within an accreditation program, we were able to realize other benefits of being an accredited testing laboratory.

Some of these benefits include the following:

1. The best aspect was the fact that we built a sound quality management system and we function within this structure. It gave our work precision and accuracy, and due to accreditor agreements with international organizations, we received a form of international recognition that allowed us to have the data and results be more readily accepted in overseas markets. We were also able to bring each piece of equipment to the market on time, and not have to worry about any delays that might be encountered during evaluation by a third-party laboratory.
2. Accredited laboratories usually issue test or calibration reports bearing the accreditation body's symbol or endorsement from that body as an indication of their accreditation. The customers of the testing laboratory are encouraged to check with the laboratory as to what specific tests or measurements they are accredited for. This information is specified in the laboratory's scope of accreditation, issued by the accreditation body, which provides the customers seeking laboratory services with clear information about the range of testing or calibration services that the laboratory can provide under their accreditation. At the end of the day, it gives a higher degree of confidence in the testing performed by that testing lab.
3. A recognition of testing competence gains global recognition under an existing mutual recognition arrangement (MRA). This system of international

MRAs among accreditation bodies has enabled accredited laboratories to achieve a form of international recognition, and thus allowed the data accompanying exported goods to be more readily accepted in overseas markets. This aspect effectively reduces costs for the involved parties, the manufacturer and/or the importer, by reducing or eliminating the need for equipment to be retested in another country.

4. Laboratory accreditation is highly regarded both nationally and internationally as a reliable indicator of technical competence. Many industries, from clinical, chemical, construction, forensic science, electrical, and food sectors routinely specify laboratory accreditation for suppliers of testing or calibration services.
5. Unlike certification to the ISO 9001, which uses general criteria and procedures that do not make any statement about the technical competence of a company (in our case a laboratory), the accreditation uses criteria and procedures specifically developed to determine technical competence, and thus assures customers that the test (or calibration or measurement) data supplied by the testing /calibration laboratory or inspection services are precise, accurate, and reliable. In this context, a mandatory proficiency testing program is a definite plus.

We found that over time, accreditation has many advantages and contributes effectively to reducing the cost for testing and evaluation. From our experience, we can state that having the testing laboratory accredited was a huge benefit for our organization. Working very closely and with ILAC, Asia Pacific Accreditation Co-operation (APAC), and others, our accreditor was able to guide us in a very constructive manner. In turn, we realized that the accreditation bodies that are members of ILAC are required to comply with appropriate international standards and the ILAC documents for consistent application of those standards (e.g., ISO 17025). It gave us confidence in the products and services that our accreditor offered us.

ILAC manages all the international accreditation arrangements in the fields of testing, calibration, medical laboratory testing, inspection bodies, proficiency testing providers, and reference material producers. The International Accreditation Forum (IAF) deals with the international accreditation arrangements in the fields of management systems, products, services, personnel, and other similar programs of conformity assessment. Both organizations, ILAC and IAF, work together and coordinate their efforts to enhance the accreditation and conformity assessment worldwide.

Regional arrangements are managed by recognized regional cooperation bodies that work in harmony with ILAC and IAF. Such bodies are EA in Europe, APAC in the Asia-Pacific, Inter American Accreditation Cooperation (IAAC) in the Americas, African Accreditation Cooperation (AFRAC) in Africa, Southern African Development Community Cooperation in Accreditation (SADCA) in Southern Africa, and Arab Accreditation Cooperation (ARAC) in the Arab region. All of these organizations are working together, and the benefit of their ties is visible to any of the accredited testing laboratory. Their reports will be acceptable to the local authorities that have jurisdiction.

The membership of ILAC consists of accreditation bodies and stakeholder organizations throughout the world (101 MRA signatories as of August 2020) and is the representative organization that is involved with

- The development of accreditation practices and procedures;
- The promotion of accreditation as a trade facilitation tool;
- Supporting the provision of local and national services;
- Assisting in developing accreditation systems;
- The recognition of competent testing (including medical laboratory) and calibration laboratories, inspection bodies, proficiency testing providers, and reference material producers around the world.

ILAC became an international organization for accreditation bodies operating under the ISO/IEC 17011 and involved in the accreditation of conformity assessment bodies including calibration laboratories (using the ISO/IEC 17025), testing laboratories (using the ISO/IEC 17025), medical testing laboratories (using the ISO 15189), inspection bodies (using the ISO/IEC 17020), proficiency testing providers (using the ISO/IEC 17043), and reference material producers (using the ISO 17034). It assisted in defining the accreditation (the independent evaluation of conformity assessment bodies against recognized standards to carry out specific activities to ensure their impartiality and competence). Through the application of national and international standards, government, procurers, and consumers can have confidence in the calibration and test results, inspection reports, and certifications provided [4].

Becoming a part of this global accreditation system represents another reason to take pride in obtaining accreditation within this elite group of testing laboratories.

7.6 Post-Production Surveillance

Once equipment is placed on the market, a manufacturer has at least some obligation to follow up on the evolution and behavior of the equipment that is in the hands of the end user by performing post-market surveillance.

By doing this on a voluntary basis for each electrical product handed over for use, a manufacturer will be able to show their customers a proactive constructive attitude and, at the same time, will be able to be the first to discover solutions to mitigate any risk that could be involved during the operation of the equipment.

An attentive post-production surveillance will be able to offer at least the following very useful information to the manufacturer for the next revisions of the equipment:

- Equipment failures and electrical safety aspects have not been detected before or during the type tests;
- Identify the risks that may occur in the field by using the equipment;
- Detailed monitoring of the performance of the equipment in real life;

- Performing adequate risk assessments in progress that will be able to design measures to mitigate any effects of possible failure and/or to make the recalls in time.

7.6.1 Follow-Up Services

In situations where an NRTL/certification body was involved in the approval process, it will be implied for the duration of the equipment life in the field to supervise by performing field inspections at the manufacturing location(s) that were approved to do so.

Once compliance is determined, the manufacturer is authorized to use the certification/listing mark at agreed-upon manufacturing locations field engineering service inspections. Some NRTLs are starting the follow-up services (FUS) program even before the testing and evaluation of the equipment for which a manufacturer asked for listing/certification. In our opinion, it is a standard way to advise the manufacturer about possible nonconformances within their QMS under whose umbrella the potential listed/certified equipment will be manufactured. As well, the NRTL has to document and then understand the ability of a certain manufacturing location of the equipment that will go under investigation.

The FUS is a required program to be followed by both parties, NRTL and the manufacturer, and the program is designed to verify that the manufacturer continues to produce certified equipment in compliance with the applicable standards requirements (standards used during the evaluation) and the applicable standards that prescribes the routine tests (production line testing) in production.

The FUS generates inspections/audits of locations where listed/certified products are manufactured. These are conducted by field engineers from the NRTL at a frequency appropriately determined for the subject listed/certified product or system. Normally inspections must be unannounced during each factory visit, since the field engineer is looking for equipment that bears the NRTL mark. However, there may be occasions when no NRTL marks are being used when the inspector is present. In these cases, they will conduct a production-ready visit. In this visit, they will inspect how the critical components mentioned within the NRTL procedures and/or may select samples from the factory (production or warehouse), and will ask to check the test records that are performed during manufacturing on those samples. Likewise, they will follow up on how accurately the QMS procedures and work instructions are applied, how people from production are trained and thus are able to follow those instructions, and the field engineer may also ask for records of routine tests and observe them as well. Health and safety aspects may come under observations during the visits.

Throughout the lifetime of the valid certification/listing, products will undergo regular inspections at the manufacturing facility to verify continued compliance with all of the agreed-upon requirements specified within written procedures and regulations that govern the certification. The certification body will follow up the steps stated in those procedures assuring in this manner that the manufacturer will use only approved components to manufacture the equipment, as well as for intermediate checks and final routine tests in production that were considered necessary. They will also check that the manufacturer is using adequate test equipment in relation to accuracy and that is properly calibrated. In the EU, especially for medical electrical products, inspections can be conducted at subcontractors that manufacture critical parts of the certified product.

Because the EU directive requests the presence of technical documentation at the manufacturer, many certification bodies request the manufacturer to have a follow-up file consisting of:

- Certification license;
- List of critical components and relevant licenses for components;
- Copies of labels;
- Schematics, drawings, and photos (external, internal) for the certified product;
- Installation and user instructions;
- Reports of previous inspections;
- CAPAs for previous deficiencies;
- Complaints from customers.

As was specified earlier, the follow-up inspections must be unannounced visits and are scheduled at 4, 6, or 12 months (depending on the certification body policy).

A follow-up inspection, a process of assessing the quality and safety of certified products by representatives of certification bodies, is a near-worldwide practice. In recent years, many certification bodies start to reexamine their follow-up inspection systems in the face of demands that the manufacture of certified products should be made more transparently accountable for the outcomes and standards that they achieve.

In many situations, the date of the inspection is announced and the manufacturer will be ready to look good on the day of the follow-up, the inspector will be well impressed, and the invoice from the certified body will be paid by the inspected manufacturer. Is this truly the purpose of the follow-up? In our opinion, certainly not. We believe the entire philosophy and methodology of follow-up inspections needs to be changed. The inspections need to be free of charge, the manufacturer responsibility and transparency need to increase, the cooperation between certification body and the manufacturer must be continuous and based on mutual trust, and the visit of the representative of the certification body at the manufacturer does not need to be conducted as an exam. Rather, this should be a work meeting in which the certification body should support the manufacturer to solve any existing problems. This is the only way that quality and safe products can be provided to the market.

7.7 Market Surveillance, Post-Market Surveillance, and Vigilance

It is expected that manufacturers will guarantee that their electrical products perform within the promised parameters and do it without generating any safety hazard and/or unacceptable risks for the end user. But unfortunately, the reality is not so. That is why monitoring product performance and safety of both certified (marked) or noncertified products on the market has become crucial to systematically identifying unacceptable risks in the practical use of the product. Therefore, the market surveillance (post-market surveillance (PMS)) system was implemented, more or less, worldwide. The post-market surveillance system aims to actively and systematically gather, record, and analyze relevant data on the quality, performance, and safety of a product throughout its entire lifetime. We need to regard the PMS from two points of view: that of the manufacturer, who needs to know how their product meets customers' requirements, and that of local regulatory authorities, who will protect users from being affected by unexpected product noncompliance. In many countries the local authorities do a random survey on shops, wholesalers, and customs, taking samples of electrical products and sending them to labs for testing according to the local requirements (standards, codes, etc.). The negative results of such evaluation can compromise the current and future presence of a specific manufacturer in this market.

Post-production surveillance is important for medical electrical equipment even if all of the electrical equipment is totally safe for the involved personnel during use. Medical electrical equipment has even more intimate contact with the operators of the equipment as well as with the patients who are subjected to diagnostic testing, treatment, or rehabilitation, and have direct contact with the equipment via applied parts.

Each piece of equipment must be used with the utmost of care around patients and users, and there must not be any unacceptable risks under normal and single-fault conditions.

Both the MDR (regulation (EU) 2017/745) and the FDA define the term post-market surveillance.

The MDR defines it as all activities carried out by the manufacturers in co-operation with other economic operators to institute and keep up to date a systematic procedure to proactively collect and review experience gained from their devices placed on the market, made available or put into service for the purpose of identifying any need to immediately apply any necessary corrective or preventive actions [5].

A similar definition comes from the FDA, which is the active, systematic, scientifically valid collection, analysis, and interpretation of data or other information about a marketed device [6].

Compared with the FDA's definition, the MDR's definition provides a more detailed explanation, and thus, it may be more helpful for those who will assure the process of surveillance, as it not only describes actions but also states the objectives of those post-market activities.

What is the difference between post-market surveillance and vigilance?

Vigilance is only one part of the post-market surveillance system, as it refers to the reporting of (serious) incidents, field safety corrective actions (FSCAs), and recalls. It is a reactive system that deals with incidents rather than the proactive collection of PMS data. Section 2 of Chapter 7 of the MDR on vigilance defines the incidents that manufacturers have to report to the relevant competent authorities and how to submit these reports. Also, it requires manufacturers to analyze their vigilance data.

Additional guidance in relation to the vigilance system currently in operation under the MDD (to complement the MEDDEV 2.12-1 rev 8 from 2013) was published in July 2019. This document refers to the new manufacturer incident report form, which already includes the International Medical Device Regulators Forum (IMDRF) [7] adverse event terminology, UDI, and the single registration number as prescribed in the MDR [8].

In our opinion, even nonmedical equipment will benefit from attentive post-market surveillance. More often than not, recalls "discover" hidden defects present from back in the design stage, and that were not detected during the type tests performed or during the routine tests in production. The most frequent examples are the luminaires, pumps, and fans, but the list can continue with other categories of equipment.

We are not suggesting that each type of electrical equipment will have to follow a similar surveillance route that became mandatory for medical equipment by incorporating elements from the MR regulations within the quality management system of the manufacturer. We just want to emphasize that attentive surveillance will bring many benefits to any manufacturer who is placing electrical products on the market.

With the entry of the new EU MDR and of the In Vitro Diagnostic Regulations (IVDR) in 2017, these new requirements (located in Chapter VII of both regulations) had a significant impact on the processes cited in these documents and named the PMS, and put even more pressure on medical electrical equipment manufacturers. These new regulations introduce more incisive and prescriptive measures based on each equipment risk level for both the MDR and IVDR.

These regulations describe what kind of information/data must be included and provided based on the PMS.

There must be a PMS plan for each device and the plan must be an integral part of the manufacturer's QMS and the device technical documentation file. This means that a PMS for each device must be planned, maintained, documented, and updated in the QMS.

The PMS requires to document within the process information:

- Serious incidents, including information from the periodic safety update reports (PSUR) and FSCAs;
- Records referring to nonserious incidents and data on any undesirable side effects;
- Data from trend reporting;
- Relevant specialist or technical literature, databases, and/or registers;
- Information, including feedbacks and complaints, provided by users, distributors, and importers;
- Publicly available information about similar medical devices.

Annex III of the MDR has another list of subjects that must also be covered by the PMS (Section 7.6.1):

- A proactive and systematic process to collect information;
- Effective and appropriate methods and processes to assess the collected data;
- Suitable indicators and threshold values that must be used in the continuous reassessment of the benefit-risk analysis and of the risk management;

- Effective and appropriate methods and tools to investigate complaints and analyze market-related experience collected in the field;
- Methods and protocols to manage the events subject to the trend report;
- Methods and protocols to communicate effectively with competent authorities (CAs), notified bodies, economic operators, and users;
- Reference to procedures to fulfill the manufacturers› obligations for PMS system, PMS plan, and PSUR;
- Systematic procedures to identify and initiate appropriate measures including corrective actions;
- Effective tools to trace and identify devices for which corrective actions might be necessary.

One of the PMS will have as a final output specific content depending on the class of the involved medical device. The second document will be the PSUR. The PSUR must be part of the technical documentation and made available to notified bodies and competent authorities.

As for the post-market surveillance report, the periodic safety update report must present results and conclusion of data gathered from the post-market surveillance plan, which includes the rationale and a description of the CAPA implemented. These CAPA may refer to the equipment and at the same time, perhaps to the design/manufacturing process that generated that equipment, and it may impose actions as follows:

- Informing the involved authorities and users;
- Providing updates and changes including clinical evaluations;
- Review and evaluation of the results that were already issued and, in some situations, initiating recalls.

7.7.1 Recalls

The idea of recalls represents a nightmare for any manufacturer, supplier, and/or distributor. Authorities rarely officially or formally force manufacturers to take back any dangerous products they have sold. Recall itself is almost as slippery a term as the adjectives that modify it.

Sometimes, due to errors in design and/or manufacturing, unforeseen behavior of an electrical product will force a manufacturer to withdraw from the market equipment that failed to behave as it was intended. They must spend money on this measure, but even an expensive recall is understood to be much cheaper than the liability involved by keeping any faulty equipment on the market.

It is not an easy task to categorize recalls: they may happen in different moments of the equipment life on the market. It is not uncommon that recalls may have their class of severity changed due to the evolution in time.

Around the world, different organizations are in charge of overseeing markets. Manufacturers must have an understanding of the rules that have to be followed in the realm of commerce, always keeping at the forefront that customer safety is paramount. Any manufacturer will have in mind the possibility that sooner or later, a recall of their equipment may happen.

What is a recall? It represents the action of removing a product from the market. Recalls may be classified (according to the FDA) as follows, and this classification includes warnings [9]:

- *Class I recall:* A situation in which there is a reasonable probability that the use of or exposure to a violative product will cause serious adverse health consequences or death. This is the worst type of recall.
- *Class II recall:* A situation in which use of or exposure to a violative product may cause temporary or medically reversible adverse health consequences or where the probability of serious adverse health consequences is remote.
- *Class III recall:* A situation in which use of or exposure to a violative product is not likely to cause adverse health consequences.
- *Market withdrawal:* Occurs when a product has a minor violation that would not be subject to FDA legal action. The firm removes the product from the market or corrects the violation. For example, a product removed from the market due to tampering, without evidence of manufacturing or distribution problems, would be a market withdrawal.
- *Medical device safety alert:* Issued in situations where a medical device may present an unreasonable risk of substantial harm. In some cases, these situations also are considered recalls.

We want to mention an aspect that is common in North America. Here, due to elements that we do not fully understand, it seems to be part of the North American culture, we are living in a very litigious society. Here, a manufacturer may face a lawsuit for injury much like in the infamous case that involved a hot coffee a long time ago.

For example, if a child's toy powered by a coin-type battery turns out to pose a choking hazard to children under three, the parents can sue for monetary damages. But as long as the product doesn't harm the child, there can be no recovery of the so-called personal injury damages. The parents may potentially sue to get their money back, but most toys don't cost enough to warrant a lawsuit for a refund. These types of economic harm cases affecting many people commonly exist as a class action lawsuit and may hurt a manufacturer if the recalls are not properly prepared. This scenario happens only, for example, when many people bought that electric toy and proceed to look for compensation. A manufacturer/supplier/distributor must be aware of the moment that initiates a recall, as it can make it harder for a group of affected consumers to sue as a class. That is because some recalls establish a procedure through which consumers can request a full refund, and if the purchase price represents the extent of the consumer's damages, there's no reason left to sue.

In a product liability lawsuit that involves personal injuries, while a recall doesn't automatically immunize the defendant, it can make it harder for a plaintiff to succeed. This is especially true if the plaintiff purchased the equipment or continued to use it after the recall occurred. But simply recalling a product does not guarantee a defendant wins the case. The defendant must also show that they provided adequate notice of the recall to consumers [10].

The globalization of the economy has forced manufacturers, suppliers, and distributors with the need to be even more proactive and to coordinate and synchronize their actions when a recall is involved. Government consumer protection authorities are helping such recall efforts by developing sophisticated online product recall portals and other automated reporting tools.

In 2012, the Organization for Economic Co-operation and Development (OECD) launched its global product recall portal to coordinate recall notifications between the U.S. CPSC, the EU's member states, which coordinate such efforts through the EC's online dangerous product notification portal, rapid alert system (RAPEX), launched in 2004, and other countries around the globe. In addition to facilitating cross-border recalls, the data captured by the OECD, CPSC, and RAPEX have led to statistical and economic reports that may benefit manufacturers and distributors in product development, regulatory compliance, and recall efficiencies.

According to the OECD website, in 2018 alone, over 3,700 product recall notifications were submitted by 39 jurisdictions on the OECD global recalls portal. Consumers' reactions to recalls remain low in most jurisdictions, some achieving response rates as low as 3%. As a result, a large proportion of products that have been the subject of a recall remain in consumers' homes, exposing them to injuries or even risk of death [11].

In 2019, OECD launched the Global Awareness Campaign on Product Recalls, which aims to assist consumers in bridging the gap between being aware of a recall and reacting to it and to enhance business understanding of how to effectively communicate recalls to consumers.

The OECD global recalls portal represents, in our opinion, a very useful tool for all the involved parties in recalls; and, as OECD states on their website, it paves the way to safer products.

References

[1] LCIE, https://www.lcie.fr/129-nous-connaitre/lcie-bureau-veritas-130-ans-d-histoire.html.

[2] "Global Market Access Program," Intertek, 2019.

[3] UKAS, https://www.ukas.com/services/accreditation-services/laboratory-accreditation-isoiec-17025/.

[4] ILAC, https://ilac.org/.

[5] EU Regulation 2017/745, "Medical Devices," Brussels, 2017, https://eur-lex.europa.eu/legal-content/EN/TXT/PDF/?uri=CELEX:32017R0745&from=EN.

[6] FDA, "Postmarketing Surveillance Programs," https://www.fda.gov/drugs/surveillance/postmarketing-surveillance-programs.

[7] IMDRF, http://www.imdrf.org/.

[8] EU, "Additional Guidance Regarding the Vigilance System as Outlined in MEDDEV 2.12-1 rev. 8," https://ec.europa.eu/docsroom/documents/32305/attachments/1/translations.

[9] FDA, "Recalls, Market Withdrawals, and Safety Alerts," https://www.fda.gov/safety/recalls-market-withdrawals-safety-alerts.

[10] *A Guide to Product Recalls: United States & European Union*, Squire Patton Boggs, 2015, www.squirepattonboggs.com.

[11] OECD, http://www.oecd.org/sti/consumer/product-recalls/.

Selected Bibliography

https://anab.ansi.org/.

https://ul.com/wp-content/uploads/sites/4/2016/08/UL_SE_DATA_Client-Test-Data-Program_2016_V4.pdf.

https://www.a2la.org/.

https://www.johner-institute.com/articles/regulatory-affairs/and-more/post-market-surveillance/.

https://www.nist.gov/nvlap.

ISO 13485, "Medical Devices–Quality Management Ssystems–Requirements for Regulatory Purposes," Geneva, 2016.

ISO 17025, "General Requirements for Competence of Testing and Calibration Laboratories," Geneva, 2017.

www.nist.gov.

About the Authors

Steli Loznen, born in 1952 in Romania, is the cofounder of Cert-Global Ltd. and is a consultant in the area of international standardization and product compliance and safety engineering. He received his M.Sc. in electronics from the Polytechnic University of Bucharest in 1974. Starting in 1988 he worked in the Tel Aviv University Department of Biomedical Engineering as a lecturer for clinical engineering and medical ethics, and was managing director of Stelco-Bioengineering Ltd., a lecturer for standardization of electrical and electronic equipment at the Center for Technological Education Holon, the chief engineer at Telematics Laboratory at the Standards Institution of Israel, and the QA and certification manager of Israel Testing Laboratories Ltd. He was also the coordinator of UL, ETL, and TUVRNA Inspections Center Tel Aviv. His work focuses on international regulation and safety of medical and laboratory electrical equipment. His main areas of expertise are:

- International standardization (ISO, IEC, UL, EN, ANSI) for medical devices and laboratory and information technology equipment;
- Product safety, international regulatory procedures, and quality assurance;
- Risk management programs (hazard identification, risk analysis, risk evaluation, risk control);
- CE Mark (MDR, MDD, IVDD, LVD, EMC, RoHS, REACH);
- FDA procedures, HIPAA;
- ISO 9001 (Quality Management Systems), ISO 14001 (Environmental), ISO 18001 (Occupational Health and Safety Management Systems), ISO/IEC 17025 (Testing and Calibration), ISO/IEC 17065 (Certification), ISO 13485 (Medical Devices Manufacturing), ISO 14155-1 (Clinical Trials).

He is also the convener for IEC Sub-Committee 62A/MT62354–General Testing Procedures for Medical Electrical Equipment and convener for SC62A/MT 29–Mechanical Hazards, and the project leader for the IEC/TR 62354 General Testing Procedures for Medical Electrical Equipment. He was also a member of IECEE-CTL (Committee for Testing Laboratories).

In 2018, he became an honorary member of the Israeli Society for Medical and Biological Engineering (ISMBE).

He has received the following awards:

- IEEE-ICCE-Berlin 2015, First Place for Best Paper
- IEC–"1906 Award," 2017

He holds the following certification: Lead and Technical Assessor for IECEE-CB Scheme.

He is one of the Charter Members of the IEEE-Product Safety Engineering Society (2005), from 2012 a member of the Board of Governors, and between 2019–2022 Vice-President for Technical Activities of IEEE-PSES. He is also a member of the A2LA Accreditation Council.

He has written seven books, including Electrical Product Compliance and Safety Engineering, Volume I (the companion to this book, Volume II), coauthored with Constantin Bolintineanu and Jan Swart and published in 2017, and more than 65 papers, and has presented at conferences more than 95 times on various topics related to product compliance and safety engineering.

Constantin Bolintineanu is currently working within the DSC Testing Laboratory, an A2LA 17025 Accredited Testing Laboratory for IT equipment. Mr. Bolintineanu has been working in the compliance regulatory area since 1977. He specializes in testing and equipment evaluations, global regulatory approvals, product requirements, and standards (electrical safety).

Mr. Bolintineanu worked as technical manager for the Electrical Division of Intertek Testing Services, providing services including testing, evaluation, and certification of diverse equipment in the area of EMC, electrical safety, and telecom, from 1995 to 1998. Since 1998 he has been the direct approval liaison engineer for the BABT for all technical related issues within Digital Security Controls Ltd.

He has been the supervisor of the DSC Testing Laboratory since 2001.

As a technical supervisor of the A2LA Accredited Testing Laboratory, Mr. Bolintineanu is responsible for developing and managing all testing and evaluation activities related to electrical safety, telecom, and environmental testing performed under the accreditation of the laboratory.

Mr. Bolintineanu has over 40 years of extensive experience in electrical safety testing and evaluation in accordance with the European, North American, South American, South African, and/or Australia/New Zealand requirements and standards for IT equipment.

Mr. Bolintineanu received an M.S. degree in electronics engineering from the Polytechnic University of Bucharest Romania in 1972. He graduated the ASQ accredited program at Ryerson Polytechnic University in Toronto in 1997, obtaining a degree in quality assurance. He has been licensed as a professional engineer in Ontario, Canada, since 1997, and he has been certified as a NARTE electrical safety engineer since 2004. Mr. Bolintineanu is a member of the A2LA Accreditation Council.

From 1972 to 1994, Mr. Bolintineanu worked as a researcher at the Center of Research of Electro-technical Engineering in Bucharest, Romania, under the direction of Prof Dr. Florin Teodor Tanasescu.

Mr. Bolintineanu is the author of several technical books and over 40 scientific papers published worldwide related to medical electrical equipment and electrical safety. He holds several patents in this field. His patents are recognized in Romania, the United States, and Japan. Along with Steli Loznen, he wrote the chapter "Product Safety and Third Party Certification," in The Electronic Packaging Handbook, edited by Glenn R. Blackwell, 1999, and along with Steli Loznen and Jan Swart wrote the first volume of this book, which was published in 2017.

Index

B

C

D

N

Artech House Technology Management and Professional Development Library

B. Michael Aucoin, Series Editor

Actionable Strategies Through Integrated Performance, Process, Project, and Risk Management, Stephen S. Bonham

Advanced Systems Thinking, Engineering, and Management, Derek K. Hitchins

Applying Total Quality Management to Systems Engineering, Joe Kasser

Building Successful Virtual Teams, Francine Gignac

Critical Chain Project Management, Third Edition, Lawrence P. Leach

Decision Making for Technology Executives: Using Multiple Perspectives to Improve Performance, Harold A. Linstone

Designing the Networked Enterprise, Igor Hawryszkiewycz

Electrical Product Compliance and Safety Engineering, Volume 2 Steli Loznen, and Constantin Bolintineanu

Engineer's and Manager's Guide to Winning Proposals, Donald Helgeson

Engineering and Technology Management Tools and Applications, B. S. Dhillon

Enterprise Release Management: Agile Delivery of a Strategic Change Portfolio, Louis Taborda

The Entrepreneurial Engineer: Starting Your Own High-Tech Company, R. Wayne Fields

Evaluation of R&D Processes: Effectiveness Through Measurements, Lynn W. Ellis

From Engineer to Manager: Mastering the Transition, Second Edition, B. Michael Aucoin

Global High-Tech Marketing: An Introduction for Technical Managers and Engineers, Jules Kadish

How to Become an IT Architect, Cristian Bojinca

Integrated IT Project Management, Kenneth R. Bainey

Introduction to Information-Based High-Tech Services, Eric Viardot

Introduction to Innovation and Technology Transfer, Ian Cooke and Paul Mayes

ISO 9001:2000 Quality Management System Design, Jay Schlickman

IT Project Portfolio Management, Stephen S. Bonham

Managing Complex Technical Projects: A Systems Engineering Approach, R. Ian Faulconbridge and Michael J. Ryan

Managing Successful High-Tech Product Introduction, Brian P. Senese

Managing Virtual Teams: Practical Techniques for High-Technology Project Managers, Martha Haywood

Mastering Technical Sales: The Sales Engineer's Handbook, Third Edition, John Care and Aron Bohlig

The New High-Tech Manager: Six Rules for Success in Changing Times, Kenneth Durham and Bruce Kennedy

The Parameter Space Investigation Method Toolkit, Roman Statnikov and Alexander Statnikov

Planning and Design for High-Tech Web-Based Training, David E. Stone and Constance L. Koskinen

A Practical Guide to Managing Information Security, Steve Purser

Practical Model-Based Systems Engineering, Jose L. Fernandez and Carlos Hernandez

Practical Reliability Data Analysis for Non-Reliability Engineers, Darcy Brooker with Mark Gerrand

The Project Management Communications Toolkit, Second Edition, Carl Pritchard

Project Managment Process Improvement, Robert K. Wysocki

Reengineering Yourself and Your Company: From Engineer to Manager to Leader, Howard Eisner

The Requirements Engineering Handbook, Ralph R. Young

Running the Successful Hi-Tech Project Office, Eduardo Miranda

Successful Marketing Strategy for High-Tech Firms, Second Edition, Eric Viardot

Successful Proposal Strategies for Small Businesses: Using Knowledge Management to Win Government, Private Sector, and International Contracts, Sixth Edition, Robert S. Frey

Systems Approach to Engineering Design, Peter H. Sydenham

Systems Engineering Principles and Practice, H. Robert Westerman

Systems Reliability and Failure Prevention, Herbert Hecht

Team Development for High-Tech Project Managers, James Williams

For further information on these and other Artech House titles, including previously considered out-of-print books now available through our In-Print-Forever® (IPF®) program, contact:

Artech House
685 Canton Street
Norwood, MA 02062
Phone: 781-769-9750
Fax: 781-769-6334
e-mail: artech@artechhouse.com

Artech House
16 Sussex Street
London SW1V 4RW UK
Phone: +44 (0)20 7596-8750
Fax: +44 (0)20 7630-0166
e-mail: artech-uk@artechhouse.com

Find us on the World Wide Web at: www.artechhouse.com